FIRST EDITION

Printed and bound in the United States of America

Published by
O'DONNELL PUBLICATIONS
7217 Popkins Farm Road
Alexandria, VA 22306

ISBN 0-9670731-0-3

Library of Congress Catalog Card Number: 99-61353

FRONT COVER (left to right): *Microscope-p. 231, Toy Ball-p. 249, Colored Whiskey Flask-p. 56, Grained Whiskey Flask-p. 243, Shaving Soap Box-p. 238, Powder Flask-p. 190, Smoking Pipe-p. 240.*

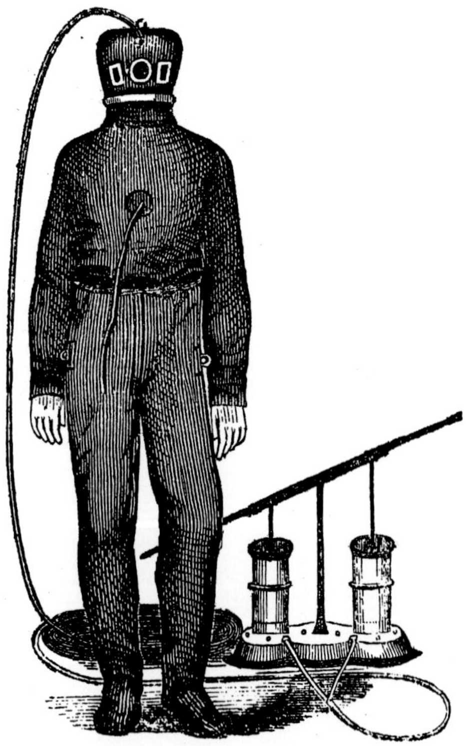

SUB MARINE ARMOR.

India-Rubber and Gutta-Percha In the Civil War Era

An Illustrated History of Rubber & Pre-Plastic Antiques and Militaria

by Mike Woshner

Photography and Design by Mike O'Donnell

O'Donnell Publications
Alexandria, Virginia

ACKNOWLEDGMENTS

The author is indebted to a number of individuals who have unselfishly contributed their time and knowledge towards the successful compilation and completion of this work. The following comments represent my sincere thanks for all who have shared so much with me and with you, the reader.

It was my good friend, Mike O'Donnell, that persuaded me to finally pursue my dream of putting together the first major collection of information and photographs on this fascinating subject. In addition to his exceptional skills in artifact photography and layout, Mike has been a tremendous resource for additional artifacts and information from individuals and sources unknown to me. This work would not have been so complete but for his ardent love of history and education.

Much of my education in military contracts and the National Archives' records came from Fred Gaede, who was always willing to share his vast experience and notes on military equipage. Fred's work with The Company of Military Historians has been long recognized and I am fortunate to be a recipient of his talented direction and support.

My awareness of overall military processes and organization was heightened by Don Kloster, retired Curator of the Division of Armed Forces History, American Military Museum, Smithsonian Institution. Don educated me on the information and accoutrements in the Smithsonian and provided considerable editing support, especially for the military sections of this work. His coaching on the presentation of several difficult points is sincerely appreciated.

When I started my humble collection of Civil War relics in 1968, I wrote to the author of *The Civil War Collector's Encyclopedia,* Francis Lord. Frank was never too busy to answer my letters and provide guidance in starting and expanding my collection. Down through the years, he has continued to support my collecting and writing efforts. I am grateful to him as a friend and a fellow collector.

The past and present staff at *North South Trader's Civil War* have always given me a printed forum for my unique areas of interest. The comments and suggestions of Bill Mussenden, Steve Sylvia, and Nançy Dearing-Rossbacher have given me the confidence I needed to follow through on this work.

My special thanks go to my family and friends that have supported and encouraged me along the way. High honors go to my wife, Janis, and my children, Craig and Lynn, for putting up with years of work, fun, excitement, and anxiety over a subject so little recognized or appreciated. I recognize and appreciate their ever-present patience, dedication, and love.

I am ever grateful to the many others who have shared information, photographs, and illustrations from their collections and archives. Every attempt has been made to cite appropriate individuals and institutions in notes and captions. Where no caption credit is provided, the item is most likely from my personal collection.

I would like to personally thank the following individuals and organizations, whose support played a critical role in the completion of this work:

Credits

Bruce Bazelon
Dan Binder
Dave Bush
Rick Carlile
Paul Chance
Richard Cies
Jonathon Kearns Cooper-Wiele
Mike Cunningham
Gordon Dammann
Jim Davie
Henry Deeks
Ray Darida
James Edmonson
Vern Eklund
Richard Ferry
Norm Flayderman
Jim Frasca
Charles Harris
Bedford Hayes
Janet Hayes
Steve Heinstrom
Frank Higginson
Keith Holland
Kim Hughes
Rance Hulshart
James Hutchins
Paul Johnson
Paul Jung
George Kegerreis
Sydney Kerksis
Clifford Krainik
Michele Krainik
Martin Lane
Paul Loane
H. Michael Madaus
Lee Manley
Mary Manley
Art Marchand
Dave Mark
Mike McAfee
John V. Miller, Jr.
Paul Millikan
Marianne Moe
Mary Frances O'Brien
Steve Osman
Ronn Palm
Alex Peck
Marius Peladeau
Stanley Phillips
Denis Reen
Ben Resnick
Brian Riel
Ray Rieser
Joe Riling
Julie Robinson
Mike Semegran
Maia Sornson
George Sparacio
Jim Stamatelos
Tony Stamatelos
Don Troiani
John Wideman
C. Young

Bertrand Conservation Laboratory, U.S. Department of the Interior, Missouri Valley, IA

Boston Public Library, Boston, MA

Buffalo Bill Historical Center, Cody, WY

Carnegie Library of Pittsburgh, Pittsburgh, PA

Dittrick Museum of Medical History, Cleveland, OH

Goodyear Tire & Rubber Company, Akron, OH

Hillman Library, University of Pittsburgh, Pittsburgh, PA

Library of Congress, Washington, DC

National Museum of American History, Division of Armed Forces History, Smithsonian Institution

Newark Public Library, Newark, NJ

S. S. White Company, Philadelphia, PA

St. Johns Archeological Society, Inc., Jacksonville, FL

U.S. National Archives and Records Administration, Washington, DC

University of Akron Archives, Akron, OH

University of Pittsburgh Center for Cultural Resource Research, Pittsburgh, PA

West Point Museum, West Point, NY

TABLE OF CONTENTS

TOPIC ARRANGEMENT

The format and topic arrangement of this work are intended to satisfy those who may want to read the entire volume as well as those who merely want to refer to a specific material, application, or patent. Some necessary overlap and duplication is present between sections and has been cross-referenced where practical.

The glossary that follows is a primer that will help the reader understand terminology applied to various rubber, rubber-like, and plastic-like substances by inventors, historians, and the public in general. With the exception of direct quotes, the preferred terms used by the author to identify the specific substances are "India-rubber" or "rubber," "hard rubber," "gutta-percha," and "hard gutta-percha" or "hard vulcanized gutta-percha.," (as opposed to "unvulcanized gutta-percha" that has hardened.) When mentioned, various compositions and waterproofing materials are identified as such or by their proper names, to distinguish them from India-rubber and gutta-percha.

The chronology is also a primer and reference source. The histories of the various substances, combined with their related patents, applications, and manufacturers, overlap and run concurrent in many phases. Along with the glossary, the chronology will help to minimize any confusion experienced throughout this work.

Part One provides an overall history of the early rubber pioneers, manufacturers, and processes, as well as early military uses. The history and general applications of India-rubber, hard rubber, gutta-percha, and other materials and compositions comprise the initial five sections. Section 6 highlights early military trials of India-rubber and gutta-percha articles from the 1830s through the 1850s. Section 7 covers their expanded use by the military during the Civil War.

Part Two consists of only Section 8—Products and Applications. This large section is an alphabetical reference embracing many unique categories and uses. Included in this section are many military, household, industrial, and general subjects.

Part Three begins with Section 9, which provides some basic guidelines to dating, cleaning, and preserving artifacts. Section 10 is an informative listing of early manufacturers and pioneers of the industry. Section 11 begins with a short overview of United States patents, followed by a comprehensive index of related patents, many of which are referenced in this work.

An index is provided to assist the reader in locating specific areas of interest.

GLOSSARY OF TERMS

Bakelite	*(Trademark)* The first totally synthetic thermoplastic. Introduced in 1909. Also used to denote any of various synthetic resins or plastics.
caoutchouc	Rubber (1775). Named by the French from the Amazon native "*cahuchu*," or "tree that weeps."
celluloid	Also "Celluloid" *(Trademark)*. A thermoplastic material patented in the United States in 1870 as an ivory substitute. Considered by some to have provided the impetus for the modern plastics industry.
composition	A term generally applied to any of several compounds that are mixed, heated, and moulded into ornate articles. Most compositions are based in gum-shellac and contain no rubber or gutta-percha.
ebonite	Hard rubber (1860).
enameled	Charles Goodyear's term for fabrics coated with India-rubber and united with the rubber in the vulcanization process. Also used by Goodyear to reference items coated with hard rubber.
gum-elastic	Rubber (1840).
gutta-percha	A substance from the latex of several Malaysian trees that quickly hardens when left untreated but resembles rubber or hard rubber when properly vulcanized (1842). Also the native name for the tree itself. May be seen as "gutta percha." The term "gutta-percha" is also generally used to denote hard unvulcanized or hard vulcanized gutta-percha.
hard rubber	A product of vulcanized rubber, produced by additional measures of sulphur and heat. The process was described in England within Thomas Hancock's vulcanization patent of 1844 and in the United States by Nelson Goodyear in his patent of May 6, 1851.
India-rubber	Rubber (1770). May be seen as "india-rubber" or "India rubber." Also used in the 1800s to denote hard rubber (1851).
latex	A milky, white fluid of various plants that is the source of rubber, gutta-percha, chicle, etc. (1835).
metallic gum-elastic	Rubber (1843). Charles Goodyear's preferred name for vulcanized rubber.
painted	Fabrics coated with unvulcanized India-rubber, tar compounds, or other waterproofing substance.

1837	June 17 - Charles Goodyear receives Patent #240 on the "Acid-gas Process" for India-rubber and effects a surface cure useful for thin sheets of India-rubber. Capes, table covers, etc. are sold by licensees.
1838	February 24 - Nathaniel Hayward receives Patent # 1,090 for the use of sulphur in the processing of India-rubber. He sells that patent to Charles Goodyear.
1839	Charles Goodyear discovers the components of the cure for India-rubber's deficiencies—sulphur, white lead, and heat.
1841	December 6 - A caveat for patent is issued to Goodyear for his process.
1842	Dr. Montgomery, an assistant surgeon in Singapore, notes the substance "gutta-percha" being used for axe handles. Samples are sent to England.
1844	May 21 - Thomas Hancock is issued a British patent for vulcanizing India-rubber. Samples of Goodyear's India-rubber had revealed the use of the key ingredient, sulphur, to Hancock. The British patent also notes the formation of hard rubber, but it is not deemed a valuable substance by Hancock.
	June 15 - Patent #3,633 is issued to Goodyear for "Improvement in India-Rubber Fabrics," which defines the process known as "vulcanization."
	L. Candee and Company, which began making India-rubber shoes in 1843, becomes the first plant to manufacture under a Goodyear license.
1845	Richard Archibald Brooman obtains British patents for moulding, stamping, embossing, and making thread of gutta-percha.
1847	Dr. Montgomery receives a gold medal for his work with gutta-percha bandages. His other suggested uses include "stopping tooth decay."
	Edwin Truman introduces gutta-percha as a denture base.
	The American Gutta Percha Company is formed by Samuel Armstrong.
1848	Licenses for the manufacture of India-rubber army and navy equipment are transferred to the newly-formed Union India Rubber Company. The company would retain these rights through the expiration of the Goodyear patent in 1865.
	April 25 - Charles Goodyear patents a process for making hollow rubber balls.
	Richard Archibald Brooman issued U. S. patents for gutta-percha processes.
	Gutta-percha golf balls are introduced.
1849	December 25 - Charles Goodyear's vulcanization patent is reissued for technical changes.
	December 31 - Nelson Goodyear, brother of Charles, receives a caveat for patent on a process for making hard rubber.

1851 May 6 - Nelson Goodyear is issued Patent #8,075 for the "Manufacture of India Rubber," which details the hard rubber process.

Charles Goodyear spends $30,000 to create a three-room display at the Great Exhibition at London's Crystal Palace. The walls, furniture, draperies, rugs, and fixtures of the display are made of India-rubber. He is one of only three Americans to win Grand Council Medals.

Thomas Hancock also erects an impressive display of India-rubber goods at the Great Exhibition. In addition, ornate gutta-percha articles are displayed by several individuals and manufacturers.

1852 June 1 - John Rider is issued U.S. Patent #8,992 for the vulcanization of gutta-percha. His patent also embraces the manufacture of hard gutta-percha.

The first license under Nelson Goodyear's hard rubber patent is issued to Meyer and Poppenhusen Company. Later offspring include the American Hard Rubber Company and the India Rubber Comb Company.

1854 May 30 - John Murphy is issued U.S. Patent #10,977 for processes for treating gutta-percha.

November 7 - William Rider and John Murphy are issued U.S. Patent #11,906 for a process of desulphurizing gutta-percha.

1855 Charles Goodyear spends $50,000 on an elaborate India-rubber display at the Exhibition Universelle in Paris. He wins the Cross of the Legion of Honor at the Exhibition but, due to an unfavorable French court ruling and loss of royalties, he receives the award in debtor's prison.

Charles Goodyear, Jr. receives a British patent for hard rubber denture plates.

1858 June 14 - Charles Goodyear's vulcanization patent is extended for 7 years.

1860 November 12 - Charles Goodyear's vulcanization patent is reissued for additional technical changes.

1861 The Civil War provides an impetus to the growing India-rubber and gutta-percha industries, as orders are placed for blankets, ponchos, and other military equipage.

1862 Hard rubber is employed extensively in reconstructive surgery on the mouth and teeth of a badly wounded soldier.

1865 May 6 - Nelson Goodyear's hard rubber patent is extended for 7 years.

June 15 - Charles Goodyear's vulcanization patent expires.

1872 May 6 - Nelson Goodyear's hard rubber patent expires.

PART ONE

BIRTH OF AN INDUSTRY

The Evolution to Rubber and Plastics

Portrait of Charles Goodyear *by G. P. A. Healy, one of the most renowned portrait artists of the mid-19th century. The painting, signed and dated 1855, also bears the inscription: "The first picture ever painted on India Rubber Panel." (Courtesy The Goodyear Tire & Rubber Company)*

SECTION 1

EARLY RUBBER HISTORY AND CHARLES GOODYEAR

"Few are aware of the extent of the applications of vulcanized india rubber; and no intelligent person can view the great variety of the fabrics and manufactures into which it is made, except with surprise and admiration..."

From an advertisement for a display of India-rubber goods, hosted by the Patent Office in 1858

The use of rubber has been documented in the earliest years following the discovery of the New World. Antonio de Herrera y Tordesillas, in his *General History of the Vast Continents and Islands of America* published in the early 1600s, noted that Columbus, on his second voyage between 1493 and 1496, saw natives of Hispaniola (now Haiti) playing with a ball made from the *"goma de un arbol"* or "gum of the tree."[1] These balls, Tordesillas wrote, ..."tho' heavy would fly and bound better than those fill'd with Wind in Spain."[2] In *Decades of the New World,* Pietro Martyre d'Anghiera records in 1511 the testimony of a witness who saw Aztecs playing with balls which were "...made of the juice of a certain herbe" hardened by heat and which, "...beeing stricken upon the ground but softly would rebound incredibly into the ayer."[3] A third Spanish writer, Juan de Torquemada, wrote *De la Monarquia Indiana* in 1615. In this work, he noted several applications of the elastic tree milk in Mexico, including the waterproofing of capes and raincoats. He also became the first to record a problem that would plague the substance for two and a half centuries when he observed, "...the items melted in the rays of the sun."[4] This shortcoming, along with the brittleness that occurred in the cold weather, presented serious obstacles to the acceptance of the material for general use.

***Illustration from* The Scientific American** of Sept. 10, 1859 *shows "Natives collecting East India rubber or tintawan."*

Throughout the sixteenth, seventeenth, and eighteenth centuries, Europeans observed South American Indian use of the milky juice not only as a drink and an aliment, but also as a means of waterproofing shoes and clothing and for fabricating bottles and toys. While soldiers in this New World also copied the Indians' applications and sent items home, Europeans generally regarded crude rubber as a novelty more than a useful substance.[5]

The most commonly used name for the gum in the late eighteenth and early nineteenth century was *caoutchouc*, named by the French from the Amazon native term *cahuchu*, meaning "tree that weeps."[6] Around 1770, the evolution towards the modern-day name for the substance began. Joseph Priestly, the discoverer of oxygen, noted that a London mathematical instrument maker by the name of Mr. Nairne sold a half-inch square piece of the material to "rub out" the marks of the black lead pencil; hence, the term "rubber" was coined. Popular notion was that it hailed from the West Indies. Early specimens came from India and, in South America, where the majority of it originated, the latex was gathered by Indians. For one of these reasons, the word "India" was added and the name "India-rubber" created. Before the modern abbreviated name of rubber became permanent, it would also be called gum-elastic, metallic-gum, and vegetable leather.[7]

***Early India-rubber eraser** in cloth bag with walnut inkwell. In the late 1700s, small blocks of what was then called caoutchouc were sold by stationers to "rub out" lead pencil marks, leading to the use of the word "rubber."*

***African Rubber,** as shown in a museum display of the early 1900s. While most rubber used in the United States came from South America, rubber-like latex or gum is found in thousands of plants around the world.*

Early Processes for Curing

In 1791, Samuel Peal was granted the first English rubber patent for "...rendering waterproof all kinds of leather, cotton, linen, etc. for the purpose of being worked into shoes, boots and other wearing apparel" through treatment with a turpentine solution.[8] Into the nineteenth century, no cure for the problems of India-rubber's instability was yet discovered. Around the same year, a Swedish professor of medicine and pharmacy, Petter Jonas Bergius, was experimenting with rubber and may have utilized sulphur in his process.[9] Various other methods employing ether, turpentine, petroleum, and other solvents met with limited success. A contemporary publication, *Ure's Dictionary of Arts*, states that "...caoutchouc has been employed very extensively in making elastic bands or braces. The original manufacturer of these webs is a major in the Austrian service, who erected a great factory at St. Deny's, near Paris, 1803."[10]

Charles Macintosh, a Scottish chemist working at the Glasgow Gas Works in 1819, combined rubber with naphtha to form a substance with which he coated fabrics to make them waterproof. Garments made from this process, patented June 17, 1823, were first called "Water-

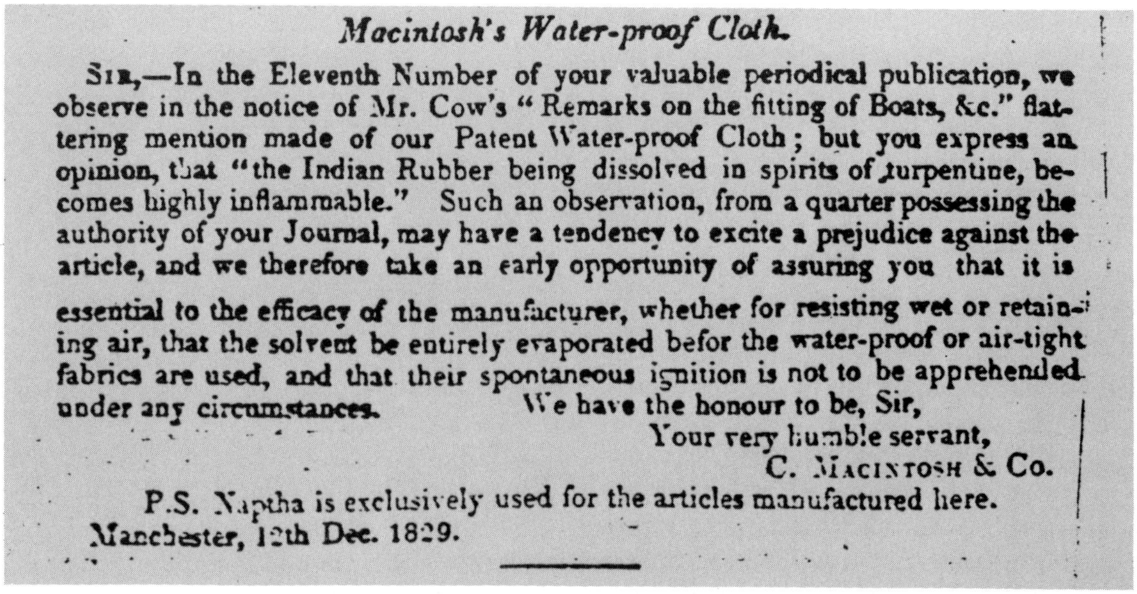

Macintosh's Water-proof Cloth.

SIR,—In the Eleventh Number of your valuable periodical publication, we observe in the notice of Mr. Cow's "Remarks on the fitting of Boats, &c." flattering mention made of our Patent Water-proof Cloth; but you express an opinion, that "the Indian Rubber being dissolved in spirits of turpentine, becomes highly inflammable." Such an observation, from a quarter possessing the authority of your Journal, may have a tendency to excite a prejudice against the article, and we therefore take an early opportunity of assuring you that it is essential to the efficacy of the manufacturer, whether for resisting wet or retaining air, that the solvent be entirely evaporated befor the water-proof or air-tight fabrics are used, and that their spontaneous ignition is not to be apprehended under any circumstances. We have the honour to be, Sir,

Your very humble servant,

C. MACINTOSH & Co.

P.S. Naptha is exclusively used for the articles manufactured here.

Manchester, 12th Dec. 1829.

Letter written by Scottish chemist Charles Macintosh *on December 12, 1829 to* The United States Journal and Naval and Military Magazine *regarding his waterproof cloth.*

proof Double Textures" and later "Macintoshes," the term that sticks to this day in England.[11] It was likely this firm that made the India-rubber diving suit used by John Deane in the late 1820s. Deane's quests for sunken treasures lead him to depths of up to ninety feet.[12]

Thomas Hancock, the dominant figure in English rubber history, became interested in rubber about the year 1815. In 1820, he patented a process for the application of gum-elastic to "springs" for trousers, gloves, shoes, stockings, etc. These springs, actually elastic stays or suspenders, would keep the garments in place without the use of laces, buttons, etc.[13] In 1825, after additional research and experiments, Hancock obtained a license from Macintosh to use his process. Hancock's products were superior to those of Macintosh, as he employed the use of a masticator to cut and mix the rubber, reducing by one-half the amount of solvent used.[14] In 1833, the firms were operating together and by 1835 Hancock was running the world's largest rubber factory, producing boots, coats, cushions, pillows, air-beds, life-preservers, etc.[15] Hancock improved manufacturing processes significantly and was responsible for the start of the French rubber industry.

Problems in America

The growth of the industry was not going quite as well in America. In the early 1820s, items like rubber shoes, made by the natives of the Amazon by repeatedly dipping wood lasts into crude rubber and smoking each layer, were imported into the United States. For several years the imports reached 500,000 pairs per year.[16] But America's seasonal temperature variations continued to cause radical changes in the condition of the rubber. Even in England, where the climate was more consistent, the garments would begin to melt if one stood by a stove for too long a time.

While several rubber companies were started in America in the 1830s, only one survived the ongoing problems of the substance. The relative success of the country's first rubber company, the Roxbury India Rubber Company of Roxbury, Massachusetts, can be attributed to the efforts of one of its founders, Edwin Chaffee. The company was founded in 1832, the same year Chaffee had invented a spreading machine for coating cloth with India-rubber dissolved in turpentine. This process had formerly been done by hand with mason's trowels. The company sold India-rubber goods for several years, until the return of too many shoes, caps, wagon covers, and other goods that had been congealed and ruined by the warm weather caused financial problems. In 1836, Chaffee invented and employed a 30 ton spreading machine known as the "mammoth" or "monster" that avoided the use of solvents and produced a slightly better product than other companies. The monster saw ongoing use with succeeding companies until the early 1890s.[17] The Roxbury company continued manufacturing shoes, life preservers, coats, caps, wagon covers, etc. until the early 1840s.[18]

Enter Charles Goodyear

It was here that Charles Goodyear was introduced to the problems of rubber. In 1834, on his first visit to the company's store in New York, he inspected a life preserver and was convinced he could make a better valve for the piece. When he returned to the store with the improved valve, the manager was impressed with its design but advised Goodyear of the demise of the rubber industry and its ultimate doom without a resolution to the problems. From that time to the end of his life in 1860, Charles Goodyear would labor obsessively on the two things he was convinced his Creator had entrusted to him; namely, the application of improved rubber for every conceivable purpose and the development of life preservers so that no person would ever drown again.[19]

While Goodyear was religious and a devoted family man, it sometimes seems that even wife and children did not come before his obsession with India-rubber. Over his years of research and experiments, he was oftentimes in debtor's prisons in America and abroad. He and his family, living in and out of poverty and despair at residences that included Philadelphia, Boston, Roxbury, New Haven, and New York, frequently went hungry for months. Even his children's school books were not spared when he once sold everything to finance his experiments.

Charles Goodyear was surrounded by inventive and industrial influence. His father, Amasa Goodyear, made America's first pearl buttons as well as military buttons for the United States during the War of 1812. He patented many improvements in farm implements including spring steel hay and manure forks. In addition to these products, his company also manufactured spoons, scythes, and clocks. In the early 1830s, Charles himself secured patents for buttons, forks, spoons, faucets, and air pumps. He had also thought of making rubber fabrics and attempted to use a rubber washer in a faucet prior to his serious rubber ventures. His brother Nelson held several patents including the 1851 process for hard rubber manufacture, and his brother Henry invented rubber shoe tips and other rubber-related articles. Cousins Ellsworth and Austin Goodyear Day patented processes associated with rubber, and Charles Goodyear, Jr. invented hard rubber dentures in 1855 and the still famous "Goodyear Welt" shoe improvement after his father's death.[20] The elder Charles held many patents in the United States and Europe. However, having investigated only United States patents, the author cannot substantiate claims that he held over 500 patents.[21]

Several cures for the rubber problem, most utilizing turpentine as a solvent, were tried in the early 1830s and met with little success. The

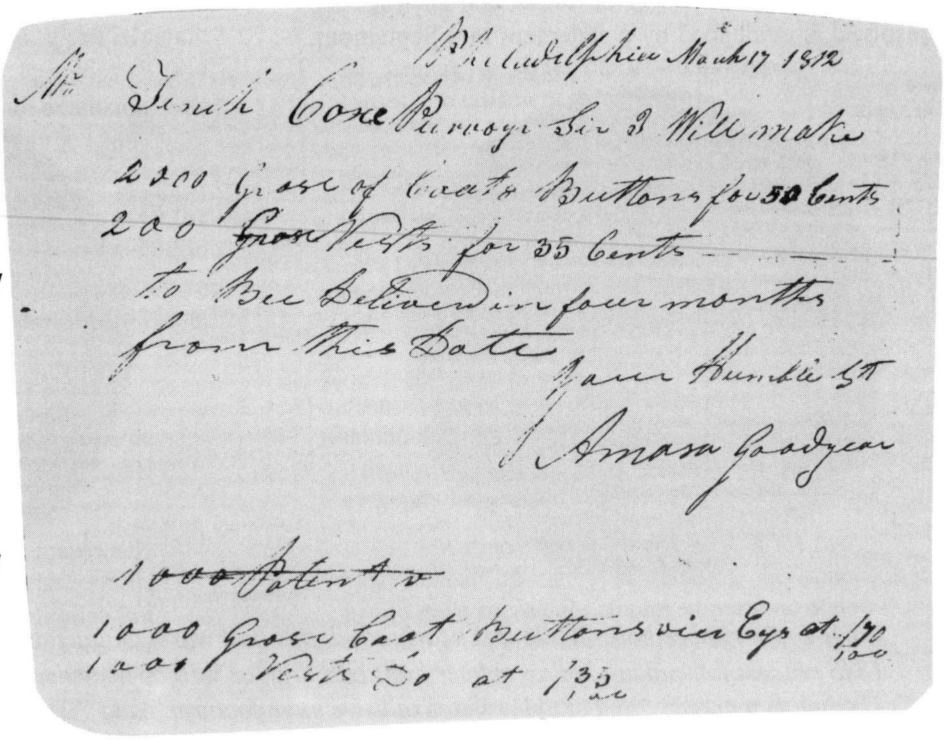
Philadelphia March 17 1812
Mr Tench Coxe Purveyor Sir I Will make
2000 Gross of Coats Buttons for 50 Cents
200 Gross Vests for 35 Cents
to Bee Delivered in four months
from This Date
Your Humble Svt
Amasa Goodyear

1000 Gross Coat Buttons

Amasa Goodyear Proposal Letter. *Handwritten proposal letter, dated March 17, 1812, from Charles Goodyear's father, Amasa Goodyear, for the production of coat and vest buttons. It was written to Trench Coxe, the government purchaser in Philadelphia at the start of the War of 1812.*[59] *(Courtesy National Archives)*

found it touched by vermin, nor altered in the slightest degree.

In conclusion, we cannot but draw the inference from our own experiments and from examination of the different kinds of fabric made from this singular composition, that it far surpasses all other attempts made to manufacture gum-elastic goods.[34]

Charles Goodyear's brother, Nelson, worked with him and held several patents, including an 1845 process for mixing sand and rubber to make a slip-resistant covering for steps and rail car roofs (U.S. Patent #4,047) and an 1849 improvement in elastic cords for suspenders (U.S. Patent #6,786). While Charles noted the formation of hard rubber, the process was perfected and patented by Nelson (U.S. Patent #8,075). Nelson Goodyear's hard rubber patent was reissued in 1858 and extended in 1865. The patent expired in 1872. The issue date of the hard rubber patent, May 6, 1851, probably appears on more items of the era than any other date and is frequently confused with Charles Goodyear's vulcanization patent.

India-rubber Pirates

After Charles Goodyear's perfection of the vulcanization process, he would have done well to become a serious manager of the licenses, manufacturers, and royalties that could have made him a millionaire. Instead, he relentlessly pursued novel and unique applications for rubber, often spending exorbitant amounts on prototype items. With no one minding the store, several manufacturers began making illegal items without a license.

The most infamous of all these rubber pirates was one Horace Day, himself an inventor, experimenter, and rubber pioneer. He patented several rubber products and manufacturing processes and even filed a patent on a vulcanization process in 1844, which was rejected because of Goodyear's pending application. Day blatantly used the vulcanization process without a license. Goodyear filed several successful lawsuits, resulting in royalty agreements which Day ignored. Day unsuccessfully attempted to have Congress overturn the Goodyear patent in 1850 on the grounds it was fraudulent.[35]

In 1851, several shoe manufacturers holding Goodyear licenses made a move to set a precedent against the pirates and filed suit in Goodyear's name against Day. They hired the seventy-one year old Daniel Webster, who took leave from his position as Secretary of State, to defend the patent and the rights of all licensees. Horace Day retained Rufus Choate, the greatest trial lawyer of the time. The case, dubbed the "Great India Rubber Suit," came before the United States Circuit Court on March 23, 1852 at Trenton, New Jersey. The decision was a total victory for the Goodyear forces and Daniel Webster. An accounting of illegal articles made and sold was ordered and a permanent injunction issued against Day. Daniel Webster was paid $15,000, the highest ever paid to a lawyer, for what was to be his final court appearance.[36] But Horace Day and other infringers would never disappear from the inventor's life, and would continue to plague him and the licensed rubber companies even after Goodyear passed away. Of thirty-two cases that Goodyear and his licensees were forced to carry to the Supreme

Patent drawing of Henry Goodyear's geographic globe. *These globes, with details in relief, could also be used as a ball or as a tool for teaching the blind. Note the 1861 reference to "Russian Territory," five years prior to the United States' purchase of Alaska.*

Court in his lifetime, not one was returned against them.[37]

Goodyear was continuously involved in patents for India-rubber and related applications. He had attempted to file British and French patents for his vulcanization process, but had been upstaged by the Englishman Thomas Hancock. Ironically, Hancock's discovery of the process was a result of studying cured rubber samples passed to him by William Brockedon, who received them from Stephen Moulton, whom Goodyear had met and entrusted the pieces. It was Brockedon who suggested the name *vulcanized rubber*, after the god Vulcan.[38] Goodyear had used and preferred the name *metallic gum-elastic* for the processed product.

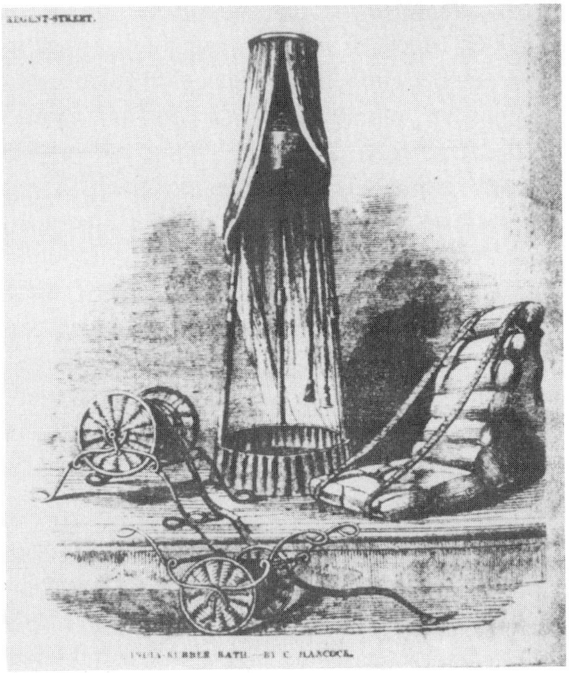

India-rubber bath. *From Thomas Hancock's display at the 1851 Great Exhibition in London, as pictured in* The Illustrated London News *of October 4, 1851.*

A World of India-rubber

In addition to monies lost to the rubber pirates and legal fees, the inventor's relentless pursuit for new applications also cost him dearly. Be it $5,000 for a set of rubber sails or a small fortune for a set of

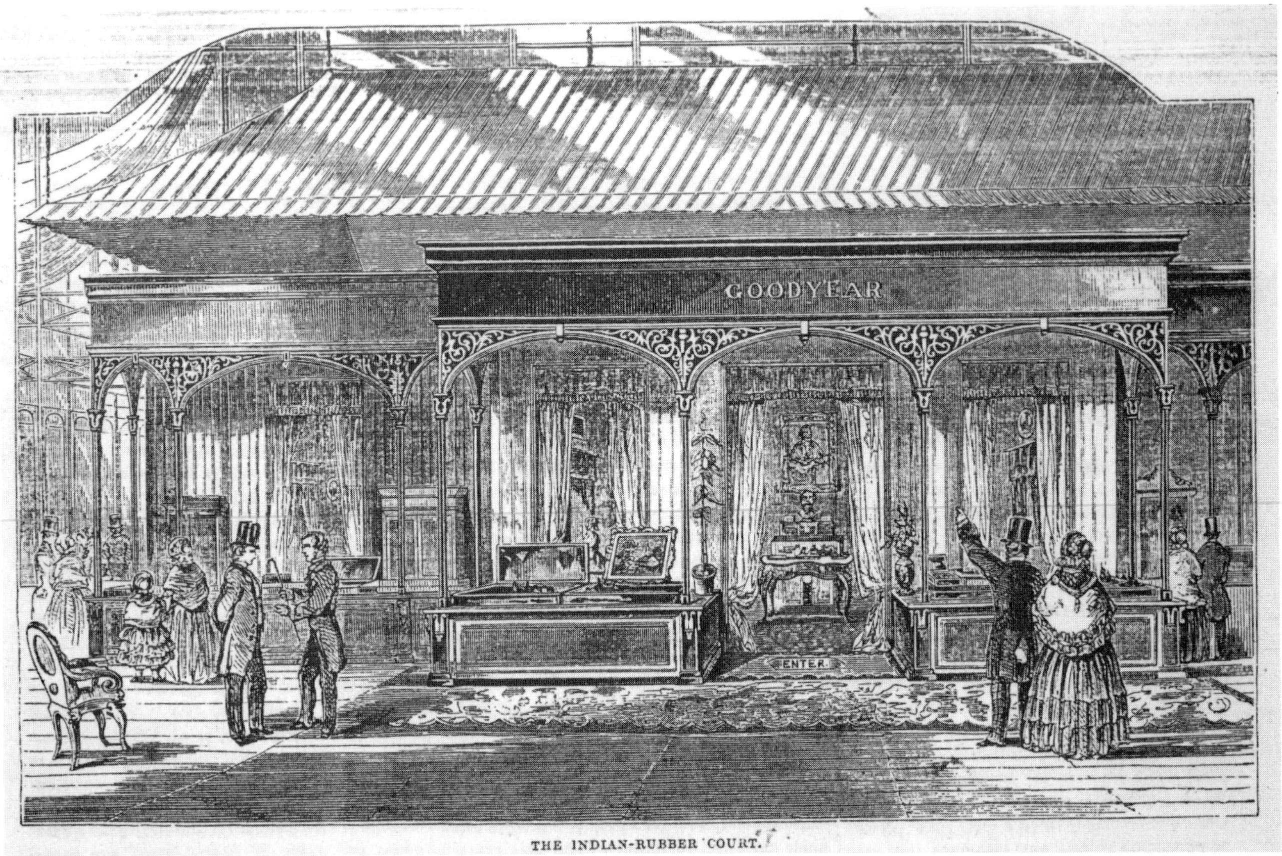

Goodyear's Vulcanite Court at the Crystal Palace Exhibition of 1851 in London's Hyde Park. *The exhibit, consisting of rooms of walls, rugs, furniture, musical instruments, and other household articles, was made entirely of India-rubber and hard rubber. Goodyear was awarded a Grand Council Medal, one of only three presented to Americans. (Courtesy of University of Akron Archives)*

hard rubber combs, the constant drain on profits prevented Goodyear from ever becoming a wealthy man. On one occasion, he and Hayward filled a large jug with powder, submerged it in a pond and, igniting a fuze that had been placed in a long rubber tube, successfully exploded a submarine mine.[39] On another brainstorm, he set out to test a theory that ships bottoms would be impervious to the fouling caused by sea life if they were encased in hard rubber. He bought several large sheets of steel, which were coated with hard rubber and submerged for several months in the sea. After inspecting the recovered pieces and noting the normal amount of barnacles and sea vegetation, Goodyear permanently abandoned this idea.[40] But not hundreds of others.

In 1851, at the Great Exhibition in the spectacular Crystal Palace in London's Hyde Park, the inventor stopped the show with a $30,000 display that was made totally of India-rubber. Called "Goodyear's Vulcanite Court," the three-room suite had rubber walls, furniture, draperies and rugs. Hydrogen-filled balloons floated about and hard rubber combs, musical instruments, and ornate carvings were admired by visitors. Despite the British disdain for American inventors, Goodyear received rave local reviews, his exhibit far surpassing Hancock's. He was one of only three Americans to win the Grand Council Medal at the Exhibition.[41]

Goodyear spent over $50,000 to create a similar exhibit at the Exposition Universelle in Paris in 1855, where every known rubber item of the era was displayed. The emperor of France, Louis Napoleon (nephew of Napoleon Bonaparte) visited the display and was so overwhelmed, that he awarded Goodyear the Grand Medal of Honor and the Cross of the Legion of Honor. In the course of the event, the French courts reversed a previous patent ruling in Goodyear's favor, voiding his French licenses and leaving the inventor once again in heavy debt. He was in Clichy Prison when his son delivered the Cross of the Legion of Honor to him. On a later occasion, Goodyear accompanied Louis Napoleon in the royal carriage to the *Bois de Boulogne*, where he gave the emperor a demonstration of his rubber pontoons.[42]

The United States Patent Office hosted an "Exhibition of India Rubber Fabrics" in 1858 at the Model Hall of the Patent Office. A promotional advertisement for the display stated:

> Few are aware of the extent of the applications of vulcanized india rubber; and no intelligent person can view the great variety of the fabrics and manufactures into which it is made, except with surprise and admiration, and without feeling that the genius, artistic skill, and enterprise, which have originated and made successful this great branch of national industry, are honorable to our country.[43]

While commendations like those voiced by the Patent Office were indirectly complimentary to Charles Goodyear, the most lasting tribute to his labors, creativity and imagination is his autobiography, *Gum-Elastic and Its Varieties*, published in 1853 and 1855. Original volumes done by the inventor contained plates and maps done on

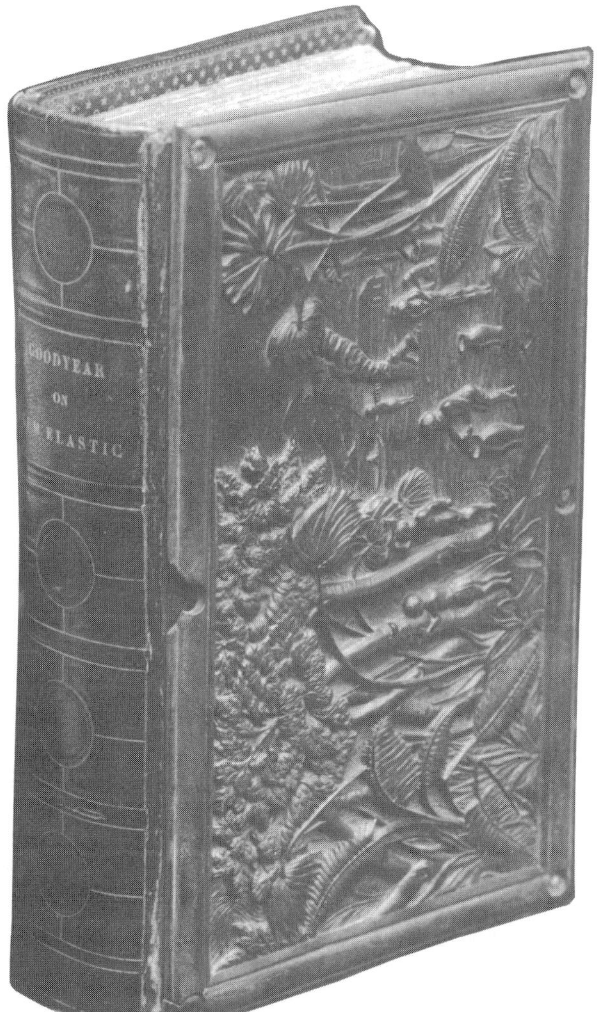

Goodyear on Gum-Elastic*. Rare personal notebook used by Charles Goodyear for outlining applications. Like a few special editions of his book, Gum-Elastic and Its Varieties, the hand-carved hard rubber covers of the notebook depict scenes of natives in various stages of the rubber collection process. (Courtesy The Goodyear Tire & Rubber Co.)*

India-rubber fabrics.[44] A few original volumes, one of which is maintained in the Library of Congress, were printed on rubber tissue. At least one surviving copy is bound in hard rubber covers with carvings of natives tapping rubber trees and processing the latex.[45]

In Volume I (interestingly dated 1855, while Volume 2 is dated 1853), Goodyear traces the history of the substance, its properties, and his own trials and tribulations while experimenting on the road to the discovery of vulcanization. He lists the properties of vulcanized India-rubber as "...elasticity, pliability, durability, insolubility, unalterability by climate or artificial heat or cold, inadhesiveness, impermeability to air, gases, liquids, plasticity, facility of receiving every kind of printing, facility of being ornamented by painting, bronzing, gilding, japanning and mixing with colors, non-electric quality and odor."[46] Each quality is described and the benefits to society outlined, most with specific examples. Goodyear related:

> An anecdote is told of a professor who, having highly charged an electric machine while wearing India rubber shoes and standing on the wire, upon taking them off and resuming his position, was convinced of their non-electric property by being knocked down without them.[47]

The only negative quality, odor, is treated cautiously by an explanation of how most of the natural offensive odor is gone, how the remaining odor will dissipate over time, and the fact that so many other articles in daily use bear offensive odors.[48]

Volume II is a detailed list of all the possible uses and applications of India-rubber—a veritable encyclopedia of contemporary civilian, industrial, and military items, with the relative advantages of making them from rubber or hard rubber clearly stated. Clothing, boots, tents, horse trappings, camp-ware, musket covers, pontoons, boats, medical supplies, toys, carriage covers, air cushions and mats, luggage, musical instruments, sails, and fire buckets are among the hundreds of articles depicted and described. Life preservers receive special attention, with variations that include life preserver pillows, cushions, jackets, self-inflating life-boats, and musket covers that doubled as life preservers. Novel uses that parallel modern-day products are air-filled safety jackets for boxers and skaters, baby jumpers, and a folding dress to wear when taking a close look at Niagara Falls.[49]

CHARLES GOODYEAR.—[From a Photograph.]

Image of Charles Goodyear. *From the first biography of Goodyear,* Trials of an Inventor, *written by Rev. Bradford Peirce in 1866.*

Goodyear mentions, "...a nautical term, that of 'squeal-gee,' " being applied to a rubber sponge fastened to a block with a long handle "fast getting into favor for scouring vessel's decks."[50]

Goodyear showed an interest in creative education by promoting various styles of globes that would be cheaper than contemporary globes made of plaster that have "...heretofore been so expensive as to be found in schools only of the higher class." He suggested inflated hydrogen-filled globes, illuminated globes, embossed colored globes that could be used as bat or foot balls, and blank globes on which students could draw or paint the earth's details and either wash them off or varnish over them for permanent use. Credit for the latter invention is given to Mr. Henry Barnard, whom Goodyear cites as the superintendent of the State schools in Massachusetts and Rhode Island.[51] Goodyear's brother, Henry, would later patent "terrestrial or celestial globes," made of India-rubber or gutta-percha with 3-dimensional geographical relief that was designed to be flush with the surface (U.S. Patent #31,311). Henry noted that his globes were useful in teaching blind children, more educational, and could also be used as toys.

Charles ventured into the medical field by suggesting that a very light, delicate fabric called

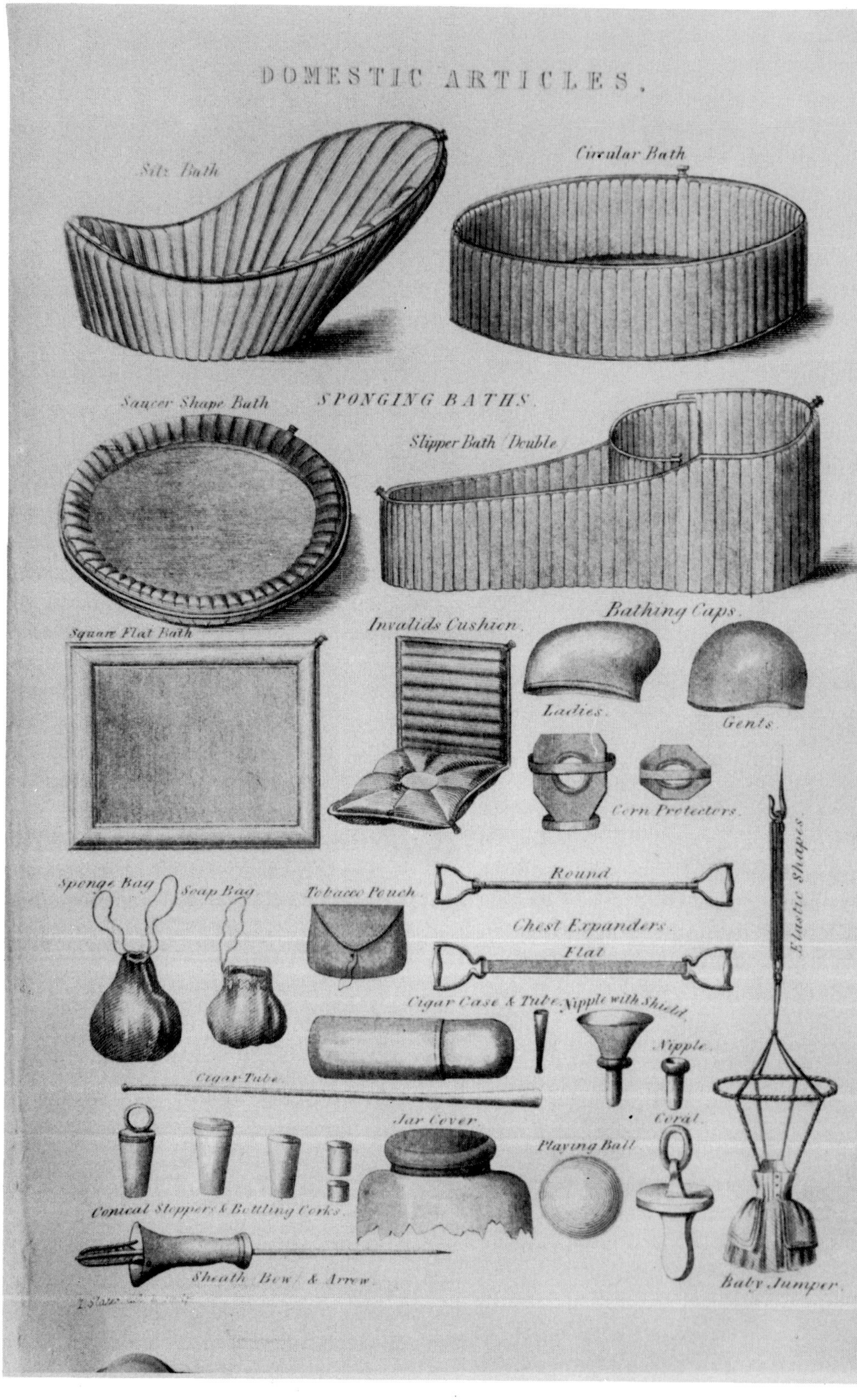

***Illustrations from* Personal Narrative of the Origin and Processes of the Caoutchouc and India-Rubber Manufacture in England**, *published by the English rubber pioneer, Thomas Hancock, in 1857.*

drapery, produced by dissolving India-rubber in camphene, could be used for bandages or to stop hemorrhage. Medicated drapery, containing opium, camphor, or perfumes prescribed by a doctor, was deemed a cure for rheumatism and "As a remedy in the case of very painful burns and scalds, it may be highly recommended."[52]

Among the unique ideas that likely never made it to production were India-rubber bank notes, which Goodyear claimed would be "exceedingly difficult to counterfeit" and could be "...cleaned as easily as glass, by washing or boiling, without injury to the tissue or the engraving." This the inventor saw as a must because, in their lifetime, bank notes "...pass through hands of every hue, and through pockets of all grades, until they acquire an odor that would be quite insufferable in any other articles."[53]

Thomas Hancock published a similar book in 1857, entitled *Personal Narrative of the Origin and Process of the Caoutchouc and India-Rubber Manufacture in England.* While not as lengthy and detailed as Goodyear's, the volume provides an excellent overview of the manufacture and applications of rubber. The works of both pioneers were reprinted in 1939, the 100th anniversary of the Goodyear vulcanization discovery, under the title *A Centennial Volume of the Writings of Charles Goodyear and Thomas Hancock.*[54] In the introduction to this work, reference is made to the hundreds of novel "modern" uses for rubber that were suggested by the two inventors long ago. An innovative rubber band for fruit trees, suggested around 1938, was found to have been completely described by Goodyear under the heading "Grafting Bandages."[55] Readers are thus warned, "It is therefore recommended that any hopeful inventor of a new use for rubber first consult Goodyear and Hancock to avoid later disillusionment and embarrassment."[56]

A Famous Name...A Dream Fulfilled

The 1850s saw the number of rubber and hard rubber manufacturers and products increase dramatically. The Goodyear name appeared on many items, first because a license requirement for rubber and likely hard rubber as well, was that goods be clearly marked with the patent dates and Goodyear name. A second reason was that many companies used the name of the inventor as a tribute or because owners were related to the Goodyears.[57] Goodyear's Metallic Shoe Company and Goodyear's India-Rubber Glove Company—both of Naugatuck, Connecticut—were among the first licensed manufacturers. (The Goodyear Tire and Rubber Company was formed in 1898, thirty-eight years after Charles Goodyear's death.) But the appearance of the Goodyear name on so many articles created the popular opinion that Charles Goodyear himself was the owner of most of the rubber companies and enormously wealthy. The fact is that he died in 1860, by some accounts in debt and by others leaving a complicated estate worth several hundred thousand dollars. This difference can be explained, as royalties were paid to the estate until the vulcanization patent expired in 1865.[58]

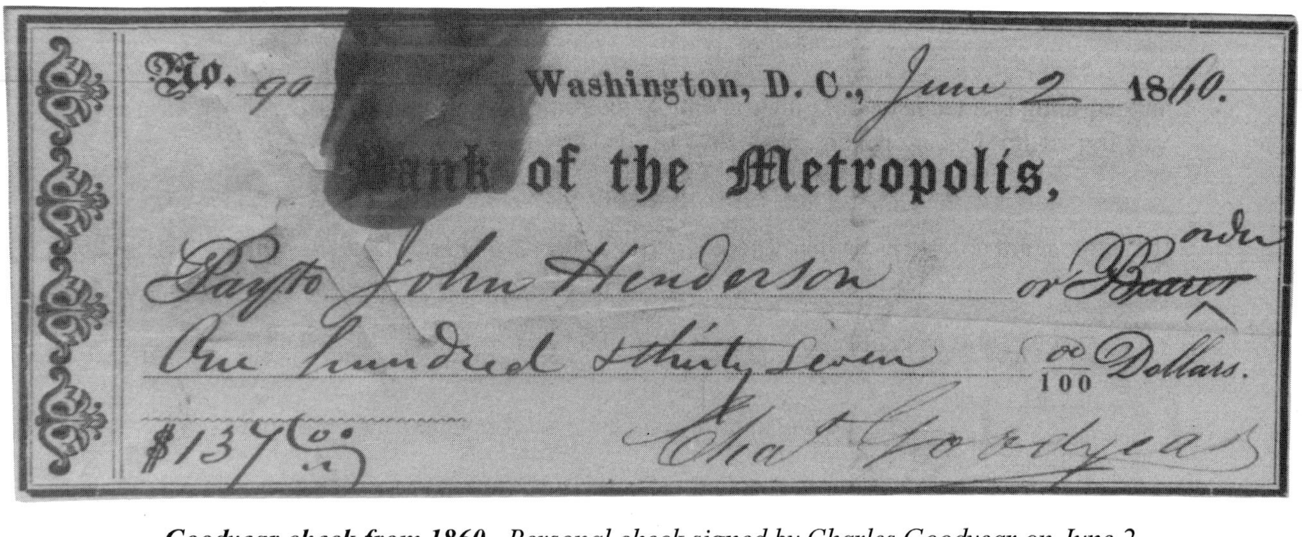

Goodyear check from 1860. *Personal check signed by Charles Goodyear on June 2, 1860, less than a month before his death.*

By any account, the inventor never personally received monetary compensation for the dedication and time he devoted to his favorite substance. It is unfortunate that he did not witness the growth of the industry during the Civil War years and the use of so many of the articles proposed by him. He would have experienced great satisfaction seeing the life of the soldier made a bit easier through the use of rubber blankets, ponchos, boots, and ambulance springs, as well as hard rubber buttons, combs, flasks, drinking cups, and medical instruments. One recognizes what Charles Goodyear's ultimate satisfaction might have been, however, from his words, "Life should not be estimated exclusively by the standard of dollars and cents. I am not disposed to complain that I have planted and others have gathered the fruits. A man has cause for regret only when he sows and no one reaps."

NOTES - Section 1

---

[1] Howard and Ralph Wolf, *Rubber* (New York: Covici-Friede, 1936) p. 23; Institution of the Rubber Industry, *History of the Rubber Industry* (Cambridge, England: W. Heffer, 1952), p. 250.
[2] Howard and Ralph Wolf, *Rubber,* p. 23.
[3] *Ibid.*
[4] *Ibid.,* p. 25.
[5] Ralph F. Wolf, *India Rubber Man* (Caldwell, Idaho: The Caxton Printers, Ltd., 1939) p. 42; Charles Goodyear, *Gum-Elastic and Its Varieties* (New Haven, CT: Goodyear, 1853 and 1855) Vol. I, pp. 16 & 45.
[6] Adolph C. Regli, *Rubber's Goodyear* (New York: Julian Messner, Inc., 1941) p. 20.
[7] Wolf, *India Rubber Man*, p. 44; Goodyear, *Gum-Elastic and Its Varieties*, Vol. I, pp. 43-45.
[8] *The Scientific American*, 22 Jan. 1860.
[9] Wolf, *India Rubber Man*, p. 131-132.
[10] Goodyear, *Gum-Elastic and Its Varieties*, Vol. I, p. 17.
[11] Wolf, *Rubber,* p. 292. The inventor's name and the garment is frequently misspelled as "Mackintosh," especially in American writings.
[12] Alexander McKee, *How We Found the Mary Rose* (New York: St. Martin's Press, 1982)
[13] Thomas Hancock, *Personal Narrative of the Origin and Process of the Caoutchouc or India-Rubber Manufacture in England* (London: Longman, Brown Green, Longman's & Roberts, 1857) p. 3.
[14] *Ibid.*, pp. 22-23.
[15] *Ibid.*, pp. 72-73.
[16] *The Scientific American*, 26 Jan. 1860.
[17] Barker, *Charles Goodyear*, pp. 19-20.
[18] Wolf, *India Rubber Man*, pp. 50-54; Goodyear, *Gum-Elastic and Its Varieties*, Vol. I, p. 54.
[19] Wolf, *India Rubber Man*, p. 57.
[20] *Ibid.* pp. 29-39 and 274-279.
[21] Wolf, *Rubber,* p. 322.
[22] James A. McLaughlin, *Charles Goodyear of Woburn, Mass* (Woburn: 1931) No page numbers.
[23] Barker, *Charles Goodyear*, p. 29.
[24] *Ibid.*, p. 32.
[25] *Ibid.*, pp. 31-32.
[26] National Archives, The Records of the Office of the Quartermaster General, hereinafter referred to as RG 92; Entry 225, Consolidated Correspondence File, Letters received by the Office of the Quartermaster General and other Offices, 1794-1890, "Rubber."

[27] Barker, *Charles Goodyear*, pp. 35-36.
[28] Goodyear, *Gum-Elastic and Its Varieties*, Vol. I, p. 164.
[29] *Ibid.*, pp. 72-74.
[30] Barker, *Charles Goodyear*, p. 38.
[31] Goodyear, *Gum-Elastic and Its Varieties*, Vol. I, p. 69.
[32] Wolf, *India Rubber Man*, p. 260.
[33] *Ibid.,* p. 274.
[34] NA, RG 92, Entry 225, CCF, "Goodyear," "Goodyear Metallic Rubber Co."
[35] Barker, *Charles Goodyear*, pp. 50-59.
[36] Wolf, *India Rubber Man*, p. 175.
[37] Regli, *Rubber's Goodyear*, p. 125.
[38] Barker, *Charles Goodyear*, p. 66.
[39] Wolf, *India Rubber Man*, p. 119.
[40] Barker, *Charles Goodyear*, p. 77.
[41] Wolf, *India Rubber Man*, pp. 239-240.
[42] *Ibid.*, pp. 248-251.
[43] NA, RG 92, Entry 225, CCF, "India Rubber."
[44] Goodyear, *Gum-Elastic and Its Varieties*, Vol. I, p. 9.
[45] Glenn D. Babcock, *History of the United States Rubber Company,* (Bloomington: Bureau of Business Research, Graduate School of Business, Indiana University, 1966) p. 18.
[46] Goodyear, *Gum-Elastic and Its Varieties*, Vol. I, p. 138.
[47] *Ibid.*, p. 143.
[48] *Ibid.*, pp. 144-145.
[49] Wolf, *India Rubber Man*, pp. 219-238.
[50] *Ibid.*, p. 221.
[51] Goodyear, *Gum-Elastic and Its Varieties*, Vol. II, pp. 35-36.
[52] *Ibid.*, Vol. I, pp. 184-185.
[53] *Ibid.*, pp. 277-278.
[54] The American Chemical Society, *A Centennial Volume of the Writings of Charles Goodyear and Thomas Hancock* (Boston: The American Chemical Society, 1939).
[55] *Ibid.*, p. 7; *Gum-Elastic and Its Varieties*, Vol. I, p. 283.
[56] *Ibid.* p. 7.
[57] Wolf, *India Rubber Man*, p. 167.
[58] Regli, *Rubber's Goodyear*, p. 205.
[59] NA, RG 92, Entry 225, CCF, "Goodyear Metallic Rubber Co."

SECTION 2

INDIA-RUBBER

"In the vulcanized form, the American specimens embrace a range which will not yield in variety to that of the mother country; for besides such articles as balls, whips, mail-bags, swimming apparatus, cushions, saddles, bags, gas bags, etc., we have no less a curiosity than a 'vulcanized india-rubber flute.'"

From *Dodd's Curiosities of Industry*,
published in England in 1851

Ambrotype of a soldier wearing an India-rubber cap
or cap cover and cape or poncho. In the 1840s and 1850s, waterproof items of clothing were becoming popular with the military, miners, travelers, and anyone with a desire to stay dry in inclement weather.

The vulcanization patent and items produced using the process eventually convinced a wary public that the problems with India-rubber had finally been overcome. Mass production techniques and man's ingenuity combined to produce a multitude of unique articles. Another look at "India Rubber at the Great Exhibition," as reported by *Dodd's Curiosities of Industry* published in England in 1851, shows the scope of novel applications.[1]

Dodd's first reviews various British contributions, including elastic webs, braces, garters, firearms, and tackle; India-rubber saddles, umbrella tents, and inflatable "cloak-boats" and paddles. From the famous firm of Macintosh were displayed life-belts, cushions, pillows, beds, waterproof garments, door-springs, noiseless wheels, maps, thread, ornamental work, and industrial components. From France, India-rubber braces, bandages, and surgical apparatus; from Holland, boots and shoes and, from Russia, India-rubber clogs.

But the most lavish praise is saved for the American contingent of articles, of which is stated:

> Every quality which india-rubber can possess seems to have been brought into requisition: impermeability, elasticity, toughness, smoothness—all are made to bear a part. We have the Hayward Rubber Company and the Goodyear Company, both contributing their boots and shoes, cloaks and capes, hats and caps,

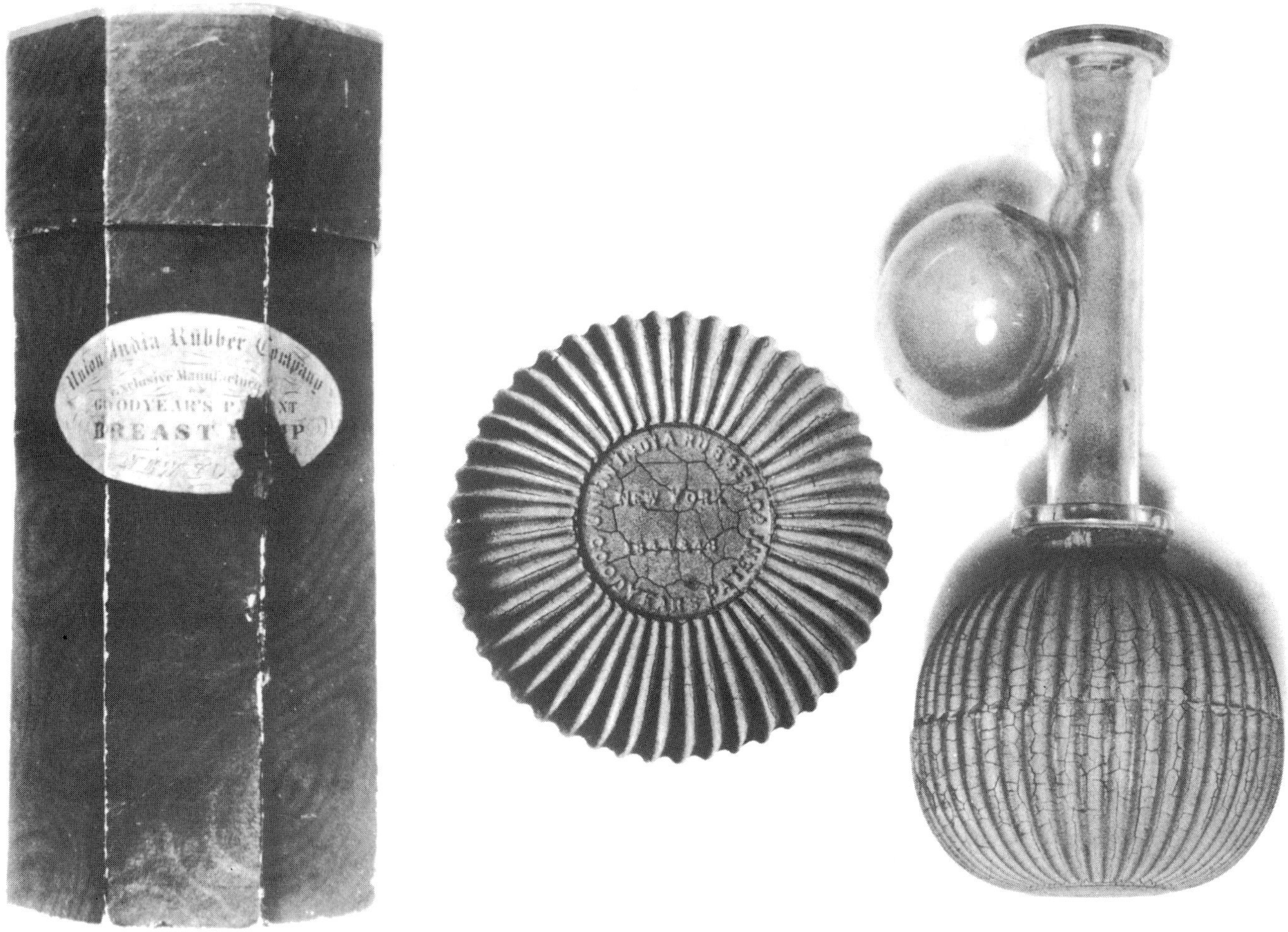

India-Rubber breast pump made by the Union India Rubber Company. *This early example of vulcanized India-rubber is combined with a blown glass bulb and reservoir. The vulcanized India-rubber suction bulb, which has hardened with age, is clearly marked: "Union India Rubber Co.-New York-Goodyear's Patent 1844 & 48." The 1844 date refers to the vulcanization patent; the 1848 date is for Charles Goodyear's process of making hollow rubber balls. The original octagonal box and cover has a label that reads: "Union India Rubber Company - New York -Exclusive Manufacturers of Goodyear's Patent Breast Pump."*

> leggings and gaiters, belts and gloves, and other water resisting garments.[2]

While the British publication never mentions Charles Goodyear by name, they compliment him indirectly by referencing the large variety of life preservers in the booth and adding:

> In the vulcanized form, the American specimens embrace a range which will not yield in variety to that of the mother country; for besides such articles as balls, whips, mail-bags, swimming apparatus, cushions, saddles, bags, gas bags, etc., we have no less a curiosity than a 'vulcanized india-rubber flute.'[3]

The flute would have been made of hard rubber, at times also called India-rubber.

The Scientific American reported in October of 1856 that the united India-rubber companies of the United States maintained an excellent display of goods in the Crystal Palace. They attempted to list all the items displayed and included:

> ...coats, pants, carriage cloths, piano and table covers (beautifully printed in various colors), blankets, saddle and gas bags, aprons, beds, pillows, boots and shoes, hose and tubeing, life preservers, bath mats, water buckets, hats and caps, bottles, drinking cups, diving or submarine dresses, breast pumps, nursing bottles, cupping cups, water bags, gloves, all kinds of toys, balls, combs, packing for steams engines, belting for machinery, pencil cases, pen holders, pulleys, insulated telegraph wire and valises in imitation of morocco leather.[4]

Early Licensees

In America, the manufacture and sale of these articles had been well underway for several years. The first plant to manufacture under Goodyear's vulcanization process was L. Candee and Company of New Haven, Connecticut, which began making rubber shoes in 1842, two years before the patent was issued. To generate interest and confidence in his products, one salesman gave away rubbers to cautious prospective customers. The following year the company was selling all they could produce—two or three dozen pairs a day.[5] Samuel J. Lewis and Company, later Goodyear's Metallic Rubber Shoe Company and the Litchfield Rubber Company, later Goodyear's India-Rubber Glove Company, were other early licensees.[6] The Newark India Rubber Company was licensed to make rubber whips, for which Goodyear noted, "The artificial gum-elastic ivory [hard rubber] is certainly a very suitable article for the mountings."[7] Other licenses were issued for mail bags, clothing, army and navy equipment, druggists goods, and elastic bands.[8] By 1851, there were about twenty establishments operating under a Goodyear license. A stipulation of each license was that the stamp of the patent be put upon all articles, as the patent law required.[9] Because his vulcanization process was still somewhat imperfect, due primarily to the inability to evenly regulate the required heat, very little royalty was paid to the inventor until he perfected the process between 1848 and 1850.

A large portion of the profitable rubber manufacture at this time was in the form of "shirred" or "corrugated" goods. This category, which today would encompass elastic portions of our clothing, etc., consisted of shirred ribbons and various kinds of "springs" used in suspenders, shoes, and corsets.[10] Early licensees of 1844 and 1845 included David Suydam, John R. Ford & Company, Onderdonk & Letson, and Hutchinson & Runyon. As part of a settlement of an infringement suit, a license was also issued to Horace Day, who flooded the market with inferior goods, creating diverse problems for the other manufacturers.[11]

Later processes for making a "…superior article of woven suspenders...sold at a much lower price"[12] diminished the demand for shirred goods. These newer articles were referred to as "elastic woven india-rubber goods."[13] Elastic bandages of India-rubber were advertised in the 1852 surgical instrument catalog of Arnold Company in England.[14]

Several early licenses for general rubber products, army and navy equipment, wearing apparel, and mail bags were transferred to the newly-formed Union India Rubber Company in 1848.[15] The principals in the firm included Jonathon Trotter, Elisha Townsend, Nicholas Dean, and William and Emory Rider. The August 22, 1848 filing for the company stated the intention to "…establish the manufacture of India-rubber and gutta-percha goods

Patent model of Horace Day's life boat. *The India-rubber has deteriorated and is falling apart, possibly due to poor materials or an inferior vulcanization process.*

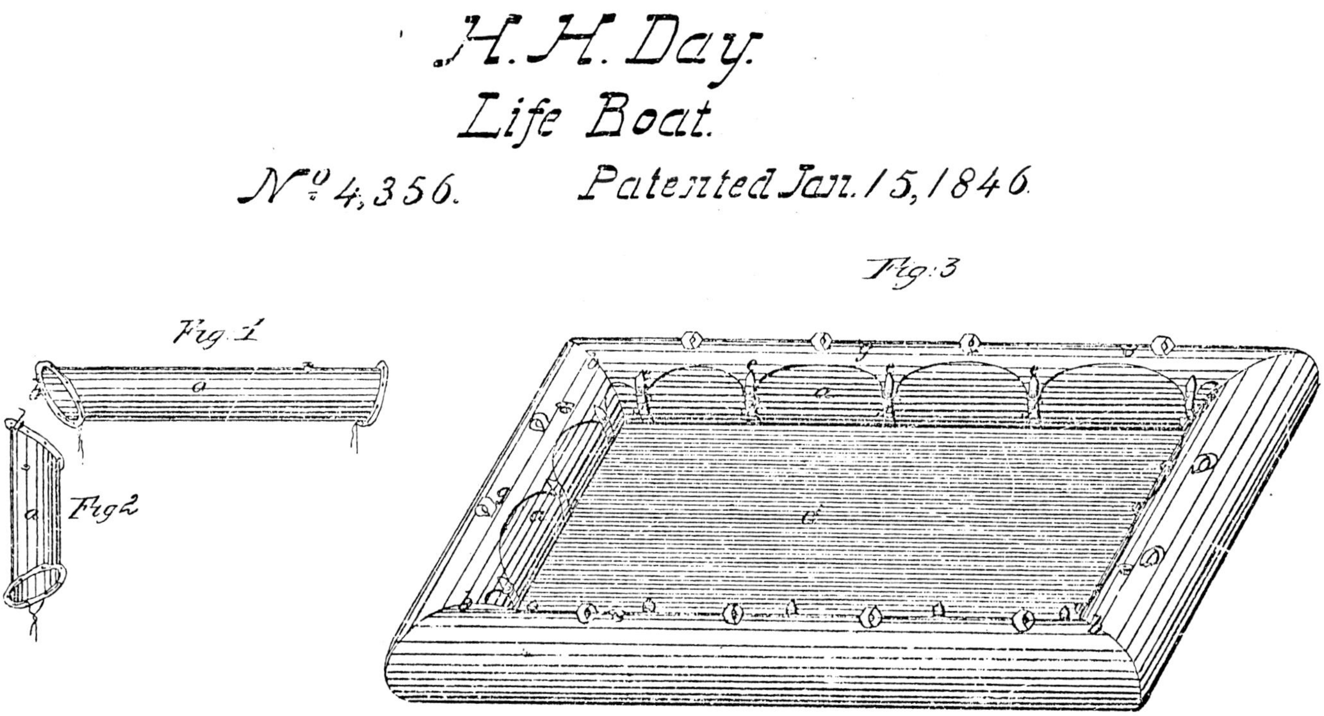

Patent drawing of Horace Day's "India-rubber Life Boat." *This life boat, described in U.S. Patent #4,356 issued January 15, 1846, featured straps that allowed the inflatable boat to be folded and carried or worn as a knapsack.*

and all kinds of chemical preparations and fibrous fabrics used in the manufacture of India-rubber and gutta-percha goods of all description known or used and to sell the same and to purchase India-rubber and gutta-percha, etc."[16] This company, with factories at Naugatuck, Connecticut and Harlem, New York and a retail store in New York City, would maintain a healthy business relationship with Charles Goodyear and later become the largest manufacturer of licensed rubber goods for the Union Army.

The Connecticut State Agricultural Society included a glowing endorsement of the company's products in an 1855 report on the state's India-rubber companies:

> The other goods manufactured at Naugatuck, are also perfect in their kind. We saw while there a rubber coat which was made by the Union Company of that place. It had been worn about two years in hot weather and cold, in wet seasons and dry. It had some of the time been taken care of, and some of the time it had been thrown under the seat of a buggy, and had experienced all the rough usage of an out-door wagon coat. In short, it had been subjected to all sorts of treatment, and yet it was perfectly whole. Not a particle of the gum was worn off, and it was just as water-tight as when it was first put on. To the farmer and to those whose occupation calls them out in all weathers, such a coat is invaluable, and as the cost is but trifling, no person, so employed, should be without one.[17]

The Gold Rush And Beyond

The Gold Rush provided an impetus to the embryonic industry. The use of rubber increased significantly as the material was further accepted by manufacturers and the public. In 1848 and 1849, dozens of New York stores sold India-rubber clothing, hats, footwear, and air-cushions, along with other supplies of varying types and materials, to Easterners headed for California with the hope of striking it rich. An advertisement in the *New York Daily Tribune* of April 11, 1850 read in part:

> Californians and Others, Attention
>
> Seely's patent life preserving hammock bed and raft, which can be converted at a moment's notice into a boat for crossing rivers. Testimonials from Californians and others can be seen at the California Depot, 11 Park-row, opposite the Astor House.

All kinds of Goodyear's Patent Metallic India Rubber Goods and California Outfits for sale.

Daniel Hodgman advertised overshoes "...manufactured from Goodyear's metallic rubber" and guaranteed that they would "...retain their brilliancy of gloss, will not stiffen in cold weather, and are extremely light and durable." And Goodyear's nemesis, Horace Day, offered his "Patent India Rubber Boat" that could be converted into a tent, at prices ranging from $45 to $75.[18] General John C. Fremont, who had utilized two of these boats in his surveys of the American West with Kit Carson, reported later that they "...went to pieces and had to be abandoned."[19]

Through the 1850s and into the 1860s, the use of rubber had expanded to a myriad of industrial and household uses in addition to many military applications that will be addressed in later sections. In the April 5, 1864 edition, the *New York Tribune's* report on the Metropolitan Sanitary Fair highlighted a display of contemporary India-rubber goods:

> In front of this stand, on the north side of the house, is a portrait of Charles Goodyear, painted on hard rubber nearly ten years ago. At the back of the stand is a splendid likeness of Daniel Webster, painted on the same material. Upward of $10,000.00 worth of rubber goods have been donated by manufacturers and dealers, comprising every article into which this material is manufactured.

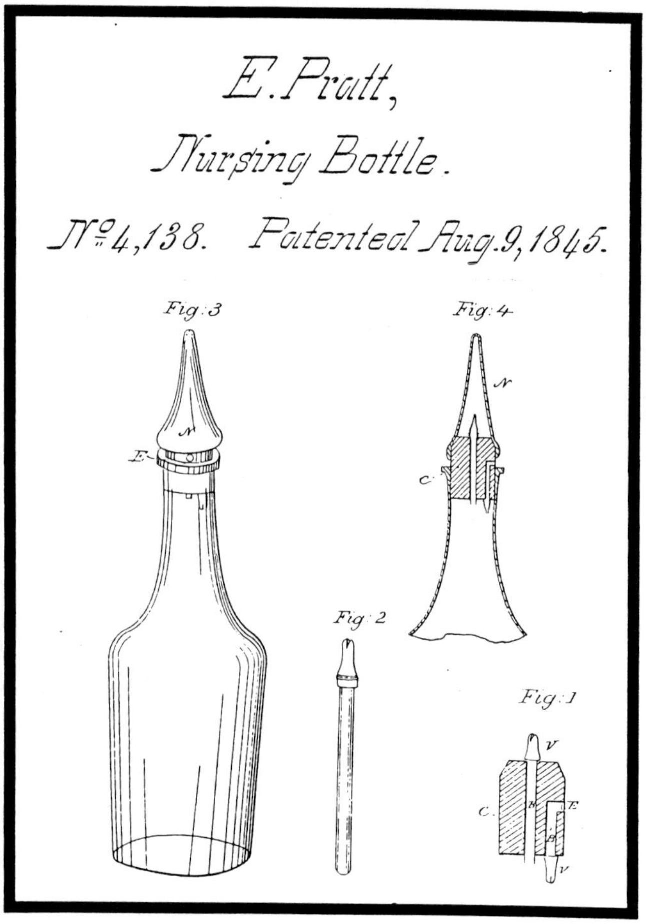

Patent drawing of a nursing bottle patented in 1845. *The use of an India-rubber valves prevented spillage when the baby was not nursing. An unvulcanized rubber nipple was considered the best substitute for nature but those made of India-rubber, vulcanized with oxides of lead or zinc, caused lead poisoning.*

The multitude of applications attracted additional manufacturers. Nathaniel Hayward, who had guided Charles Goodyear to the use of sulphur for the curing of India-rubber, stayed actively involved in the rubber industry. He operated the Hayward Rubber Company and the Haywardville Rubber Company. In 1860, the Haywardville company's output included 14,000 India-rubber boots and 50,200 pairs of shoes,[20] which may have been among those manufactured for the British market under an agreement with Charles Macintosh and Company.

One source shows that, in 1860, there were 27 rubber establishments in the United States—nine in Connecticut, five in Massachusetts, five in New Jersey, five in New York, one in Pennsylvania, and two in Rhode Island. The output for that year in rubber goods is listed as $5,642,700 and the number of direct employees was 1,795, of which 973 were women.[21] *The Scientific American* of July 1, 1865 quoted 150 rubber manufacturers in America and Europe, each employing between 400 and 500 workers, and consuming more than 10 million pounds of crude rubber per year.

Health and Safety Issues

Occupational health problems among workers in the rubber factories were receiving some press in the 1860s. *The Scientific American* of February 1, 1862 noted that tests on animals confirmed that the sulphuret of carbon vapors created by the vulcanization process were responsible for the many maladies afflicting rubber workers. Symptoms included irritability, sleeplessness, heart, circulation and respiratory problems, and loss of memory and sexual powers. In the Industrial Age that preceded OSHA by a century, the recommended correction was:

> ...that a glass screen should intervene between them and the caoutchouc under preparation, their arms being passed through aperatures properly stuffed in order to prevent the entry of vapor, much benefit would accrue, in a hygienic point of view, to the employees in this branch of trade.[22]

It is doubtful that any serious attempts were made to implement this recommendation.

At a time when the effects of lead poisoning were being confirmed, at least one specific application of India-rubber raised a stir in the medical community. A medical report issued in 1862 stated that teats, nipple-shields, and the stoppers of feeding bottles made of vulcanized India-rubber caused vomiting and disturbances of the digestive organs in infants. According to the report, nothing but crude rubber and sulphur was necessary to produce vulcanized India-rubber but, since the processed rubber was sold by weight, manufacturers added compounds which included lead, zinc, antimony and arsenic, creating "adulterated rubber." The Royal Medical College of the Rhenish provinces denounced the use of vulcanized India-rubber containing oxides of lead and zinc as extremely dangerous. At Dusseldorf and Berlin, authorities threatened anyone making or selling these articles with severe penalties. The report suggested using only "...teats made of pure caoutchouc" (i.e. unvulcanized), further suggesting ways to identify these, the simplest being that they were brownish, shiny and floated in water.[23]

Origin of the Gum

While many varieties of gum or latex were being experimented with at this time, India-rubber and gutta-percha were the most commonly used in manufacturing. In *Gum-Elastic and Its Varieties*, Charles Goodyear described the origin of several types of rubber.[24] In 1899, author Henry Pearson notes that the earth contains "...thousands of species of trees, vines, bushes, weeds, roots, and tubers which contain rubber."[25] Both authors indicate that most of the India-rubber of the mid-eighteenth century came from huge rubber-producing forests of hevea trees in the Amazon valley of South America. The area of Para in Brazil produced the best quality of India-rubber and the city of Para was the major export point, so this product became known to many as "Para rubber."[26] The specific variety of Hevea, from which the true American or "genuine India-rubber" was produced, was *Siphonia Elastica* or, as Goodyear calls it, the Seringa tree.[27] Since the first accurate account of Para rubber given by C. M. la Condamine while in the Amazon in 1835 on an astronomical mission included the common use of elastic syringes or "squirts" in social and religious activities, the source was known to the Indians as the Seringa tree, from the Portuguese *Pao di Xirringa* or the "Syringe tree."[28]

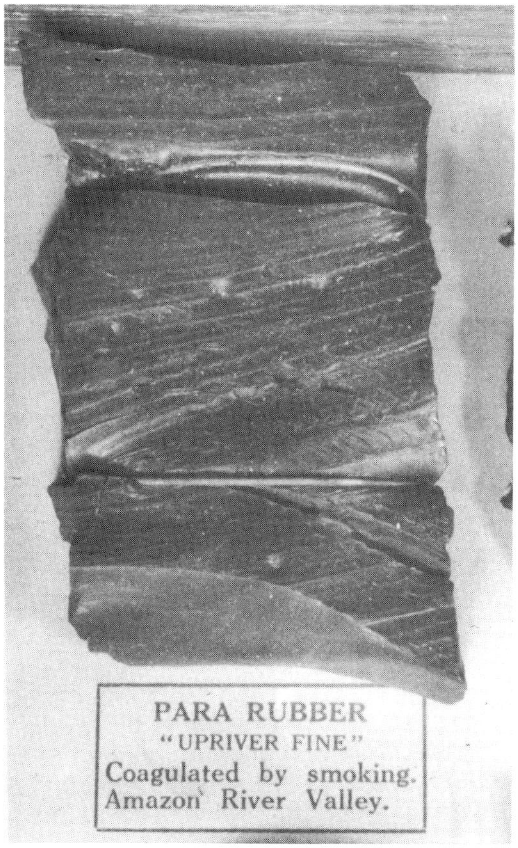

***"Para Rubber,"** produced from the latex of the Hevea tree in the Para region of Brazil, was the highest quality India-rubber available. The milky juice was "smoked" and dried in layers on clay forms.*

When the tree trunk is tapped, it yields a milky juice that would stay liquid for some time if bottled. However, when the natives collected the crude gum, they would spread a layer on clay forms and dry it over the heat of a smoky fire. Next, another layer would be applied, smoked, and so on until a large mass was formed. The clay would then be shattered, leaving a shell of India-rubber for export. Early exports of the gum were in smaller forms called "bottles."

MANUFACTORY OF THE N. Y. BELTING AND PACKING CO.

Tour of the New York Belting and Packing *Company. The series of illustrations on this page and the following are from* The Scientific American *of September 10, 1859, showing the various stages of the vulcanization process.*

THE GREAT CALENDER MACHINE.

THE BELT ROOM.

The Vulcanization Patent and Process

Charles Goodyear's vulcanization patent opened the door for an ever-increasing demand for this crude India-rubber. The process for converting the unstable, crude substance into a stable, consistent product is described in Goodyear's introductory description in United States Patent #3,633, dated June 15, 1844:

> My principal improvement consists in the combining of sulphur and white lead with the india-rubber, and in the submitting of the compound thus formed to the action of heat at a regulated temperature, by which combination and exposure to heat it will be so far altered in its qualities as not to become softened by the action of the solar ray or of artificial heat at a temperature below that to which it was submitted in its preparation—say to a heat of 270° of Fahrenheit's scale—nor will it be injuriously affected by exposure to cold. It will also resist the action of the expressed oils, and that likewise of spirit of turpentine, or of the other essential oils at common temperatures, which oils are its usual solvents.

Goodyear then described several variations of rubber fabric created by mixing the compound with fabrics, applying it to fabrics or leather with various kinds of machinery made use of by India-rubber manufacturers. An all-rubber fabric is made by spreading the compound on a glazed cloth and later removing it.

Tour of an India-rubber Factory

An overview of the process as it was used by manufacturers could best be observed by taking a tour of a leading rubber company of the period. In May of 1859, writers for *The Scientific American* visited the New York Belting and Packing Company and published a detailed report on the manufacturing process. The following excerpts highlight the steps along the way:

INDIA-RUBBER GRINDING MILL.

Machine for Grinding Rubber.

MACHINE FOR WASHING INDIA-RUBBER.

STEAM HEATERS FOR VULCANIZING.

As the goods manufactured in this establishment are principally for mechanical purposes, the proprietors are at great pains to use the rubber which has the strongest fiber, and the East India gum, the product of the beautiful "ficus elastica," which they import from Penang, Singapore, has been found to be the best. It is imported in rude masses, most of them being about two feet long and one foot thick, and covered with a rude matting, woven in wide meshes, through which the dark rubber is easily seen.

Cleansing

From the store-houses we passed into the factory, and there we were lead [sic] to see the huge water-wheel which drives the powerful machinery that is required in the various processes. The first process which the rubber undergoes is cleansing it of foreign matter, for the masses of native rubber are by no means pure...in the process the rubber loses over 20% of its weight. The rubber is first placed in a large vat, filled with hot water, where it remains for some time, until the exterior is partially softened and the workmen are enabled to strip off the basket-work which is woven around the rubber.

The "Crackers"

The masses of rubber are then cut into slabs of about an inch in thickness, by means of a large circular knife, between three and four feet in diameter. The slabs of rubber are then taken to the "crackers," as they are called. These crackers are large, deeply-grooved iron cylinders, invented for this purpose, which revolve in pairs, slowly and heavily, grinding the rubber between and driving out much of the bark and dust.

The "Washing Machine"

From the crackers the rubber is taken to the "washing machine," a large vat, where it is cut into small pieces by numerous sharp knives which revolve under the water and where it undergoes a kneading and washing process. By this process all dirt and foreign substances are perfectly expelled, and the pure rubber alone is left.

Grinding

From the washing-machine the rubber is taken to powerful grinding-machines...and here the rubber...is pressed and kneaded into thick sheets or mats. At this stage...the mats are suspended in long drying-rooms, where they are allowed to hang for many months before they are thought fit for use.

Mixing

The rubber thus cleansed and dried is taken to the mixing-machines. The mixing-machines, like most of the machines employed in the factory, are hollow iron cylinders, and it is necessary that they should be kept at high but regulated degrees of heat, as the tough masses of rubber would otherwise resist the action of the machinery, however powerful. Two are placed near together, which, as they revolve towards each other, knead the substances placed between them like dough.

Compounding

When the rubber is somewhat softened, the workman mixes slowly the various

***Balata**, a variety of gum first noted in 1857, was frequently mixed with India-rubber or gutta-percha for industrial applications. One of several rubber products made from South American tree sap. Specimen displayed in circa 1900 museum case.*

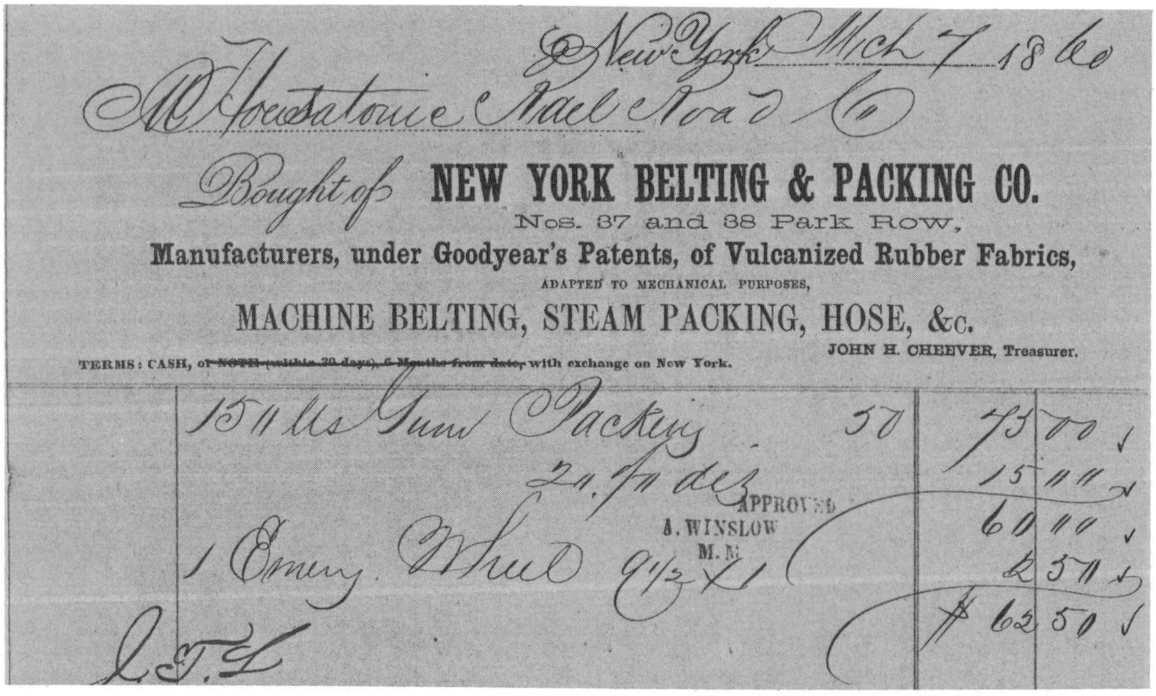

New York Mch 7 1860

The Housatonic Rail Road Co

Bought of NEW YORK BELTING & PACKING CO.
Nos. 37 and 38 Park Row,
Manufacturers, under Goodyear's Patents, of Vulcanized Rubber Fabrics,
ADAPTED TO MECHANICAL PURPOSES,
MACHINE BELTING, STEAM PACKING, HOSE, &c.
JOHN H. CHEEVER, Treasurer.
TERMS: CASH, ~~or NOTE (within 30 days), 6 Months from date,~~ with exchange on New York.

150 lbs Gum Packing	50	75 00
20 pr dis		15 00
		60 00
1 Emery Wheel 9 1/2 x 1		2 50
		$ 62 50

APPROVED
J. WINSLOW
M. M.

New York Belting & Packing Company receipt for "Gum Packing." *Writers for* The Scientific American *marveled at a one-piece rubber belt, 300 feet long and 3 feet wide, that had been manufactured by this company in 1859.*

substances which are to be incorporated with it; these consist principally of sulphur and of the oxyds [sic-contemporary]of various metals, zinc, lead, iron, etc., which are combined in various proportions, according to the uses for which the rubber is destined. It is in this department that the greatest science and experience are required, for different qualities of rubber require different compounds, and every difference in the compound makes a different treatment necessary in the subsequent stages of the manufacture. When the rubber is thus prepared it is ready to be molded and shaped into the various forms in which it is to be finally perfected and used.

The "Calendering Machines"

The rubber...is then taken to another department to the 'calendering-machines.' These somewhat represent the other machines, but they are composed of more cylinders, and are of much larger size, and of a perfectly polished surface. Upon these calendars the prepared rubber is placed, and after passing between the cylinders is rolled out in a perfect and even sheet, upon a web of powerful cotton duck, which has previously been coated with rubber, driven through and through its meshes by powerful machinery.

Cutting

The 'bolts' of duck covered with rubber, after this process is completed, are taken to the belt-room; here the long webs are taken by the skillful workmen and unrolled upon tables 100 feet long, and in an incredibly short time, are cut into strips and folded together into machine-belting. In order to give the required strength to the belt, folds upon folds of the heavy duck are placed one upon the other, and then forced together by the tremendous power of the rolling-machines, until a belt is formed, more tough and solid than the best sole-leather.

The Magic of Vulcanization

From this room the belts are taken to the heaters; these are immense steam-boilers, with a long iron frame or railway, which can be thrust in or drawn out at pleasure; the goods are placed on the railway and...into the boilers, which are then closed and steam is admitted. This part of the process is the most remarkable of all; for the rubber, which, when placed in the heaters, is like a tough, unelastic dough spread upon the various fabrics for which it is used, becomes wholly changed into the new and peculiar substance called metallic or vulcanized rubber. Heat and cold, which destroy the value of native rubber, have no effect upon it; the solvents in which the native rubber dissolved like gum have no influence upon it whatever; in fact, it becomes, and it has been well called, an 'elastic metal.'[29]

The article goes on to say that the writers examined:

...the most striking specimen...a belt of seven

DAVID CULVER, (Inventor and Patentee) 32 Cliff street, manufacturer and dealer, wholesale and retail, in Furnaces, Registers, Ventilators, &c.

INDIA RUBBER GOODS.

THE NEWARK INDIA RUBBER CO., 59 Maiden Lane, N. Y., Manufacturers and Wholesale Dealers in Goodyear's Patent India Rubber Boots, Shoes, Whips, Clothing, Balls, Toys, &c., &c. fe6-1yp

HENRY DAVENPORT, 86 John street, Manufacturer and Dealer in every description of Goodyear's Patent Rubber Fabrics, Fancy Rubber Goods and Toys. sep13-1y

UNION INDIA RUBBER CO., 41 John st., make all kinds of Rubber Clothing, Cloths, Druggists' Articles, &c.; Coats from $2,60 to $6,00 each. fe52-1yp

LADIES' ELASTIC SKIRTS, (HOUGH'S PATENT.)

LADIES' ELASTIC SKIRTS, Hough's Patent—P. BATES, 31 Barclay street, opposite the Astor House, New York, manufacturer of the above splen-

India Rubber Goods. *Advertisements for Goodyear's patent India-rubber goods from the Pittsburgh Gazette of October 19, 1854.*

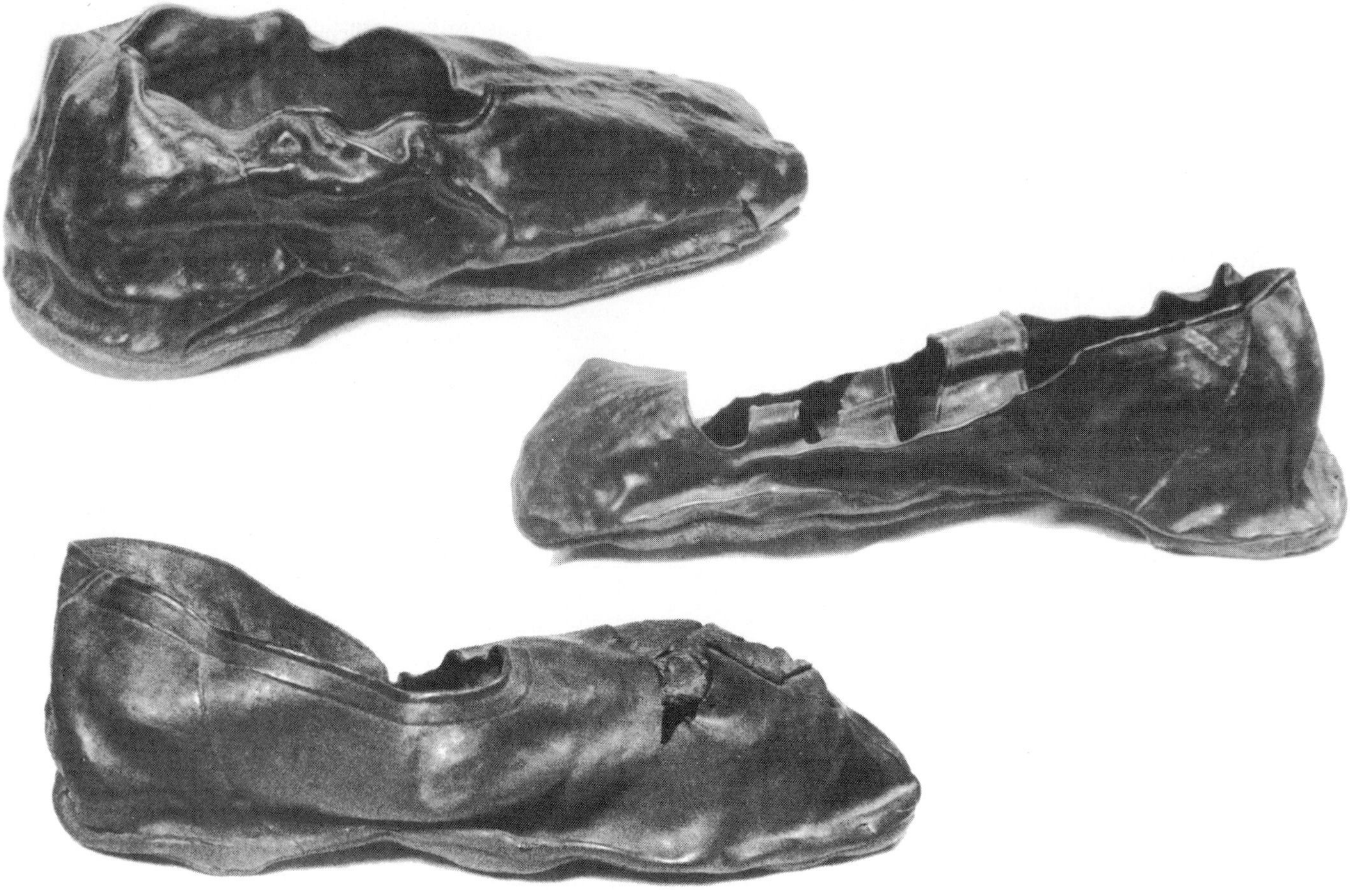

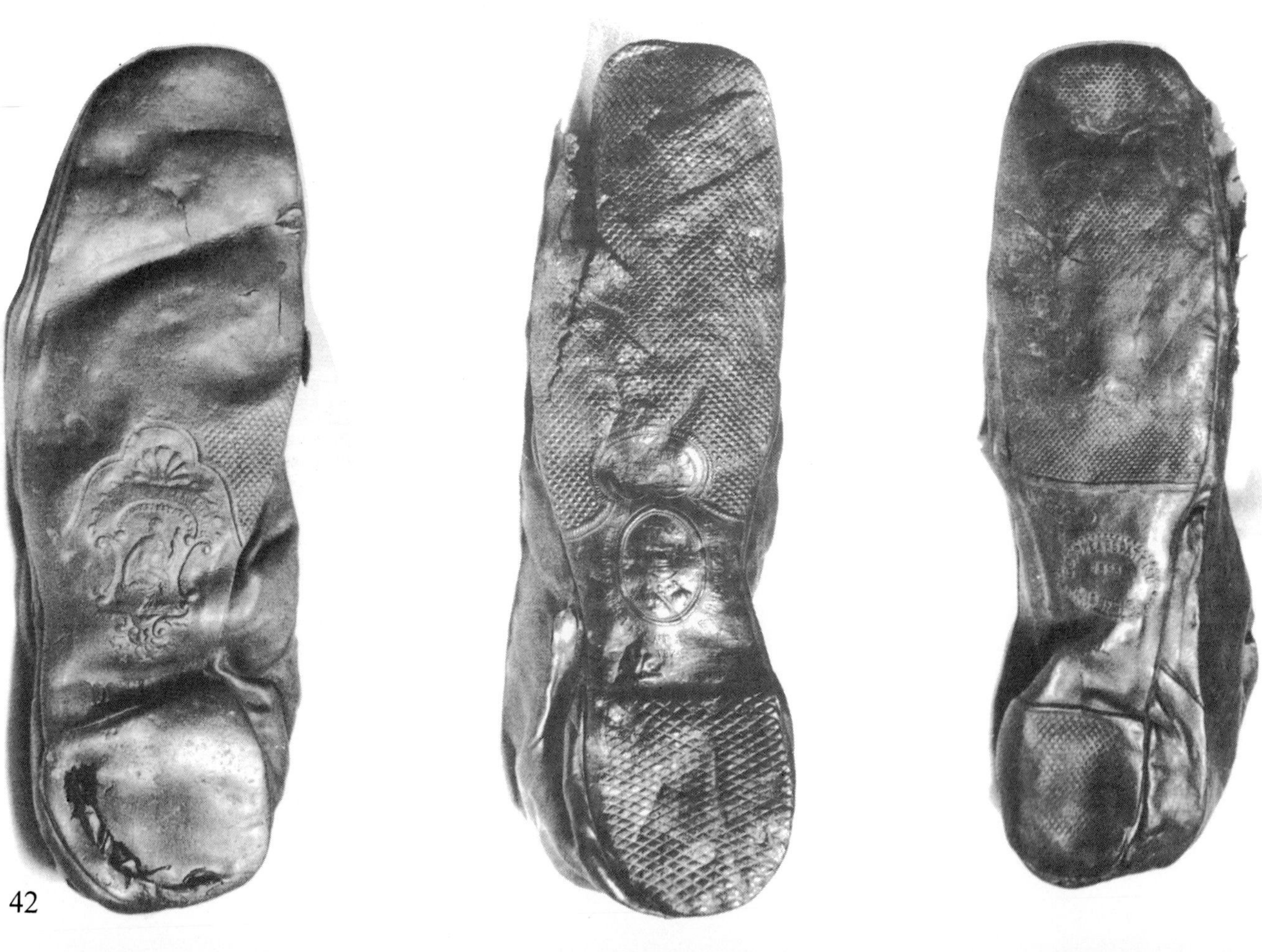

plies lately made, over a yard wide, and nearly 300 feet long. Such a belt, if made in the old-fashioned, clumsy way, from leather, would have required the hides of 120 oxen, and would have been fastened together by copper rivets; but by the triumph of mechanical art, the great rubber belt was made in one operation, without a joint or seam or imperfection.

This reference to the "...old-fashioned, clumsy way, from leather" was surely not appreciated by the large number of leather workers, whose products were also being replaced by rubber and gutta-percha in shoes, boots, soles, and heels.[30]

Terminology

Various types of India-rubber, rubber fabrics, and waterproof materials were produced and terms such as "plated," "coated," and "painted" were oftentimes used to describe rubberized garments or other articles. This created some difficulty for the buying public, government contractors, and manufacturers of true vulcanized India-rubber or gutta-percha articles, as inferior goods were sometimes associated with quality merchandise.

The terms painted, coated, or saturated generally referred to fabrics or materials manufactured by a process of merely applying liquid rubber compounds or other waterproofing substances and varnishes with a roller or brush as opposed to vulcanizing the rubber in the fabric or material using Goodyear's process.

PREVIOUS PAGE: ***Early India-rubber overshoes excavated in the early 1970s.*** *While some quality problems existed with crude India-rubber and in the vulcanization and manufacturing processes, surviving artifacts like these overshoes or "rubbers," excavated near the San Francisco coastal area, attest to the durability of the product. The rubbers, which show signs of wear and possible fire damage, were likely layered in sand with other goods or materials. Names of manufacturers and Goodyear patent markings found on the soles of several excavated rubbers include: (bottom left to right) "New Brunswick Rubber Co. - Goodyear's Patent;" "Ford Rubber Co., New Brunswick, N.J. - Goodyear's Patent - 1849;" and "Goodyear's Patent - 1858 - N. I. R. Mfg. Co." (Newark India-rubber Mfg. Co.)*

The military employed both vulcanized and painted equipage during the Civil War. A letter to Quartermaster General Meigs dated May 2, 1863 stated in reply to a request for a comparison of the relative durability of India-rubber versus painted blankets, that "...no painted blankets, as far as I know, have been used in this Department—those that were sent to this depot were sent to the Department of the Cumberland for issue."[31] Lt. John Matthews referenced both varieties in a letter of September 25, 1863, where he questioned whether his men were chargeable for the "gum or painted blankets" he had drawn the previous day.[32] Horace Day had furnished some painted poncho tents to the Union army and, as evidenced by testimony furnished in an 1865 court proceeding where Day was seeking payment, the Army's Quartermaster General was certain of the difference in quality between vulcanized articles and Day's painted goods. The records noted that "...some of the Day's ponchos...after one day's use in the sleet were...absolutely worthless, the varnish being cracked and peeled in the short time it had been used, so as to permit the water to pass through freely."[33]

Various rubber processes are discussed by Goodyear in his book, *Gum-Elastic and Its Varieties.* In describing solarization or "natural vulcanization," he speaks of the "...invention of 'plated fabrics,' by which an extremely thin coat of caoutchouc and its compounds is made to answer a better purpose in all respects than the large quantity formerly used, whereby, on account of the thinness of the 'plating' of the fabrics, solarization may be substituted in these fabrics instead of vulcanization."[34] The Union India Rubber Company is cited for the garments they produced using this method.[35]

Goodyear goes on to apply the same term, plating, to vulcanized India-rubber rubber goods and distinguishes them from painted or coated fabrics:

> The term 'plating' has been adopted for this improvement, from its resemblance to the art of plating metals, and also to distinguish it from the common method of coating India rubber cloths. The invention consists in interposing a bat or fleece of cotton wool, or other fibrous substance, between the coating of gum and the fabric. The gum is thus intermixed with the bat or fleece...the gum is by this process prevented from peeling off and abrasion...it is by this process of plating that the manufacture of porous fabrics has been made practical.[36]

The inventor's enumeration of other vulcanized India-rubber materials included embossed fabrics, ventilated goods, quilted fabrics, and "...India rubber cloths napped with woolen flocks...for wearing apparel, they are more comfortable and pleasant, and for blankets and imitation buffalo robes, they are warmer than India rubber goods that are not napped."[37]

In a printed booklet entitled *Scientific Reports of the Comparative Value of Vulcanized and Unvulcanized India-Rubber*, published in 1865, several professors and chemists reported on tests conducted on four samples of India-rubber cloth. The consistent message throughout the reports is that the quality and durability of the vulcanized samples far surpasses that of the unvulcanized cloth, which was treated with a solution of sulphur and bisulphide of carbon. The following excerpts support the findings:

> All other processes [except vulcanization] for effecting chemical change in caoutchouc, for manufacturing commercial articles, consist in the use of various mixtures in liquid form, which are either spread on the rubber with a brush, or its equivalent; or, into which the goods are immersed. When cotton or linen cloth constitutes a portion of the goods, the chlorine liquor is brushed only on the rubber side...to effectually prevent the hydrochloric acid gas from 'rotting' and destroying the fabric. The artisans...are necessitated to leave their rubber uncured on the side next the muslin and, consequently, inferior to articles which have been thoroughly cured, or, in other words, properly vulcanized by heat.

> On an hasty comparison of the samples, no very noteable difference appears, but a more careful scrutiny will demonstrate a lack of uniformity in [the unvulcanized] samples...and will pass through the degeneration to which it is well known that unvulcanized rubber is subject.

> If there were no other objection to this class of goods under consideration than that they stiffen and become hard when cold, and eventually crack, with continual warm temperatures they soften and become sticky, while moisture only hastens their decay, I think a true economy would prevent either their manufacture or use.[38]

Quality Problems with Vulcanized Goods

While the quality of India-rubber had improved overall, problems due to inconsistent crude rubber plagued even the manufacturers of vulcanized products. Fine Para rubber was not always obtainable, especially in Britain, and the use of substitutes of inferior quality hurt rubber's reputation. As late as 1888, in a patent litigation case, expert witness, consulting engineer, and patent agent William Gadd stated, "India-rubber is a very treacherous substance indeed. You cannot rely upon two pieces being the same."[39]

Defects in products were also blamed directly on manufacturers and in some cases created a lasting bias against India-rubber and new inventions. In an 1856 statement regarding a fractured casing for rubber railway springs manufactured by the premier British rubber company, an agent says:

> This is due to Macintosh, one of the cones was as good as when first put in, a clear proof that if they had been of equal quality they would have all stood. They have been working up all the rubbish in the place in many of them. It makes my work much harder than a first application as there is much antipathy to overcome.[40]

Three years later, the same agent commented:

> These too frequently occurring break-downs...take ten times the trouble to patch up than getting the first introduction...in fact, getting one rung up the ladder and being pulled down ten.[41]

In 1857, an engineer for the London and North Western Railway reported that one spring "...has no more elasticity than a block of wood."[42] An 1860 letter to the editor of *The Scientific American* from a reader in Mobile, Alabama stated, "India-rubber goods frequently stick together in this warm climate. Even fine silk coats delicately covered with India-rubber will, when hanging up in the wardrobe, stick together whenever parts come in contact."[43]

Despite these quality problems in England and in the United States, the number of manufacturers and the demand for rubber goods continued to grow. No longer did salesmen have to give away products to prove their worth. The industry was booming. India-rubber was here to stay!

NOTES - Section 2

---

1 *Dodd's Curiosities of Industry* (Published in England circa 1851).
2 *Ibid.*, p. 10.
3 *Ibid.*
4 *The Scientific American*, 25 Oct. 1856.
5 Advertisement for Candee Rubber Company; circa 1900. P. W. Barker, in *Charles Goodyear*, p. 51, gives the first year of Candee's production as 1843.
6 P. W. Barker, *Charles Goodyear* (Boston: Godfrey L. Cabot, Inc., 1940) p. 51.
7 Charles Goodyear, *Gum-Elastic and Its Varieties* (New Haven: Goodyear, 1853 & 1855) Vol. II, p. 221.
8 Barker, *Charles Goodyear*, pp. 51-53; Fisher, *Rubber and Its Use*, 1941.
9 Goodyear, *Gum-Elastic and Its Varieties*, Vol. I, p. 226.
10 *Ibid.*, p. 195.
11 Ralph F. Wolf, *India Rubber Man* (Caldwell, Idaho: The Caxton Printers, Ltd., 1939) pp. 154-155.
12 *Eighth Census of the United States*, 1860, p. lxxviii.
13 *The Scientific American*, 19 Feb. 1859.
14 Institution of the Rubber Industry, *History of the Rubber Industry* (Cambridge, England: W. Heffer, 1952) p. 266.
15 National Archives, Record Group 92, Entry 225, Consolidated Correspondence File, "Union India Rubber Co."
16 *Ibid.*, "India Rubber." Handwritten filing.
17 Executive Committee, *Transactions of the Connecticut State Agricultural Society for the Year 1855* (Hartford: Case, Tiffany and Company, 1856) p. 212.
18 Barker, *Charles Goodyear*, pp. 55-56.
19 Adolph C. Regli, *Rubber's Goodyear* (New York: Julian Messner, Inc., 1941) p. 134.
20 Zellie, Carole, & Martha Coons, *Stoneham, Massachusetts: A Shoe Town* (Stoneham: Stoneham Historical Commission, 1981) p. 57.
21 Barker, *Charles Goodyear*, p. 96.
22 *The Scientific American*, 1 Feb. 1862.
23 *Journal of Practical Medicine and Surgery*, No. 522, Article 6240 (2 July 1862).
24 Goodyear, *Gum-Elastic and Its Varieties*, Vol. I, pp. 15-27.
25 Henry Pearson, *Crude Rubber and Compounding Ingredients* (New York: The India Rubber Publishing Company, 1920) p. 7.
26 *Ibid.*, p. 9; Goodyear, Gum-Elastic and Its Varieties, Vol. I, p. 16.
27 *Ibid.*, Vol. I, p. 18.
28 R. H. Lock, Sc. D., *Rubber and Rubber Planting* (New York: G. P. Putnam's Sons, 1913) pp. 3-4.
29 *The Scientific American*, 10 Sep. 1859.
30 *Dodd's Curiosities of Industry*, p. 21; *Eighth Census of the United States*, p. lxxx.
31 NA, RG 92, Entry 225, CCF, "Blankets" Letter to Meigs. Illegible signature.
32 *Ibid.* Letter to Meigs from Lt. Matthews.
33 *Ibid.*, "Horace Day." No date noted.
34 Goodyear, *Gum-Elastic and Its Varieties*, Vol. I, p. 73.
35 *Ibid.*
36 *Ibid.*, p. 168.
37 *Ibid.,* pp. 197-200.
38 NA, RG 92, Entry 225, CCF, "India Rubber." No author or publisher shown.
39 P. L. Payne, *Rubber and Railways in the Nineteenth Century—A Study of the Spencer Papers* (Liverpool: Liverpool University Press, 1961) p. 11.
40 *Ibid.*, p. 27.
41 *Ibid.*
42 *Ibid.*
43 *The Scientific American*, 8 Sep. 1860.

Goodyear's Patent Hard Rubber Goods,

MANUFACTURED BY THE

AMERICAN HARD RUBBER COMPANY,

No. 63 Maiden Lane, New-York,

POWDER FLASKS.

	2 oz.	8 oz.	12 oz.	16 oz.
No. 50, Common Top, Unpolished, Plain, per dozen, ...	$4 00	$7 00	$9 00	$12 00
" 55, " " " Trimmed, "		8 00	10 00	13 00
" 60, Patent, " " Plain, "		9 50	11 50	14 50
" 65, " " " Trimmed, "		10 50	12 50	15 50
" 120, Common " Polished, Plain, "	5 25	9 00	12 00	15 00
" 130, " " " Trimmed, "		10 00	13 00	16 00
" 220, Screw, " " Plain, "	6 00	10 00	13 00	16 00
" 230, " " " Trimmed, "		11 00	14 00	17 00
" 320, Patent, " " Plain, "	10 00	13 00	16 00	19 00
" 330, " " " Trimmed, "		14 00	17 00	20 00

PISTOL, SWORD, DENTAL and SURGICAL INSTRUMENT HANDLES,

New-York, July 1st, 1859.

Dear Sir:

Having purchased of Mr. Goodyear, in addition to the privileges previously possessed by us, the EXCLUSIVE RIGHT *to apply the* HARD COMPOUNDS OF CAOUTCHOUC *to all new purposes, we are prepared to treat, on liberal terms, with parties who may wish to make* NOVEL *and* VALUABLE *applications of the same; and we would call the attention of manufacturers particularly to the peculiar properties of this material, feeling confident that, for many purposes, they will find it much superior to any other they may be using.*

Respectfully,

AMERICAN HARD RUBBER COMPANY.

PRICE LIST AND SALES LITERATURE OF AMERICAN HARD RUBBER COMPANY

– 196 –

Composite advertisement and sales literature. *Hard rubber powder flasks and other items made in the 1850s by the American Hard Rubber Company. (From* The Powder Flask Book *by Ray Riling; New York: Bonanza Books, 1953. Courtesy of Joe Riling)*

SECTION 3

HARD RUBBER

"The discovery of the process of making hard rubber is one of the greatest discoveries of the age, and it will not be many years before this material will supersede to a great extent, ebony and other woods, ivory and horn."

Quoted from *Transactions of the Connecticut State Agricultural Society for the Year 1855*

The formation of hard rubber was likely observed long before the formal discovery and patent applications of the 1840s. An article in a 1791 publication called *The Bee* stated:

> It is many years since Dr. Bergius, at Stockholm, made some experiments on this substance [India-rubber] in Papin's digester [a vessel for softening or breaking down by heat and moisture or chemicals or for extracting soluble ingredients]. By subjecting it in this way to an intense degree of heat it is said to have been converted into a hard elastic horn-like substance."[1]

While Nelson Goodyear obtained the United States patent (U.S. Patent #8,075 dated May 6, 1851) for hard rubber, it is certain that others, including his brother, Charles, and the British rubber pioneer Thomas Hancock, noticed the formation of the substance when the amounts of additives were varied and heating temperatures and times of the vulcanization process increased. Charles Goodyear directed the manufacture of many unique hard rubber items in the late 1840s and the early 1850s, oftentimes presenting the finished articles to friends as gifts. It is generally accepted, however, that Hancock is the true inventor of the hard rubber process. His vulcanization patent noted that rubber immersed in sulphur and heated for longer than two

Hard rubber dice cup with felt lining. *Marked: "xxx Goodyear's Patent May 6, 1851xxx." All items bearing the month as well as the year of the patent are believed to date no later than 1872.*

hours produces a material that "turns nearly black, and has something of the appearance of horn, and may be pared with a knife similarly to that substance."[2] In a patent infringement suit of the late 1850s, a certain Oscar Falke was mentioned by the New York Gutta Percha and India Rubber Vulcanite Company as having brought samples of Hancock's hard rubber to Nelson Goodyear prior to Nelson's patent.[3] In 1861, Leonard D. Gale, the Principal Examiner of the U.S. Patent Office, testified that he would not have granted the hard rubber patent to Nelson Goodyear if he had been aware of Hancock's existing British patent for the same process.[4]

After Nelson Goodyear's death in 1852, his brother, Henry Bateman Goodyear, controlled the hard rubber patent. In the same year, negotiations between Charles Goodyear and Thomas Hancock over patent and manufacturing rights in England resulted in Goodyear obtaining the exclusive right to manufacture hard rubber in that country. This agreement may have been dissolved when Goodyear lost a suit on behalf of his vulcanization patent.[5]

Several other inventors patented processes for making hard rubber, hard gutta-percha and combinations of the two. Austin Goodyear Day (no relation to Horace Day), a nephew and employee/assistant of Charles Goodyear since 1841,[6] received a patent for his own hard rubber process in 1858 (U.S. Patent #21,222). The 1854 patents of L. O. P. Meyer (U. S. Patents #10,586 and 10,741), as well as the Goodyear hard rubber patent reissues of May 26, 1858, specify "India-rubber or other vulcanizable gums."

Terminology

Within the realm of hard rubber, and aside from the somewhat erroneous name of "gutta-percha" frequently applied to the material, a number of words have been used to denote various compounds of hard rubber. Early terms for the novel substance were vegetable ivory, artificial horn, or simply India-rubber. In describing manufacturing processes and applications in his book, Hancock called the material "hard vulcanized rubber."[7] Charles Goodyear, classifying varieties of hard compounds by their relative hardness and flexibility, called it caoutchouc enamel, caoutchouc ivory, caoutchouc buck-horn, and caoutchouc whalebone.[8] He further noted

Set of 12 table knives with hard rubber handles in original box. *Each knife is marked on the blade: "India Rubber Handle Cutlery Co. C. Poppenhusen's Patent March 31, 1857." Box label has similar markings and various other patent dates, cleaning instructions, and a likeness of the factory in New York.*

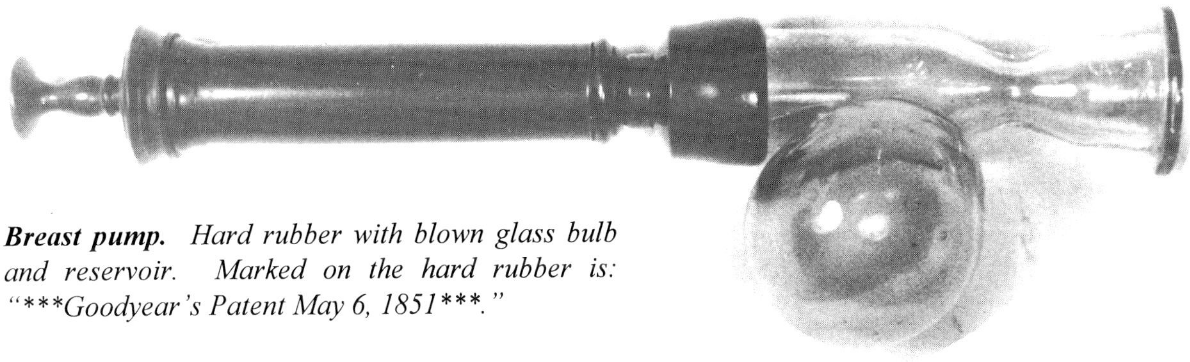

Breast pump. *Hard rubber with blown glass bulb and reservoir. Marked on the hard rubber is: "***Goodyear's Patent May 6, 1851***."*

"caoutchouc deal boards...formed by alternate layers or sheets of the compound and cloth or canvas" to create a "plastic wood or flexible board" and "enameled ware," made by plating iron, glass, ceramic, etc. with hard rubber.[9] In another instance, he referred to the material as "artificial gum-elastic ivory."[10]

Vulcanite and ebonite are the most commonly used alternative names for hard rubber. Credible rubber historians Howard and Ralph Wolf said "hard rubber...alias vulcanite or ebonite"[11] as if the words were all synonymous. And the famous display at the Crystal Palace was called "Goodyear's Vulcanite Court." Yet, *A Practical Treatise on Caoutchouc and Gutta Percha,* published in 1883,[12] and some modern-day collectors refer to vulcanite as a unique and different material. Webster's dictionary has no word vulcanite, but defines ebonite as "hard rubber," with the date of origin noted as 1861. Goodyear, Webster's, and noted authors combining to define hard rubber, vulcanite, or ebonite as the same substance in the 1850s and 1860s indicates that they were indeed synonymous in that era.

An Endless Variety Of Articles

At first, few realized the possibilities for this substance, as contemporary mind-sets and factories were accustomed to working with metals, wood, ivory, horn, or whalebone and produced articles by carving or cutting processes rather than molding. While Charles Goodyear recognized the potential of hard rubber, even he may not have fully understood the ultimate impact of the revolution in Industrial Age technology and manufacturing that was about to take place. In his book, he mentions in a trivial and matter-of-fact way, the dimension of hard rubber that would set the stage for the birth of the plastics industry:

> There is also an important economy in the manufacture of these substances for these purposes, because they are moulded into the different forms, either plain or ornamental, with trifling labor without waste of material.[13]

Thus, the mere possibility for endless uses and mass production earned hard rubber a prominent place in Goodyear's ongoing dream of a world of India-rubber. The previously-listed articles of hard rubber displayed at the Crystal Palace represent a large number of the early applications. And the variety of articles continued to increase as others thought of novel applications. In the late 1850s, John Jacob Bausch of Bausch and Lomb fame found a piece of hard rubber on the street and discovered he could form eyeglass frames from the material. By the start of the Civil War, the sale of hard rubber frames had skyrocketed and within five years the firm's name became The Vulcanite Optical Instrument Company.[14]

An 1855 report on hard rubber manufacturers in Connecticut stated:

> As an insulator for telegraph wire, it is unsurpassed. Gutta percha cannot compare with it. There is nothing like it for caster-rollers, as they roll with scarce any friction, and do not cut carpets or matting. Handles for table-knives can not be cracked by any amount of carelessness on the part of ignorant servants, and hard rubber sheaves in ship's blocks, will by want of friction and early turning, draw out as many blessings from sailors as the old lignumvitae sheaves have curses. The discovery of the process of making hard rubber is one of the greatest discoveries of the age, and it will not be many years before this material will supersede to a great extent, ebony and other woods, ivory and horn.[15]

On their tour of the New York Belting and Packing Company in 1859, the writers for *The Scientific American* observed, "Here, too, we saw sinks cast without a joint or seam, from hard india-rubber; and every fair lady whose china has been broken in the hard cast-iron sinks now in use, will appreciate the saving in china and in temper which this new invention will make."[16]

Color

When examining surviving hard rubber artifacts, the issue of color becomes an interesting one. Most items of the period are black, brown or something close to a combination of the two. A large number of artifacts from the mid-1800s appear to have originally been black and have turned brown on their exterior surfaces, possibly through oxidation. While early references to hard rubber articles indicate the color was black, some surviving items are so evenly brown on all surfaces that it appears they were manufactured that shade. Continued research on this issue has resulted in no positive conclusions. Speculation that items which have turned brown are actually hard vulcanized gutta-percha is not borne out on basic examination, which reveals the sulphurous, "rubbery" odor common to hard rubber.[38]

Goodyear said hard India-rubber could be "mixed with any color that is desired in the manufacture, and the color remains unfading."[39] While Hancock described flutes he had seen "...the colour ...jet black and it polishes like ebony," he also detailed the same techniques of coloring as Goodyear and commented, "Gutta percha and compounds of this substance with rubber are equally susceptible of improvements in the same way."[40] An 1883 publication noted that, "Hard rubber, prepared with caoutchouc and sulphur only, has a black color, and takes a very high degree of polish. Articles manufactured from this material, although black, can also be colored any desired shade."[41] It went on to say, however, that coloring additives can injure the properties of the hard rubber by their proportions or interactions with the sulphur and suggests "plating or enammelling" if color is desired.[42] So, over thirty years, several sources discussed ways of coloring, but few colored items from the period are seen today. Noted exceptions are reddish-orange and marbleized hard rubber drinking flasks bearing the "Goodyear-1865" patent extension date.

Setting a Standard

The large number of remaining examples of hard rubber items are a true testimony to the popularity and durability of the remarkable substance. Many hard rubber artifacts have survived exposure to acids, inks, and medications during their use in the mid-1800s and over a century of burial underground and underwater. Until the turn of the century, many

Glass one pint whiskey flask with red hard rubber covering. *The two-piece, smooth body has a removable cup and screw-on cap. The top of the cap is marked: "Ind. Rubber Comb Co. Goodyear's Pat. 1865." The 1865 date refers to the extension of the Goodyear hard rubber patent, so the flask was likely manufactured between 1865 and 1872, when the patent expired.*

more hard rubber products would grace catalogs, stores, and homes. And, while not considered the first true plastic due to its natural origin and inability to be reheated and remolded, hard rubber set the standard for the flexible, moldable, durable plastics that were to become a part of everyone's life.

NOTES - Section 3

[1] Howard and Ralph Wolf, *Rubber* (New York: Covici-Friede, 1936) p. 316.
[2] P. W. Barker, *Charles Goodyear* (Boston, Mass: Godfrey L. Cabot, Inc., 1940) pp. 74-75.
[3] Ralph F. Wolf, *India Rubber Man* (Caldwell, Idaho: The Caxton Printers, Ltd., 1939) p. 276.
[4] Barker, *Charles Goodyear*, p. 76.
[5] *Ibid.*, p. 75.
[6] *Ibid.*, p. 76.
[7] Thomas Hancock, *Personal Narrative of the Origin and Process of the Caoutchouc or India-Rubber Manufacture in England* (London: Longman, Brown Green, Longman's & Roberts, 1857) p. 120.
[8] Charles Goodyear, *Gum-Elastic and Its Varieties* (New Haven: Goodyear, 1853 and 1855) Vol. I, pp. 213-218.
[9] *Ibid.*, p. 217.
[10] *Ibid.*, Vol. II, p. 221.
[11] Wolf, *Rubber*, p. 316.
[12] William T. Brannt, *A Practical Treatise on Caoutchouc and Gutta Percha* (Philadelphia: Henry Carey Baird & Co., 1883) p. 113.
[13] Goodyear, *Gum-Elastic and Its Varieties*, Vol. II, p. 137.
[14] Bausch & Lomb, "About History" (http://www.bausch.com/About/chronology.html).
[15] Executive Committee, *Transactions of the Connecticut State Agricultural Society for the Year 1855* (Hartford: Case, Tiffany and Company, 1856) p. 215.
[16] *The Scientific American*, 10 Sep. 1859.
[17] *Ibid.*, 24 Sep. 1859.
[18] *Ibid.*, 23 Feb. 1863.
[19] Brannt, *A Practical Treatise on Caoutchouc and Gutta Percha*, pp. 111-112.
[20] *Ibid.*, p. 112.
[21] Goodyear, *Gum-Elastic and Its Varieties*, Vol. I, p. 170.
[22] *Ibid.*, pp. 170-171.
[23] Executive Committee, *Transactions of the Connecticut State Agricultural Society for the Year 1855*, pp. 213-214.
[24] *Ibid.*, p. 214.
[25] *Ibid.*, p. 172.
[26] Ray Riling, *The Powder Flask Book* (New York: Bonanza Books, 1953) p. 156.
[27] Barker, *Charles Goodyear*, pp. 77-78.
[28] *Ibid.*, p. 79.
[29] *Ibid.*, p. 78.
[30] Riling, *The Powder Flask Book*, p. 196.
[31] Barker, *Charles Goodyear*, pp. 78-79.
[32] *Ibid.*, p. 79.
[33] *Ibid.*
[34] *Ibid.*
[35] Wolf, *India Rubber Man*, p. 276.
[36] Barker, *Charles Goodyear*, p. 79.
[37] Author's opinion based on examination of dozens of documented artifacts and markings.
[38] Experiments conducted by the author using a heated needle to melt known examples of black and brown hard rubber and other compositions.
[39] Goodyear, *Gum-Elastic and Its Varieties*, Vol. I, p. 142.
[40] Hancock, *Personal Narrative*, pp. 120-125.
[41] Brannt, *A Practical Treatise on Caoutchouc and Gutta Percha*, p. 108.
[42] *Ibid.*, p. 109.

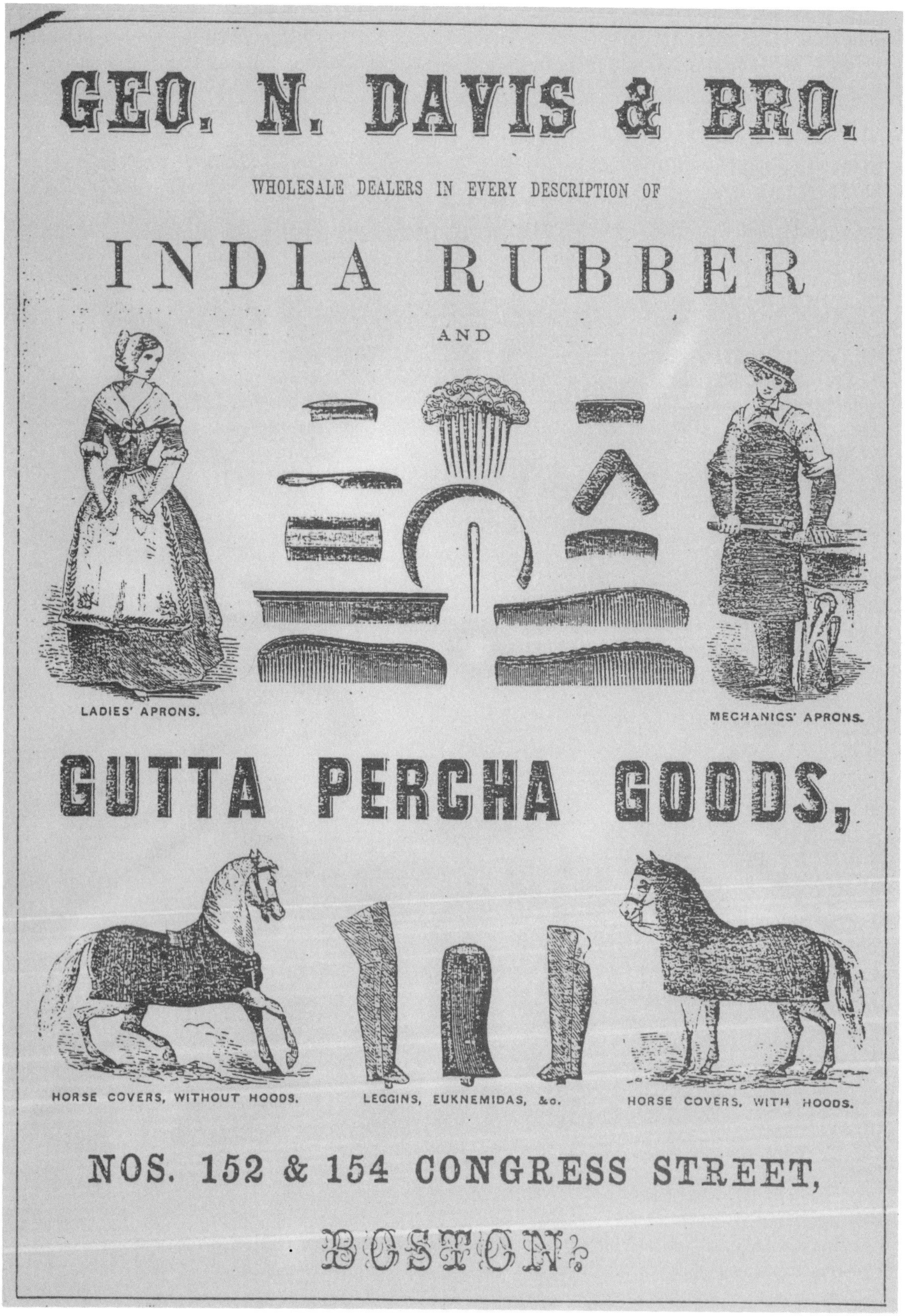

1850s advertisement for India-rubber and gutta-percha goods. *Competition was keen between early patentees and manufacturers for the lucrative business in boots, waterproof clothing, and other articles made of these substances.*

SECTION 4

GUTTA-PERCHA

"We never expected to find a substitute for India-rubber, but one has been recently found in the article of gutta-percha."

From the *Republican and Argus*, quoted in an 1854 advertising brochure for The North American Gutta Percha Company

English traveler and gardener John Tradescant and his brother are credited with bringing the first samples of gutta-percha to Europe about the middle of the 17th century. The curious material was used by the Chinese and Malays for walking sticks, knife handles, and whips.[1] While some of these articles made their way to Europe and America, two centuries passed before gutta-percha was exported from Singapore to England for commercial use (between 1842[2] and 1845).[3] The exact date varies among references, as does the date that the substance was first noted in the 19th century. One source says that, in 1842, an Assistant Surgeon in Singapore, Dr. Montgomery (also noted as "Montgomerie"), observed a native wood-cutter using an axe with a handle very different from the normal wooden handles. Upon questioning, the native revealed that the material from which the axe handle was made could be heated through in boiling water and formed to any desired shape, which it would retain as it cooled. Montgomery sent various samples to England for further study. Around the same time a botanist, Thomas Lobb, was visiting the islands and observed the tree and the unusual gum it produced.[4]

***A sample of hardened crude gutta-percha** from an early 20th century museum display. Gutta-percha was vulcanized and used for many of the same applications as India-rubber and hard rubber. From a circa 1900 Pennsylvania state museum display case.*

In an article published in 1847, Dr. Montgomery is quoted as saying that he first observed some specimens of gutta-percha in 1822, and infers that he noted the axe handle about the same time.[5] (The 1822 date was probably a misprint that was meant to read 1842.) He instituted experiments and determined that the substance was more suitable than India-rubber for certain parts of surgical instruments, as it was more stable than the unvulcanized gum in the tropical climates.[6] In 1847, the Society of Arts in London awarded him a gold medal for this work and his work with gutta-percha surgical bandages.[7] It is interesting to note that Dr. Montgomery suggested the use of gutta-percha to make raised type and embossed maps for the blind, and thought it might be "...found serviceable in stopping tooth decay."[8] Dentists still use gutta-percha in modern-day root canal therapy.

According to early references, early gutta-percha was produced from the juice or latex of the Isonandra tree, also called the gutta or gutta-percha tree, found only in the Malay Peninsula, Borneo, and Ceylon. The Isonandra gutta is a large, high tree with soft, spongy wood containing longitudinal cavities filled with brown stripes of gutta-percha.[9] The origin of the unique word that has been misused more than any other in identifying plastic-like materials of the nineteenth century is with the natives. *"Gutta"* is the Malayan word meaning gum or juice and *"percha"* the Malayan name for the tree.[10] At least one source

[Leaf and Fruit of the Gutta Percha Tree.]

[Gutta Percha Forest, with Natives gathering Gutta Percha.]

The Gutta-percha tree and natives gathering the gum. *Illustrations from the 1854 catalog of the North American Gutta Percha Company. Indiscriminate harvesting of the gum by felling the trees created the need to educate the natives on proper techniques for draining the gutta-percha without killing the tree. (National Archives)*

gives the Malayan translation as "ragged gum."[11] Another reference claims that the word "gutta-percha" as used to identify this gum, was a misnomer in itself, as the gum actually came from a different tree, the gutta taban.[12] In fact, at least nine varieties of gutta-percha plants in the natural order *Sapotaceae* were later noted as growing in the Malay Peninsula.[13] These varieties yielded gutta-percha that varied in elasticity and quality including *balata,* a substance first noted in 1857 and frequently used in mixtures with India-rubber and gutta-percha.[14] Various mispronunciations of the word "gutta-percha" were being corrected as early as 1847. According to the Commissioner of Patents' report of that year, the correct pronunciation is "gutta," as in "gut" and "percha" as in the fish, "perch."[15]

As the Malayan natives realized the value of gutta-percha to buyers in Europe and America, quick profit was the initial response as the trees were chopped down and drained of the gum. In a three and a half year period in the late 1840s, 270,000 trees were felled to quickly harvest the gum.[16] Prudent planning and education followed, as the natives were taught to tap the trees by cutting notches in the branches at certain intervals, allowing the sap to flow without killing the tree.[17]

Gutta-percha was usually shipped in large, square blocks, but it was sometimes formed into fanciful fish, birds, and other shapes.[18] Some amount of foreign matter such as dirt, leaves and bark generally became accidentally mixed in with the gutta-percha while being collected. But it was often intentionally mixed with bark or inferior gums from other trees, which adulterated the gutta-percha and produced greater profits for the exporters.[19] Since the natives sold gutta-percha by weight, it was not uncommon to find large stones rolled up with the gum when the mass was cut open at the factory.[20]

An Alternative to India-rubber

Many touted gutta-percha as superior to India-rubber, probably comparing well-manufactured gutta-percha goods to early or improperly vulcanized rubber products. While some articles in *The Scientific American* praised gutta-percha, at least one called it "...similar, though inferior" to India-rubber.[21]

While the two gums share many physical properties when vulcanized and applied to similar applications, they differ in their raw state and in some scientific respects. An interesting observation in an 1883 publication was:

> Examined in thin sections under the microscope it [gutta-percha] seems to possess a porous structure. Caoutchouc under these conditions gives little or no change of color, while gutta-percha exhibits a beautiful spectacle. It appears to be built up of prisms of every variety of hue and, as it were, fused into each other.[22]

***Gutta-jelutong**, one of many varieties of "gutta-percha." Like balata, this gum was apparently mixed with India-rubber or other forms of gutta-percha to address specific applications.*

Genetically or botanically, gutta-percha is not related to India-rubber,[23] but chemically they are similar. Both gums are comprised of about seven-eighths carbon to one-eighth hydrogen, but gutta-percha contains oxygen, which is lacking in India-rubber.[24] Early chemical analysis showed that gutta-percha contains 86.36% carbon, 12.15% hydrogen and 1.49% oxygen.[25]

Unlike raw India-rubber, raw gutta-percha solidifies and stays hard soon after exposure to air. When heated or immersed in hot water, gutta-percha becomes soft; as it cools it becomes hard and retains the shape given it when warm. This quality allowed it to be readily used for hard, horn-like applications without being vulcanized as hard India-rubber is. But

Gutta-percha water bucket of the 1850s. *Made of vulcanized gutta-percha over canvas or duck with reinforced gutta-percha handle and iron rings. Marked in orange/red stamping on bottom: "Vulcanized Gutta Percha Made By the Gutta Percha Manufacturing Co. New York - Rider's & Murphy's Patent 1852 & 1854." The gutta-percha is still flexible, with some minor flaking near the base.*

moisture, warmth, and time caused rapid deterioration of unvulcanized gutta-percha and it soon discolored, became brittle, and turned to powder.[26]

The Illustrated London News of September 6, 1851 pictured a number of beautiful items of furniture displayed at the Great Exhibition in London. Among the items was a side-board made of unvulcanized gutta-percha, of which was noted:

> The glass-frame and side-board by the Gutta Percha Company is intended as a specimen of what may be done of their material on a large scale, as a substitute for wood carving, etc. We must be excused from investigating its beauties as a work of art; as a piece of serviceable furniture, however, we have our misgivings about it or any thing of the kind composed of this treacherous material. The card affixed to this sideboard announces that it is intended to exhibit the capabilities of this material for ornamental purposes, "particularly the long-sought-for desideratum of a non-fragile pendant." Unfortunately, although this highly-decorated structure in gutta percha is carefully surrounded by a cordon, and has yet experienced no wear and tear—we discover symptoms of dislocation in part of the "pendant" foliage, and something like a "split" in a pear of no ordinary proportions.[27]

Early inventors had not yet realized that crude, unvulcanized gutta-percha was not a practical material for most applications.

Early Applications

The first gutta-percha patents in England were issued to Richard Archibald Brooman in March of 1845 for moulding, stamping, embossing and making thread of the substance. In May of 1848 he received patents in the United States. Patents were issued in England and the United States about the same time to Charles Hancock—for making bands or belts of gutta-percha; to Charles Keene—for boots, gaiters and overalls of gutta-percha combined with India-rubber; and to Henry Bewley—for making flexible syringes, tubes, etc. of gutta-percha.[28] Some of these patents and processes produced inferior goods, becoming a sticky mass after a few months, much like India-rubber goods prior to vulcanization.[29] But the experimentation and introduction of products continued on.

Literally all uses previously suggested for India-rubber were ultimately suggested by manufacturers of gutta-percha. The earliest uses for the gum were in surgical bandages, syringes, tubing, elastic bands, articles of clothing and temporary fillings in dentistry. Displays at the Great Exhibition included many gutta-percha items, such as bowls, shoe-soles, thread, whips, life-preservers with whistles, inkstands, pen-trays, picture frames, chemical bottles, buckets, boats, stethoscopes, and artificial teeth.[30] Golf balls of gutta-percha eventually replaced the traditional "featherie" (also spelled "feathery"), a golf ball made of stitched cowhide and stuffed with boiled feathers.[31] Writer and researcher Henry Pearson notes that the gutta-percha in golf balls "...should be specifically lighter than water, in order not to sink if dropped accidentally into a ditch."[32]

Early references indicate a popular early use of gutta-percha was in the manufacture of submarine and subterranean telegraph cables. In 1847, Werner von Siemens designed a screw press for the seamless covering of wires with the material. Eleven years later, the ship *Great Eastern* placed the first transatlantic cable, which was insulated with gutta-percha.

Another application being advanced in the early 1850s was the use of gutta-percha for gas and water pipes. Gas pipes, including a coiled pipe that could be uncoiled to serve as a portable lamp as one moved about the workshop or room, were being developed. For water pipes, made chiefly of lead in the smaller diameters, gutta-percha was deemed to be more efficient, strong, and sanitary than the heavy metal. In the latter recommendation, attention was directed to notable medical evidence of what we now know as lead poisoning. Dr. Thomas Smith of Cheltenham, England stated:

> Many serious and alarming disorders, such as mania, epilepsy, sudden death, nervous affection, paralysis, consumption hydrocephalus, heart disease, etc., owe their origin in some instances, their intractable character in others, to the gradual and continuous infinitesimal doses of

BIRD'S EYE VIEW OF THE NORTH AMERICAN GUTTAPERCHA COMPANY'S WORKS, ON TWENTY-FIFTH STREET, NEW YORK.

Illustration from an 1850s periodical, The United States Magazine of Science, Art, Manufactures, Agriculture, Commerce and Trade, *showing the North American Gutta Percha Company. (Courtesy of Cliff Krainik)*

Model for U. S. Patent #14,929, "Lining Metal Pipes." *The model consists of a 2 ¾" section of lead pipe with a brass fitting. The lead pipe is lined with gutta-percha to prevent corrosion of the pipe and to "...keep soda or mineral fountain liquids pure and tasteful while they were resident in the pipe." The model has the original Patent Office tag and the pipe is engraved "A. D. Puffer," who patented the lining on May 20, 1856.*

> lead, copper, etc., introduced into the system through the channel of our daily drink.[33]

While lead and zinc were used in manufacturing vulcanized gutta-percha, the amounts were apparently presumed safer than the lead water pipes.

For the hard-of-hearing, various ear trumpets and speaking tubes of gutta-percha, which concentrated sound from the large trumpet-shaped speaking end to the smaller ear tube, were developed. Portable units in various styles were popular and Exeter Hall in London, which boasted a capacity of four-thousand, utilized a "deaf pew," which was equipped with several ear pieces connected, via a singular gutta-percha tube, to an ornate speaking tube mounted on the podium. The same principle was applied at the factory of the Gutta Percha Company in London, where a gutta-percha tube connecting two buildings served as an intercom.[34]

Vulcanization of Gutta-percha

Soft and hard gutta-percha articles, produced by a type of vulcanization of the raw or crude gutta-percha, were widely used in the 1850s and 1860s. Interestingly, vulcanization of India-rubber eliminates some of its undesirable elasticity and pliability, while the vulcanization of gutta-percha adds elasticity and pliability to achieve a similar end product.[35]

Early attempts to vulcanize gutta-percha and produce fabrics using processes already developed for India-rubber were futile. In 1847, Charles Hancock was issued a British patent for a compound of refined gutta-percha, India-rubber, and a substance called gintawan. Using varying portions and degrees of heat, Hancock claimed he could produce materials

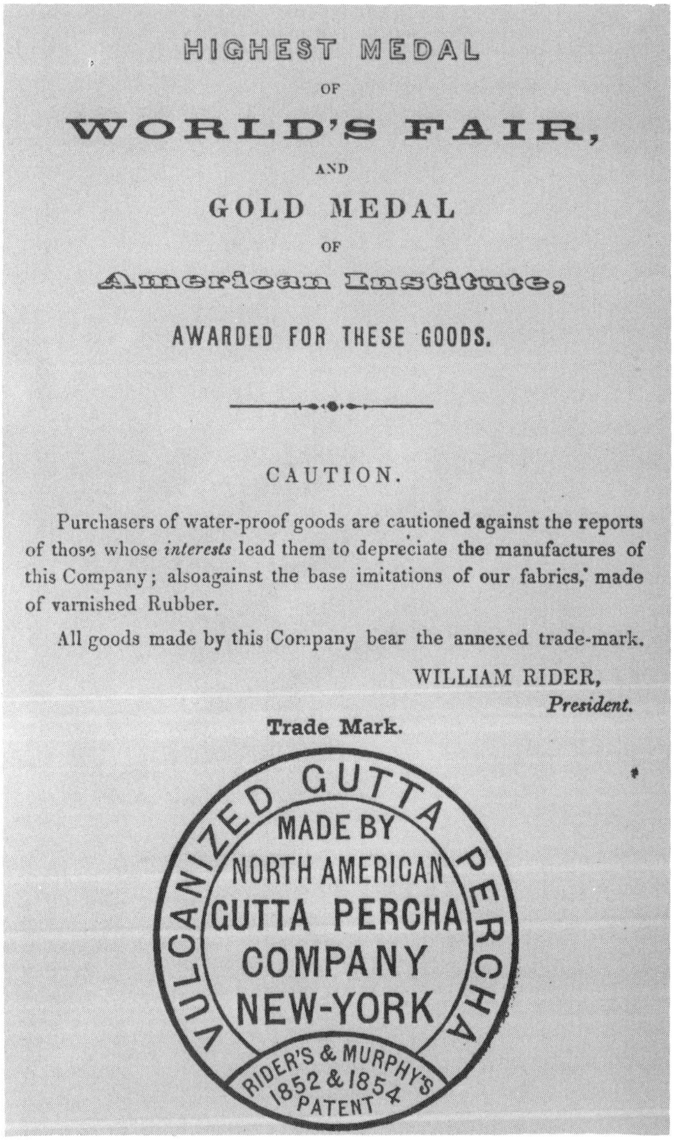

Trademark and patent markings. *From an advertising brochure of the North American Gutta Percha Company. These markings are found on surviving articles made by the company. (National Archives)*

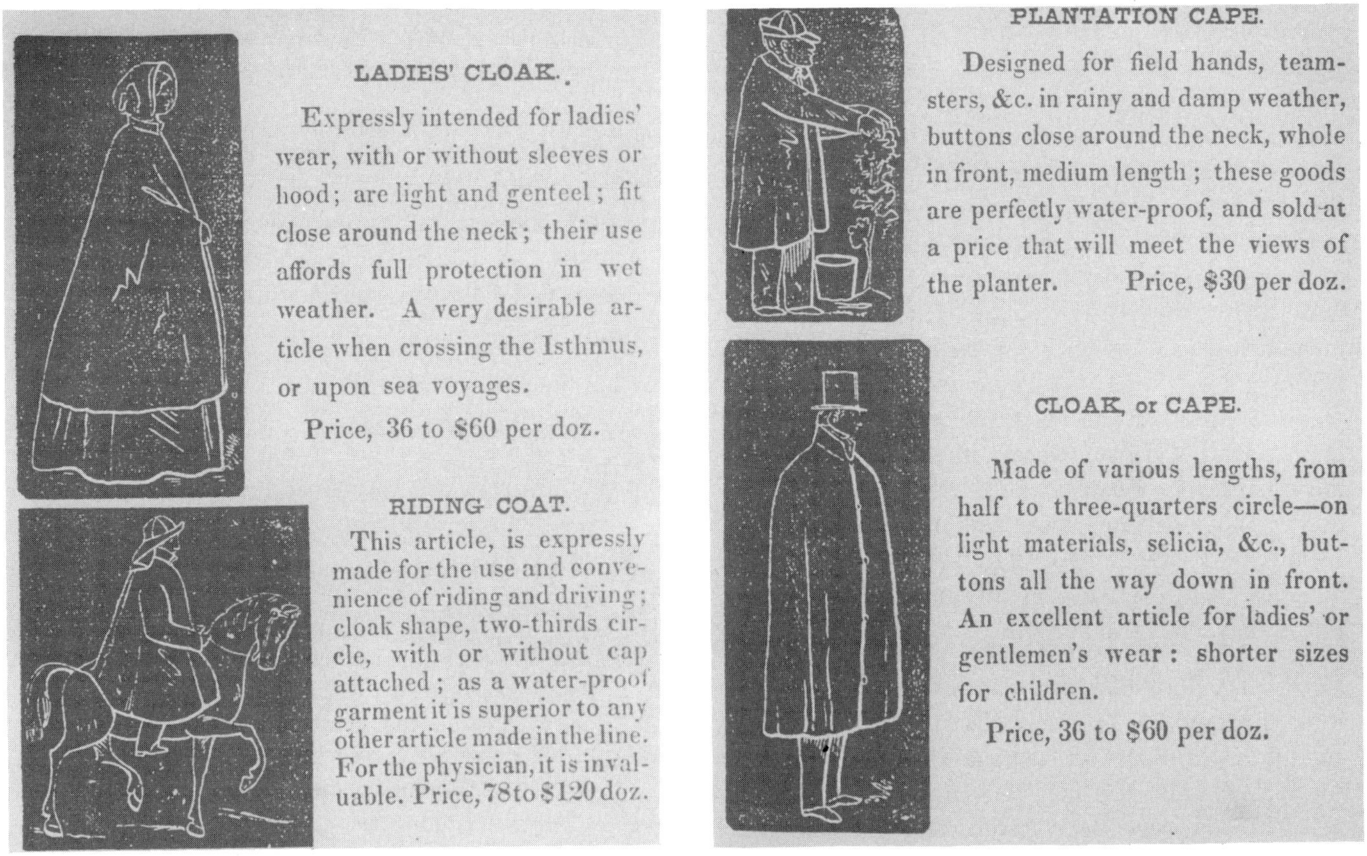

LADIES' CLOAK.

Expressly intended for ladies' wear, with or without sleeves or hood; are light and genteel; fit close around the neck; their use affords full protection in wet weather. A very desirable article when crossing the Isthmus, or upon sea voyages.

Price, 36 to $60 per doz.

RIDING COAT.

This article, is expressly made for the use and convenience of riding and driving; cloak shape, two-thirds circle, with or without cap attached; as a water-proof garment it is superior to any other article made in the line. For the physician, it is invaluable. Price, 78 to $120 doz.

PLANTATION CAPE.

Designed for field hands, teamsters, &c. in rainy and damp weather, buttons close around the neck, whole in front, medium length; these goods are perfectly water-proof, and sold at a price that will meet the views of the planter. Price, $30 per doz.

CLOAK, or CAPE.

Made of various lengths, from half to three-quarters circle—on light materials, selicia, &c., buttons all the way down in front. An excellent article for ladies' or gentlemen's wear: shorter sizes for children.

Price, 36 to $60 per doz.

***Composite of illustrations, descriptions, and prices** from the North American Gutta Percha Company's 1854 catalog, which listed almost 200 products made of soft and hard gutta-percha. (National Archives)*

that were soft and elastic, as well as hard, like wood and ivory. Elastic springs, cushions, belts, and mattresses are among the soft items listed, while the list of hard, molded items included walking sticks, buttons, combs, flutes, and handles of swords and knives.[36]

Several individuals, including Thomas Hancock, held British patents[37] and inventors in England, Ireland, and America held United States patents for improvements in the manufacture of gutta-percha using sulphides, carbonates, muriate of lime, etc. Horace Day was issued U.S. Patent #5,543 on May 2, 1848 for "Gutta-percha Fabrics—Preparing in Imitation of Patent Leather." In the late 1840s, the fabrics were termed useless and many manufacturers suspended operations in the early 1850s.[38] William and John Rider, who had been involved with Charles Goodyear, persisted in experiments and, despite opposition from Charles Goodyear and the India-rubber licensees who maintained the gums were identical,[39] were granted patents in 1852 and 1854 for preparing and vulcanizing gutta-percha.

In John Rider's U.S. Patent #8,992 of June 1, 1852, the gutta-percha was first heated to a temperature of 400 to 450 degrees F. to get rid of any woody or foreign impurities. Next hyposulphide of lead or zinc was added, sometimes in combination with metallic sulphurets, and the mixture was passed through heated rollers as in the vulcanization process for India-rubber. The mixture could then be applied to cloth in sheets or other shapes and "cured" by placing in an air-tight room or container and heating to 280-300 degrees for two to ten hours, depending on the thickness of the mass. Patents issued in 1854 to John Murphy (U.S. Patent #10,977) and William Rider and John Murphy (U.S. Patent #11,906) defined other processes for preparing and "desulphurizing" gutta-percha. While various other methods of treating gutta-percha employed sulphur, sulphur compounds, or sulphurets, some processes used no form of sulphur at all.[40]

The resultant products were similar in look and feel to those made of India-rubber. But the Riders maintained that, even though the public thought them alike, there were distinct differences. They boasted that vulcanized gutta-percha goods were much more beautiful in finish and lighter than those of vulcanized India-rubber, and that, unlike the latter, the former would not decompose, become sticky or emit a very disagreeable odor. In true advertising form, their brochure stated that, "Vulcanized gutta-percha, from its great superiority, in every respect, must soon supersede the uses of vulcanized, as well as other kinds of rubber manufacture."[41] The absence of the rubbery smell due to the reduction or elimination of sulphur in the vulcanization process is evident even on surviving gutta-percha artifacts of today. In the mid-1850s, the Union India Rubber Company manufactured lighter India-rubber products, employing a refined solarization process that utilized much less sulphur and much less actual India-rubber on the fabric.[42] This move was likely made to counter the competitive benefits stated by the gutta-percha trade. The Goodyear vulcanization patent reissues of 1860 (U.S. Patent Reissues #1,084 and 1,085), which included reduced amounts of sulphur in varying forms, may also have been linked to this issue.

The gutta-percha products of Murphy and the Rider's earned the "Highest Medal of World's Fair" and the "Gold Medal of American Institute" awards.[43] An 1855 circular for their company contains several personal and newspaper testimonials including the following from the *Republican and Argus*:

> We never expected to find a substitute for India-rubber, but one has been recently found in the article of gutta-percha. A mode of manufacture has been discovered by Mr. Rider of New York, which gives gutta-percha all the pliability and elasticity of India-rubber, without its unpleasant and objectionable qualities.[44]

The circular goes on to warn that, "Base imitations of varnished India Rubber are already being offered in the market; therefore purchasers should bear in mind that all genuine goods are stamped with the annexed Trademark of the Company."[45] Some rare surviving examples of gutta-percha articles bear the trademark and patent markings.

Manufacturers and Improved Products

An article entitled "Gutta Percha and India Rubber," which appeared in *Hunt's Merchants' Magazine* of 1856, noted:

> Very many persons, says our cotemporary [sic], the *Southern Argus*, when they first rise in the morning, take a bath in a gutta percha tub, comb their hair with a gutta percha comb, and shave with a gutta-percha-handled razor, sharpened on a gutta percha razor strop, before a mirror with a gutta percha frame; eat their breakfast on a gutta percha table cover, and over a gutta percha crumb-cloth, and after they have finished, take their gutta percha walking stick, and sally forth to their business; and if it is raining, don themselves from head to foot in gutta-percha garments. And yet many of these intelligent

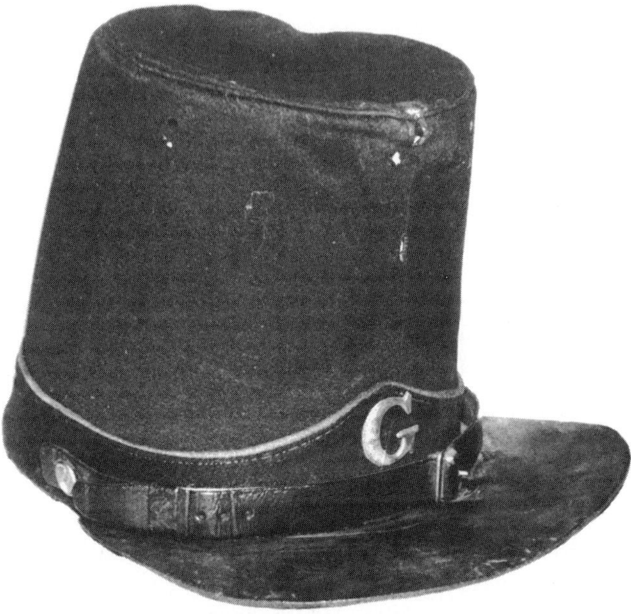

Enlisted man's 1851 pattern dress cap with gutta-percha lining *covered by blue wool. Interior view shows patent markings, trademark, and size No. 4. The patent dates and branch of service color on piping only date this item from 1854 or later. (Courtesy Smithsonian Institution)*

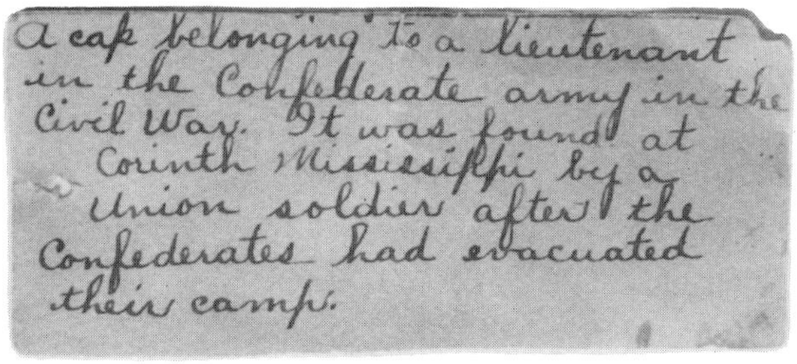

***Captured rain cap** made of vulcanized gutta-percha by the North American Gutta Percha Company and worn by a Connecticut volunteer. It exhibits a high quality appearance and could be worn today. (Courtesy Don Troiani)*

> people are totally ignorant of the article which they use so extensively, and many imagine that it is a preparation of the old fashioned India rubber.[46]

Mr. E. L. Simpson, "a gentleman of eminence in the scientific world," is quoted in the same article as saying, "Perhaps no material was ever discovered which was so extensively shipped as an article of commerce—taken up so eagerly, and manufactured at once so extensively, as has been the article of gutta percha."[47]

The North American Gutta Percha Company of New York was the largest manufacturer of gutta-percha goods in the 1850s. The firm's president, William Rider, was a financial supporter of Charles Goodyear and one of the founders of the Union India Rubber Company. His brother, John, along with John Murphy held gutta-percha vulcanization patents (1852 and 1854) and were also involved in the firm. An 1854 catalog of their merchandise depicts clothing of all types, horse covers, life preservers, boots, haversacks, canteens, etc. *The Scientific American* of November 1, 1856 reported that Rider and Murphy were exhibiting a number of their gutta-percha articles at the Crystal Palace including life-preservers, tents, coverlids, hats, coats, pants, drinking cups, valises, buckets, etc. The journal stated that they "...are similar in appearance and nature to vulcanized india rubber.....are all dark in color, and not so varied in their forms as vulcanized india rubber fabrics, but they will yet become more numerous."[48] The *United States Magazine of Science, Art, Manufactures, Agriculture, Commerce and Trade* mentioned shoes, caps, garments, army & navy equipment, and mail-bags of gutta-percha, noting, "Varnished fabrics of gutta-percha, after being vulcanized, have much the appearance of patent leather."[49] Additional circulars, contracts, and testimonials, document the use of gutta-percha for a variety of applications throughout the 1850s and 1860s.

The American Gutta Percha Company began in 1847 as the manufacturing plant for the patents of Samuel Armstrong, former agent of the Roxbury India Rubber Company. Armstrong was instrumental in introducing gutta-percha to the United States, and in 1848 proposed to Congress the placement of a submarine telegraph cable insulated with the material between America and some point on the English or Irish coast at a cost of three million dollars.[50] While some telegraph cables were placed across rivers over the next few years, it would be a decade before a transatlantic cable would be placed.

Samuel C. Bishop, an employee of Armstrong since 1848, succeeded Armstrong in 1853 and the American Gutta Percha Company became the Bishop Gutta Percha Company. Bishop's company appears to have focused on the production of bottles, pitchers, photograph dishes, and other items made of hard gutta-percha. Worthy of mention is the fact that, on New Year's Day of 1866, Bishop's employees presented him with a silver pitcher bearing the inscription: "For being the inaugurator of the Eight Hour System of Labor in the United States."[51]

Like India-rubber in the 1840s, gutta-percha was tested in the 1850s for various experimental military products such as canteens, water kegs, and pontoons. Civil War use of gutta-percha talmas and blankets is documented in letters to the Gutta Percha Manufacturing Company, which was formed in 1860.[52]

Hard Vulcanized Gutta-percha

Hard vulcanized gutta-percha was used in the manufacture of the same variety of items made of hard rubber. Raw gutta-percha can be heated, molded, then cooled many times but, when thoroughly vulcanized with the use of sulphur or sulphides, it can be made quite hard and resistant to high degrees of heat.[53] Rider's 1852 patent states that, in the second step of the vulcanization process (for soft gutta-percha), Paris white or whiting and magnesia could be added to the other ingredients when "...it is desirable to give solidity or body to the goods." The 1854 catalog of the North American Gutta Percha Company notes that the hard vulcanized gutta-percha is "...a substitute for ivory, horn, shell or pearl" and that "...in appearance it resembles buffalo horn; is jet black - receives a brilliant polish....can be moulded, while hot, into any shape or form desired, which it retains when cold." The list of available products made from this hard compound includes combs, canes, knobs, books covers, knife handles, picture frames, and fencing sticks.[54]

The "jet-black" color of vulcanized gutta-percha referred to in the catalog is not apparent in descriptions of unvulcanized gum. An 1851 report stated:

> Some of the richly-moulded [crude gutta-percha] articles just described display on their surface a diversity of brown tints...these brown tints are apt to be attributed to a painting of the surface, artificially produced; but it is due to the natural colours of the substance. Some specimens of gutta-percha are darker than others, and all have a tendency to darken by age; and the workman dexterously avails himself of these varying tints to produce a pattern.[55]

While many items were made of hard vulcanized gutta-percha as opposed to the unstable non-vulcanized gum, the vast majority of items called gutta-percha today are actually hard rubber. Furthermore, many articles made of compositions or "thermoplastic" materials are frequently called gutta-percha, even though they contain no actual gutta-percha.

At this time, the author is unaware of any hard vulcanized gutta-percha items with specific gutta-percha patent or manufacturer's markings. Comb boxes used by the India Rubber Comb Company in the 1850s-1860s contained "India Rubber and Gutta Percha Combs," according to their labels. Included on the labels are the 1858 Goodyear hard rubber patent reissues and L. O. P. Meyer's 1854 patents (U. S. Patents 10,586 and 10,741), all issued for processing "...India-rubber or other vulcanizable gums." India Rubber Comb Company packaging of the 1870s, reading "Goodyear's India Rubber and Gutta Percha Combs, contains combs that are marked "Goodyear-1851." These patents and references open the possibility that the patent dates and markings may have been used on some hard vulcanized gutta-percha articles.

Waterproof cap covers, hats, and hoods *were made of gutta-percha, India-rubber, and other materials. Variations are sometimes seen in period images. (Courtesy of Henry Deeks)*

A Few Decades of Success

While India-rubber ultimately became the public's choice for waterproof clothing and other popular products, many commercial gutta-percha products were manufactured through the remainder of the century. In his book, *India Rubber, Gutta-Percha, and Balata,* published in 1900, William T. Brannt noted:

> In many cases, gutta percha replaces leather, and for some purposes it is preferable to it. It is especially useful for articles exposed to moisture, damp colds and acids, and for that reason is employed in the form of hose for conducting cold water, beer, vinegar, wine and acids; for belts running in wet places; for buckets, ladles, shovels, bottles, siphons, funnels and spigots in chemical factories. It is largely used in galvanoplasty for the preparation of moulds and matrices. It is also employed for surgical purposes. When perfectly pure it serves for filling hollow teeth and for the manufacture of plates for artificial sets of teeth, and rolled out to the thickness of paper for different surgical bandages and compresses. In the form of very thin taffeta it is used for sweat bands in hats and caps. An important application of gutta percha is its use for coating quick-matches so they can be carried through water without injury.[56]

The author went on to discuss gutta-percha hose, picture frames made in metallic molds, and industrial belting of a gutta-percha and rubber compound. But he noted that the primary value of gutta-percha was the same as was recognized a half-century earlier, "However, the most important use of gutta-percha is as insulating material for electrical wires, especially for submarine and underground cables."[57]

While not maintaining the popularity it enjoyed in the 1850s, gutta-percha had made its mark on history, creating exciting competition and alternatives at a time when innovations flourished. And, while rare indeed, surviving items made of true gutta-percha display the quality and durability that made them popular in their infancy.

NOTES - Section 4

---

[1] *Encyclopedia Brittanica* (Chicago: Encyclopedia Brittanica, 1943), Vol. 11, p. 14.

[2] "Gutta Percha," *Annual report of the Commissioner of Patents for the Year 1847* (Washington: Wendell and Van Benthuysen, 1848) p. 22. [Quoted from *Lond. Jour., Bot.* No. LXI, Jan., 1847, p. 33.]

[3] *The Scientific American*, 18 July 1863.

[4] *Dodd's Curiosities of Industry* (Published in England circa 1851) pp. 10-11.

[5] "Gutta Percha," *Annual Report of the Commissioner of Patents for the Year 1847*, p. 22.

[6] *Ibid.*

[7] Institution of the Rubber Industry, *History of the Rubber Industry* (Cambridge, England: W. Heffer, 1952) p. 266.

[8] "Gutta Percha," *Annual Report of the Commissioner of Patents for the Year 1847*, p. 23.

[9] *The Scientific American,* 18 July 1863.

[10] *Ibid.*, 10 Oct. 1857.

[11] "Gutta Percha," *Annual report of the Commissioner of Patents for the Year 1847*, p. 22.

[12] *Ibid.*, 31 Jan. 1857.

[13] William T. Brannt, *India Rubber, Gutta-Percha, and Balata* (Philadelphia: Henry Carey Baird & Co., 1900) pp. 228-233.

[14] *Ibid.,* p. 303.

[15] *Ibid.*, p. 21. [*Lond. Jour., Bot.* No. LXI, Jan. 1847, p.33.] Common mispronunciations known to the author include "goota-percha" and "goota-perka."

[16] *Hunt's Merchant Magazine & Commercial Review,* Vol. 24 (1851): p. 517.

[17] *Dodd's Curiosities of Industry*, pp. 12-13.

[18] *The Illustrated London News*, 9 Aug. 1851.

[19] Charles Goodyear, *Gum-Elastic and Its Varieties* (New Haven: Goodyear, 1853 & 1855) Vol. I, p. 29 and *The Scientific American*, 31 Jan. 1857.
[20] *Dodd's Curiosities of Industry*, p. 13.
[21] *The Scientific American*, 22 Jan. 1860.
[22] William T. Brannt, *A Practical Treatise on Caoutchouc and Gutta Percha* (Philadelphia: Henry Carey Baird & Co., 1883) p. 154.
[23] *The Scientific American*, 10 Oct. 1857.
[24] *Dodd's Curiosities of Industry*, p. 1.
[25] Brannt, *A Practical Treatise on Caoutchouc and Gutta Percha*, p. 125.
[26] Clifford and Michele Krainik with Carl Walvoord, *Union Cases-A Collector's Guide to the Art of America's First Plastics*, p. 17. [Vander Laan, *Production of Gutta-Percha*, p. 26, quoting a 1927 Department of Commerce report.]
[27] *The Illustrated London News,* 6 Sep. 1851.
[28] *Eighth Census of the United States*, 1860, p. lxxix.
[29] *Ibid.*, p. lxxx.
[30] *The Illustrated London News,* 9 Aug. 1851.
[31] "Sports," *Science 81* (July-Aug. 1981).
[32] Henry C. Pearson, *Crude Rubber and Compounding Ingredients*, (New York: The India Rubber Publishing Company, 1920) p. 385.
[33] *Dodd's Curiosities of Industry*, pp. 15-17.
[34] *Ibid.*, pp. 17-18.
[35] North American Gutta Percha Company Catalog, 1854, pp. 7 & 8; National Archives, Records Group 92, Entry 225, Consolidated Correspondence File, "North American Gutta Percha Company."
[36] "Gutta Percha," *Annual report of the Commissioner of Patents for the Year 1847*, p. 23.
[37] Institution of the Rubber Industry, *History of the Rubber Industry*, p. xviii.
[38] North American Gutta-Percha Company Catalog, 1854, pp. 6-7.
[39] *The United States Magazine of Science, Art, Manufactures, Agriculture, Commerce and Trade,* Vol. 1 (Circa 1855): pp. 374.
[40] Pearson, *Crude Rubber and Compounding Ingredients*, pp. 387-392.
[41] *Ibid.*, p 7.
[42] Goodyear, *Gum-Elastic and Its Varieties*, Vol. I, pp. 72-74.
[43] *Ibid.*
[44] Gutta-Percha Advertising Brochure, circa 1855, p. 4; NA, RG 92, Entry 225, CCF, "Goodyear," "Goodyear Metallic Rubber Company."
[45] *Ibid.*
[46] *Hunt's Merchants' Magazine and Commercial Review* (1856): p. 123.
[47] *Ibid.*
[48] *The Scientific American*, 1 Nov. 1856.
[49] *The United States Magazine of Science, Art, Manufactures, Agriculture, Commerce and Trade,* Vol. 1 (Circa 1855): pp. 378.
[50] *Eighth Census of the United States*, 1860, p. lxxx.
[51] J. Leander Bishop, *A History of American Manufacturers from 1608 to 1860* (Philadelphia, Edward Young & Co., 1868) pp. 198-199.
[52] Gutta-Percha Advertising Brochure, Circa 1860s; NA, RG 92, Entry 225, CCF, "Gutta-Percha."
[53] Pearson, *Crude Rubber and Compounding Ingredients*, p. 389.
[54] North American Gutta-Percha Company Catalog, 1854, p. 28.
[55] *Dodd's Curiosities of Industry*, p. 24.
[56] Brannt, *India Rubber, Gutta-Percha, and Balata,* pp. 263, 264.
[57] *Ibid.,* p. 264

SECTION 5

OTHER MATERIALS

Various other soft and hard materials and compositions were introduced in the 1850s and 1860s. They are similar in appearance to vulcanized India-rubber or gutta-percha and are frequently confused with those substances.

Observations of the author

Recycled India-Rubber

Several processes were developed in the 1850s and 1860s to restore rubber waste or reprocess worn out products into new articles. One of the earliest successful attempts at recycling was implemented by Daniel Hayward, who was issued United States Patent #11,608 on August 29, 1854. The patent, entitled "Manufacture of India Rubber," covered a process for remolding worn-out India-rubber goods. Hayward produced shoe soles, spittoons, chamber pots, and various fire and water buckets under this patent at the Haywardville Rubber Company near Stoneham, Massachusetts.[1] Some of these items have survived and are in good condition today. In most cases, the recycled rubber has become hard and brittle but, like gutta-percha, can become more flexible when exposed to the sun or another heat source. All surviving items are marked with the patent date, although some markings may be very faint or worn away entirely.

After Hayward's patent, at least ten other patents were issued for reworked, restored, reprocessed, or otherwise recycled India-rubber goods. Six patents were assigned to the Beverly Rubber Company for restoring waste rubber. One of these patents, issued to Hiram L. Hall in 1858 (U.S. Patent #19,172), detailed a process of boiling waste vulcanized India-rubber in water, reducing it to a powder, then sheeting it. At least one manufacturer was still using this process in 1899.[2]

A patent was issued in 1855 to Charles Morey (U.S. Patent #12,212) for reworking hard rubber compounds by heating the scrapings, dust, powder, etc. of hard rubber.

Waterproof Materials

Tarred knapsacks, oilcloth, and non-rubber "painted" garments, cap covers, and accessories are among the more well known examples of waterproof items made of materials other than India-rubber and gutta-percha. Many materials resembling India-rubber and gutta-percha are detailed in patent records and periodicals of the era.

In 1855, Alexander Parkes combined guncotton, rendered less inflammable by the addition of biphosphate of ammonia, magnesia, talc, alum, or similar substances, with gums or resins to produce a substance like India-rubber. Guncotton had been used earlier to render fabrics waterproof, for varnishing India-rubber to make it impervious to gases, and for insulation work. It was prepared by treating cotton wool with acids, rinsing and drying, with care taken

Fire bucket made of recycled rubber. *One of the earliest processes for recycling India-rubber was patented by Daniel Hayward in 1854 (U.S. Patent #11,608). Remains of the stenciled word "FIRE" are still on the bucket. Embossed into the rubber on the underside of the bucket is "D. Hayward's Patent 1854 - Manufactured By The N. Hayward Co. Stoneham, Mass."*

to remove all traces of acid that could cause an explosion.[3]

Another waterproof fabric, patented by R. Liddell in 1864 (U.S. Patent #42,953), specified cloth treated with linseed oil, black lead, and plumbago.

Peterson's Ladies Magazine published the following article at about the start of the Civil War:

> WATERPROOF CLOTH FOR SOLDIER'S OVERCOATS
>
> Twenty thousand tunics, rendered water-proof, and yet porous were served out to the French Army during the late war with Russia. They were prepared after the following recipe:
>
> Take two pounds, four ounces of alum and dissolve it in ten gallons of water; in like manner dissolve the same quantity of sugar of lead in a similar quantity of water, and mix the two together. They form a precipitate of the sulphate of lead. The clear liquid is now withdrawn, and the cloth immersed for one hour in the solution, when it is taken out, dried in the shade, washed in clean water and dried again. This preparation enables the cloth to repel water like the feathers on a duck's back, and yet allows the perspiration to pass somewhat freely through it, which is not the case with Gutta-Percha or India-Rubber cloth.[4]

In the December 27, 1862 issue, *The Scientific American* published the same formula, for use with cotton or woolen cloth, with some additional comments:

> Persons who wear air-tight garments soon become faint, if they are undergoing severe exercise, such as that to which soldiers are exposed when on a march.
>
> The sulphate of lead is taken up in the pores of the cloth, and it is unaffected by rains or moisture, and yet it does not render the cloth air-tight. Such cloth is also partially non-inflammable. A solution of alum, itself, will render cloth, prepared as described, partially water-proof, but it is not so good as the sulphate of lead.[5]

India-rubber compounds were used in the

production of various painted or coated articles. Some patents issued for "waterproof cloth" or "waterproof fabric," contained India-rubber, which was combined with silk, fur or residual gums (U.S. Patents #31,195, 34,947, and 34,428). Gutta-percha was combined with campene or naphtha in two patents for waterproof fabrics (U.S. Patents #38,005 & 38,006). A composition for waterproofing leather consisted of India-rubber, dissolved in turpentine and petroleum, combined with fish oil, rosin pitch, beeswax, tallow, soap and lard (U.S. Patent #31,453). Samuel LaForge patented a method of waterproofing leather by applying India-rubber before vulcanization (U.S. Patent #27,337).

***Composition cover** for a soap or snuff box, featuring a raised bust of Napoleon Bonaparte enclosed within a wreath.*

Hard Compositions

The popularity of hard rubber and hard gutta-percha articles in mid-century induced creative scams in at least one case, when baby dolls advertised and sold as gutta-percha to an eager public in London were found to actually be a mixture of glue and molasses.[6] But various compositions, capable of being molded and manufactured into novel articles, were being introduced by legitimate inventors.

While not hard rubber or hard gutta-percha in the true sense, some hard compositions were created using gutta-percha, India-rubber, or rubber scraps as an ingredient. In 1854, Julius Herriet patented an "Improvement in Gutta-Percha Compositions," which consisted of "Equal parts, by weight, of gutta-percha or India-rubber and pulverized graphite or soapstone, or plaster of Paris or chloride of lime or peroxide of mangenese or other equivalent mixed together, heated and moulded while hot." (U.S. Patent #11,831) Robert Haering's patent of 1862 combined gutta-percha with linseed oil, asphalt, and sulphur to form a "composition substitute for horn, hard rubber, etc." (U.S. Patent #36,406). Denton Drake's 1864 patent for "Manufacture of Coffins and Other Articles" specified an "asphaltic compound of pitch or coal tar with plaster and sand or India-rubber" (U.S. Patent #41,284).

***Composition daguerreotype case of the 1850s.** These cases are frequently referred to as gutta-percha but were actually made of compositions of gum-shellac, fibrous materials, and other substances, that were heated and formed in metal molds. The manufacturer and patent labels are generally found under the mat inside the case.*

Early composition book covers. *(Right) Ornate covers of a composition material protect* The Miracles of Our Lord, *published in London in 1848. The author notes that "...it has been partly taken from a magnificent cover in carved ivory, executed in the 12th century." (Left) Another set of composition covers from an early religious book, probably also produced in the late 1840s.*

"Union" Composition Cases

The most popular composition articles produced in the 1850s and 1860s are the well-known daguerreotype cases, commonly and incorrectly identified as "gutta-percha." At the height of their popularity in the 19th century, they were called "Union" or "composition" cases,[7] so the mistake in terminology appears to be with later generations. The vast majority of these cases and other similar composition items were made of shellac-based compounds that contained no India-rubber or gutta-percha. Some collectors today refer to these compositions as "thermoplastics," a term generally applied to substances that can be heated, softened, re-shaped, and hardened again.

That these cases resemble hard rubber or hard gutta-percha on initial examination is unquestionable. On closer scrutiny, however, differences become more obvious. As beautiful as they may be, items made of these compositions are somewhat more prone to chipping and cracking than hard rubber. When chipping occurs, the composition's appearance is generally course and porous, unlike hard rubber, which is smoother, shinier, and chips or cracks more cleanly.

Patent markings also help easily identify some compositions. The requirement for marking items with the patent date to protect the patentee's rights is evident with daguerreotype cases, as most have retained a paper label under the inside mat containing all the patent dates and manufacturer's name. A composition case may contain multiple dates representing patents for the composition and the hinges. In the manufacturing process for composition cases and other articles, the heated material was placed into intricately designed steel molds, pressed, and hardened upon cooling to create the finished product.

The first formula for a composition that achieved success can be found in Samuel Peck's patent (U.S. Patent #11,758) of October 3, 1854, issued for an "Improvement in the Manufacture of Daguerreotype Cases." Peck had actually been making shellac-based cases since 1852. Earlier daguerreotype cases were made of leather, cloth, metal, tortoise shell or papier-mâché.[8] The patent specified a composition of shellac and sawdust or fibrous material with suitable coloring pressed between hot rollers. In 1855, Halvor Halvorson, Peck's brother-in-law, patented an "Improvement in the Manufacture of Daguerreotype Cases" that was very similar to Peck's patent and was later assigned to Peck.[9] Halvorson patented the combination of a metallic frame with velvet-covered glass as a "Miniature Case" in 1896 (U.S. Patent #14,501 assigned to Slocum & Watkinson, Boston, Massachusetts). In the same year, Peck patented hinges with a metal reinforcement (U.S. Patent #14,202) to help reduce or eliminate the "thirty percent of cases...spoiled by breaking them in riveting on the hinges."[10]

In 1858, Mark Tomlinson patented (U.S. Patent #21,285) a composition containing "equal parts, by weight, of shellac, Breckenridge or cannel-coal, and ivory black. Tomlinson claimed in his patent that this composition was more durable and less likely to fracture than others.[11]

Beginning in 1854, the A. P. Critchlow & Company, located in Florence, Mass., manufactured daguerreotype cases, buttons and photo frames using their own shellac-based composition. The firm's founder, Alfred P. Critchlow, had been making cases since 1852 and claimed to be the inventor of the composition. In 1856, he patented (U.S. Patent #15,915) an "Improved Embracing Hinge" to correct that commonly weak part of the case. Critchlow took on Daniel G. Littlefield as a partner in 1857 and sold out his interest to Littlefield in 1858. The new company, Littlefield, Parsons & Co., made daguerreotype cases, buttons, and cigar, watch and pistol cases. Records indicate that at least 14,391 of these beautiful composition pistol cases were manufactured for Allen & Wheelock, Sharps, Smith & Wesson and Warner pistols from 1858 through 1867.[12]

In 1866, the company name was changed to the Florence Manufacturing Company. Patent dates of June 19, 1866, February 4, 1873, and May 29, 1883 appear, along with the name "Florence" on many hand mirrors and brushes of the post-Civil War era. The company has undergone several name changes and still does business today as the Pro Corporation.[13]

There were several additional composition case makers of the period, including Holmes, Booth and Hayden, who entered the business in Waterbury, Mass. in 1857; Wadham's Manufacturing Company of Stoughton, Mass, 1858; and Frederick C. Key and Son of Philadelphia.[14] Other known case manufacturers are J. Lewis Baldwin, A. B. Chapman, Kinsley and Parker, Patent American, and, in England, John Smith.[15]

When ambrotypes and daguerreotypes

Decorative Composition. *George and Martha Washington form a belt buckle. The belt buckle, with stamped brass hooks on the reverse, may have been produced for the Centennial Exposition in 1876.*

became obsolete in the 1870s, both the Florence Manufacturing Company and the Scovill Manufacturing Company made other items, such as buttons, belt buckles, jewelry, combs, etc. of compositions. Scovill also made men's collar boxes with decorative composition lids.[16]

Other Compositions

Compositions were also being made into fine products in Great Britain. Several religious books of the late 1840s are bound in covers of black composition. One such book "The Miracles of Our Lord," published by Longman & Company of London in 1848, has gilded pages and beautiful color engravings on the pages. The ornate molded covers consist of twelve medallions depicting various miracles.

A scarce composition Confederate Navy button, depicting crossed cannon and anchor with the letters "CSN" is sometimes identified as hard rubber. The texture, feel and overall quality of the button is very similar to buttons made of hard rubber. The markings on the reverse of the button are "Courtney and Tennant," (Confederate contractors and importers) and "Manton's Patent."[17] This patent was filed as "No. 2,907" in England by John Shakespeare Manton and Thomas Islip, a button maker, on November 26, 1860 and sealed by the Patent Office on May 24, 1861.

Composition black oval hand mirror and brush. *Both have ornate relief of birds and sailing ship. The mirror is marked "Florence - Pat. June 19, 1866." The brush is marked "Florence - Feb. 4, '73 May 29, '83."*

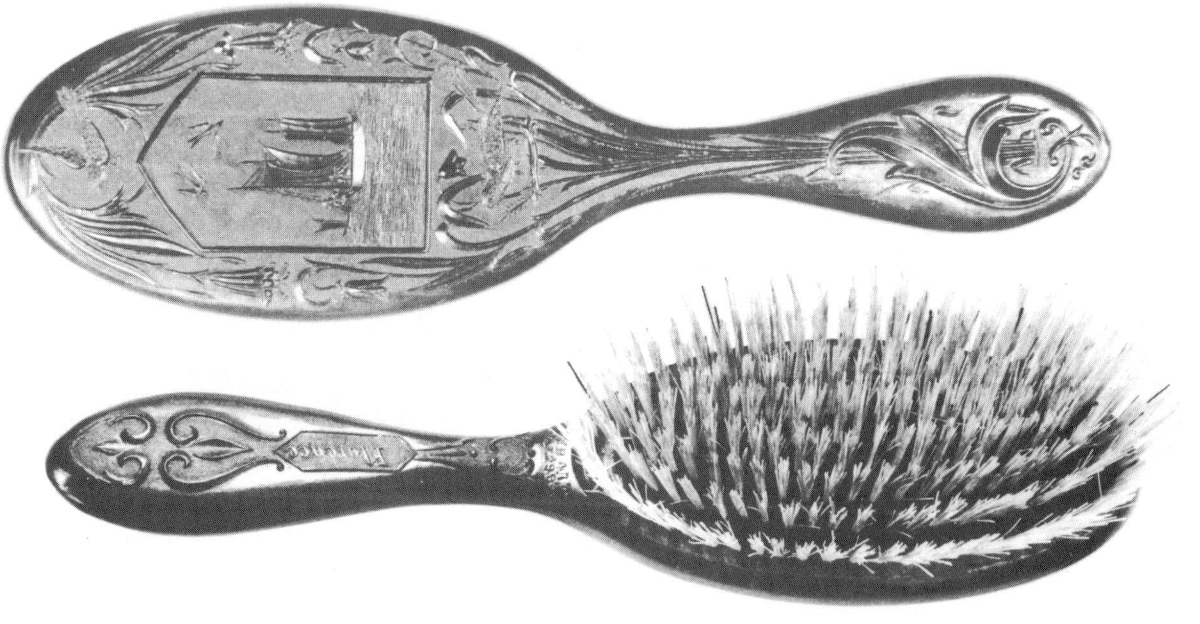

The patent is referenced under the categories of "Casting Plastic" and "Buttons, Buckles, Hooks & Eyes, and Other Dress Fastening," and includes the following specifications:

> The compositions consist of shellac mixed with mineral, earthy, arenaceous, and other substances, colouring matter being used when required. Ground porcelain or other baked earths, powdered glass or pebbles, marbles, slate, basalt, slags properly pulverized, or ivory dust, are amongst the materials we employ. These are to be amalgamated with the shellac under heat into a paste coloured according to taste and transferred to dies, by which they receive any desired impression. For some uses a solution of pyroxylene or collodion mixed with oils, gums, or resins may be employed instead of the shellac.

The specification further defines an improved machine and process for riveting together "articles of a brittle description."

"Parkesine," a composition of guncotton and camphor, was developed by the English chemist,

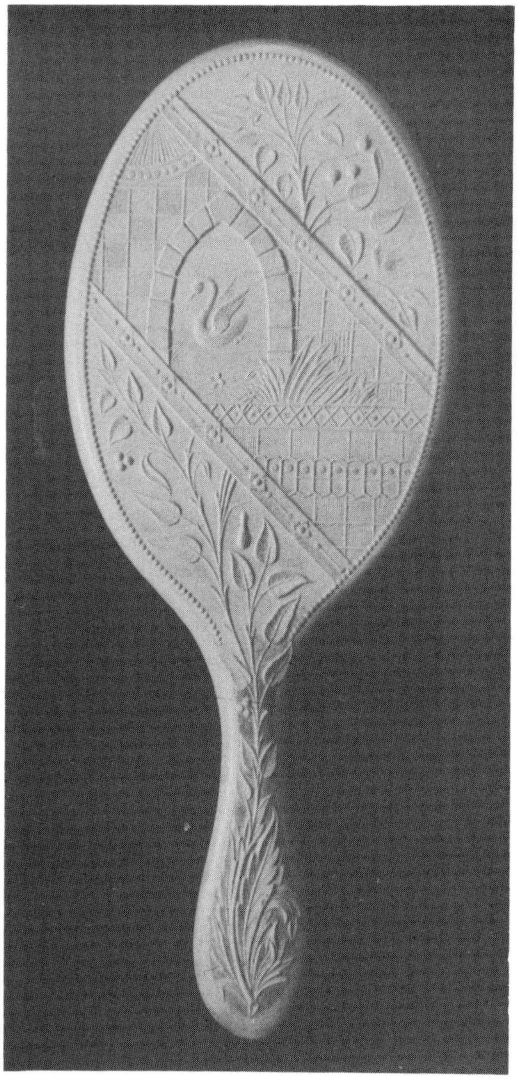

White composition mirror. *While colorings were mixed with some compositions, the majority of the articles produced were black.*

Set of 8 remaining celluloid or "patent ivory" handled table knives in their original box. *Marked "Meriden Cutlery Co. - July 12, 1870, etc." Celluloid and other hard materials gradually replaced hard rubber towards the latter part of the century.*

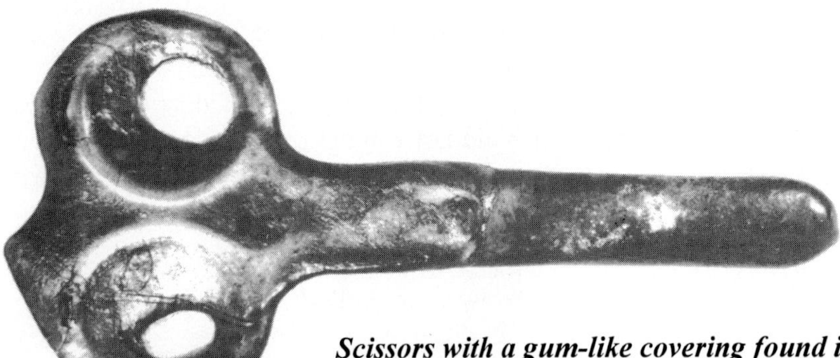

Scissors with a gum-like covering found in a Civil War campsite. *The covering was designed to prevent rust and has served well, as the shiny scissors are visible through the translucent gum.*

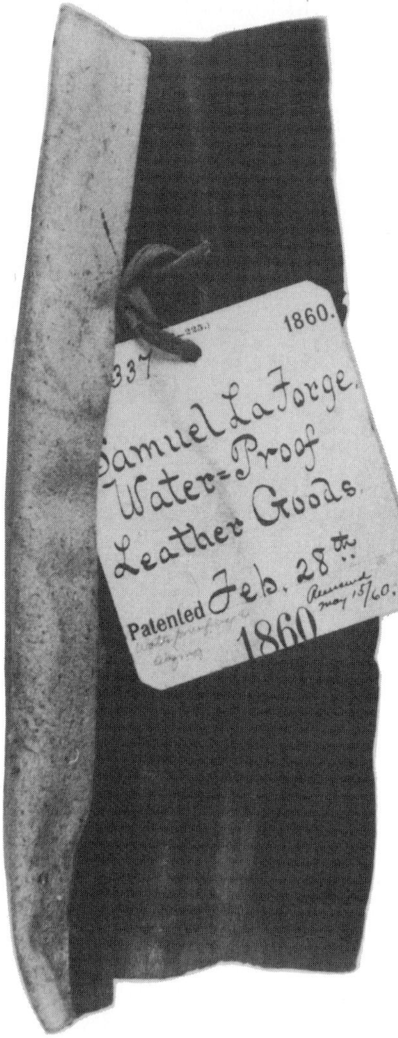

Patent model for "Improvement in Waterproof Leather Goods," *issued to Samuel LaForge as U. S. Patent #27,337 on February 28, 1860. In his patent, LaForge specified "...undressed skins, exposed to heat, coated with a charged leather solution, then subjected to the vulcanization process." After nearly 140 years, the rubberized leather is still very soft and pliable.*

Cotton or canvas cape with a waterproof coating. *This garment is very light and was likely a woman's cape.*

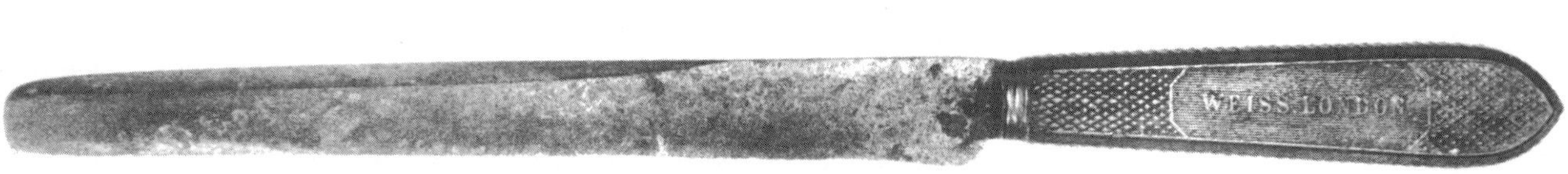

Surgical knife with composition handle. *Marked "Weiss-London."*

Alexander Parkes. In the late 1850s, he made and sold buttons, pins, medallions, book covers, bracelets, knife handles, shoe horns, and combs. Variations of this substance, called "Xylonite" and "Ivoride," were used to produce similar products from the late 1860s until 1891.[18]

The Scientific American described an "India-rubber substitute," developed by a Mr. John Root. The 1865 article claimed that it was "...made from the solid residuum that remains in the still after the more volatile vapors are driven off the well oil or petroleum. He makes beautiful picture frames, medallions, bottles and jars equal to any made of rubber."[19]

Later Compositions

A patent was issued to J. M. Merrick, Jr., of Boston, on December 15, 1868 for a composition of gum shellac and diatomaceous earth (silica) for molding. Doorknobs marked "Norwalk Lock Company - Silica - Diatite - Patented December 15, 1868" were produced under this patent. Other articles, such as hand mirrors, manufactured by the Florence Manufacturing Company are marked "Diatite" and "Patent 1868," and/or "March 19, '78," so Florence likely bought the manufacturing rights or sub-contracted this work. The name "Diatite" was no doubt created from the ingredient "diatomaceous earth."[20]

In 1864, Phelan and Collander, America's largest billiard supply company, offered a $10,000 reward for the first person to produce an imitation ivory to replace expensive elephant ivory used in the manufacture of billiard balls.[21] A patent was issued to Julius Hackert on May 31, 1864. (U.S. Patent #42,942) for a "Compound for Artificial Ivory," containing "vinegar, and nitric ether with shavings of bone, ivory, horn, etc.," that could be heated and molded. An article in *The Scientific American* of July 16, 1864, claimed that an M. Ghoulston Ghislain presented an artificial ivory substance to the French Academy that consisted of powder of marine plants, glue, and coal tar, which was boiled, dried, and bleached until white.[22] These substances may well have been developed with an eye toward the reward offered by Phelan and Collander.

Also seeking the prize (which may have never been paid) was John Wesley Hyatt, who, with his brother, Isaiah, perfected their composition and process for manufacturing a hard material. On July 12, 1870, they were issued U.S. Patent #105,338 for "Improvements in Treating and Molding Pyroxyline [sic]," which did not win the prize money, but created an industry with the birth of a new substance known as "celluloid."

The Hyatts' composition used a solid form of pyroxylin known as collodion, which had been patented in England as a hard rubber substitute by John Macintosh in 1859.[23] The collodion, with camphor added as a solvent, was mixed with cellulose fiber and molded by the application of heat with pressure. Through various additives and processes, celluloid could be colored and made very flexible and practically incombustible.[24] In addition to billiard balls, other items such as dental plates and a variety of boxes, jewelry, combs, and other ornamental articles were made from variations of the original composition. At the turn of the century, the motion picture industry became a large user of celluloid. Because it is made up of part natural and part chemical composition, celluloid is considered a semi-synthetic thermoplastic.[25]

A short article on The Fibrone-Terraloid Company of Newark, New Jersey appears in an 1894 history of that city. The article states that the company was incorporated in January, 1892,

> ...for the manufacture of the newly discovered material named 'fibrone,' and its application as a substitute for hard rubber, celluloid, vulcanized fibre, papier-mâché, and other products of like nature." While the composition is not detailed, it was claimed that it had "...the good qualities of rubber with the added attraction that 'fibrone' is cheaper.[26]

Bakelite, the first totally synthetic thermoplastic, was developed by Dr. Leo Hendrick Baekeland of Yonkers, New York, and introduced in 1909. Bakelite, a transparent, solid amber material, is formed when formaldehyde and carbolic acid are combined and cured by heat and compressed air. Originally intended for applications in the electrical field, Bakelite was also mixed with wood fibers and coloring agents and used for many commercial and household uses. Peak usage of Bakelite occurred between 1920 and 1950.[27]

NOTES - Section 5

[1] Zellie, Carole, & Martha Coons, *Stoneham, Massachusetts: A Shoe Town* (Stoneham: Stoneham Historical Commission, 1981) p. 57.
[2] Henry C. Pearson, *Crude Rubber and Compounding Ingredients* (New York: The India Rubber Publishing Company, 1920) p. 292.
[3] Pearson, *Crude Rubber and Compounding Ingredients*, p. 146.
[4] *American Review*, Vol. 1, No. 5. (N.D.); [*Peterson's Ladies Magazine*, Spring 1861].
[5] *The Scientific American*, 27 Dec 1862.
[6] Constance Eileen King, *The Collector's History of Dolls* (St. Martin's Press, Inc., 1977; New York: Bonanza Books, 1981) p. 490.
[7] Clifford and Michele Krainik with Carl Walvoord, *Union Cases: A Collector's Guide to the Art of America's First Plastics* (Grantsburg, WI: Centennial Photo Service, 1988) p. 5.
[8] *Ibid.*, p. 3.
[9] Floyd and Marion Rinhart, *American Miniature Case Art* (Cranbury, NJ: A.S. Barnes, 1969) p. 33.
[10] *Ibid.*, p. 33.
[11] Clifford and Michele Krainik with Carl Walvoord, *Union Cases*: p. 10.
[12] Vern Eklund, "'Gutta Percha' Pistol Cases," *Smith & Wesson Journal,* Vol. 21, Number 3, (Fall 1987) and "Another Visit to Littlefield, Parsons & Co.," *Smith & Wesson Journal,* Vol. 24, Number 4, (Winter 1990): p. 11.
[13] Vern Eklund, "You Just Never Know," *The Doorknob Collector,* Number 45, (Jan.-Feb. 1991): pp. 1-3.
[14] Clifford and Michele Krainik with Carl Walvoord, *Union Cases*, p. 10.
[15] Paul Berg, *Nineteenth Century Photographic Cases and Wall Frames* (Huntington Beach, CA: Huntington Valley Press, 1995) p. X.
[16] Clifford and Michele Krainik with Carl Walvoord, *Union Cases,* p. 14.
[17] Alphaeus H. Albert, *Record of American Uniform and Historical Buttons* (Alphaeus H. Albert, Bicentennial Edition, 1976) p. 362, CS 60.
[18] Barbara and Ken Beem, "Collectors May Find Their Future in Plastics," *Antique Week,* (12 Apr 1993): pp. 1-32.
[19] *The Scientific American*, 1 July 1865.
[20] Vern Eklund, "You Just Never Know," pp. 1-3.
[21] Julie Robinson, "John Wesley Hyatt and the Invention of Celluloid," *Antique Week,* (19 June 1995): pp. 1, 22.
[22] *The Scientific American*, 16 July 1864.
[23] Samuel S. White, *A History of Dental and Oral Science in America* (Philadelphia: White, 1876) pp. 71-72.
[24] Pearson, *Crude Rubber and Compounding Ingredients*, p. 143.
[25] Robinson, "John Wesley Hyatt and the Invention of Celluloid," pp. 1, 22.
[26] *Newark, New Jersey's Greatest Manufacturing Centre, Illustrated*, (Newark: The Consolidated Illustrating Company, 1894).
[27] Julie Robinson, "Bakelite: The Material of a Thousand Uses," *Antique Week,* (31 July 1995): pp. 1, 48.

SECTION 6

MILITARY TRIALS

"I am desirous of having a series of experiments made with India Rubber Cloth...I consider the matter of so much importance to the public service that I earnestly recommend that the experiments be authorized."

Quartermaster General S. Jesup to Secretary of War L. Cass, August 26, 1835

Early Military Trials

An early War Department letter to a Mr. John Harris, dated January 8, 1799 and signed by James W. Henry and Samuel Hodgson, placed an order for various stationery supplies to be delivered to Mr. John Fancin. The articles ordered include "...Twelve black lead pencils, Twelve bunches of Camels Hair, One Box of Water Colours, Two paste Boards, One piece of Indian Rubber."[1] At that time, the use of India-rubber as an eraser was the only practical use for the substance in this country.

Various applications for unvulcanized India-rubber in the 1810s and 1820s may have involved the military, but the substance hadn't been widely introduced at this point. An 1818 Army contract for 8,000 gummed knapsacks provided by William P. Gould[2] and an 1826 contract for 1,000 gummed knapsacks provided by Benjamin Morange & Son of New York[3] may reference India-rubber.

A significant request appeared in a March 26, 1833 correspondence from Commissary General of Purchases, Callender Irvine, to Lemuel Blake, agent to the Roxbury India Rubber Factory:

> Pressing public engagements have prevented my acknowledging, earlier, the receipt of your favor of the 15th instant, and the samples of India Rubber water proof cloth and leather accompanying it.

***Federal cavalryman wears a long-sleeved waterproof talma**, one of the few India-rubber items formally adopted by the United States Army after the military trials of the 1850s. Note also the waterproof kepi cover. (Courtesy Jim and Tony Stamatelos)*

Do me the favor to have made, and forwarded to this Office, a Knapsack of the India Rubber Cloth, as you have proposed, and I will cause a fair experiment to be made with it. I will in the mean time ascertain whether cloth, like any of the other samples sent, can, with advantage, be employed for the military service.[4]

The military was now interested in exploring new applications. On December 28, 1833, Colonel George Bomford, Chief of Ordnance, recommended to the Secretary of War that India-rubber covers or cases be purchased to protect small arms from dampness and moisture.[5] Three years later, Colonel Bomford again expressed interest in India-rubber. A letter sent by one Robert Dingee on September 26, 1836 stated:

I suppose you have received the two pair holsters which I handed to Capt'n Bradford as Samples. You will see that the covers are made of broad Cloath [sic] lined with ticken, also an india-rubber cloath which is put between the broad cloath and ticken to prevent water from penetrating. They will cost more than the Seal Skin.[6]

On November 27, 1836, Colonel Bomford received a suggestion to construct ammunition cases of "Strong Leather well covered inside and out with India Rubber Cloth" to keep cartridges and caps dry.[7] In the same year, Captain Mackay of the firm Mackay and Walker offered to make the Quartermaster General's Office 5,000 knapsacks with India-rubber covers for $1.12 ½ each.[8]

On August 26, 1835, Quartermaster General Jesup requested permission from Secretary of War Lewis Cass to conduct "...a series of experiments made with India Rubber Cloth, for the purpose of ascertaining whether it may not be advantageously used in the construction of Pontons, as well as employed for other military purposes." Jesup stated that he felt the India-rubber cloth could also be used as a cover for baggage wagons and as a tent. He asked that Lieutenant John F. Lane, who first suggested use of India-rubber pontoons, be assigned to special duty to carry out these experiments, which was so ordered the next day.[9] Correspondence by S. T. Armstrong, agent for the Roxbury India Rubber Company, on November 4, 1836 indicated that he had

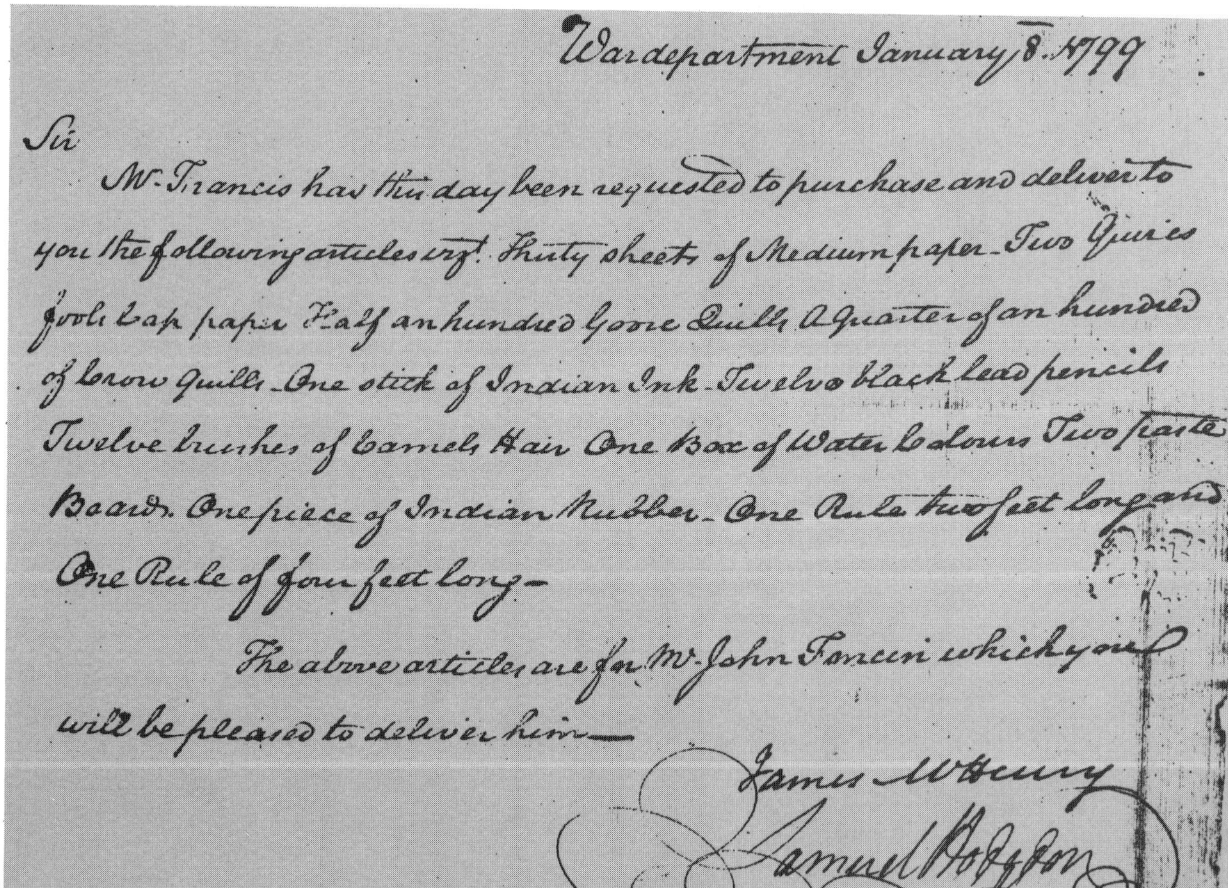
War department January 8. 1799

Sir

Mr. Francis has this day been requested to purchase and deliver to you the following articles viz. Thirty sheets of Medium paper - Two Quires foolscap paper Half an hundred Goose Quills A Quarter of an hundred of Crow Quills - One stick of Indian Ink - Twelve black lead pencils Twelve brushes of Camels Hair One Box of Water Colours Two paste Boards. One piece of Indian Rubber - One Rule two feet long and One Rule of four feet long -

The above articles are for Mr. John Foncin which you will be pleased to deliver him -

James McHenry

Samuel Hodgdon

***Original order from the War Department**, dated January 8, 1799, ordering various stationery items, including "one piece of Indian Rubber." (National Archives)*

entered into a contract with Lieutenant Lane to manufacture "...ponton sacks for air bridges, waggon [sic] covers, dragoon capes, etc.."[10]

Much time was spent developing applications with pontoons and wagon floats made of India-rubber. On August 13 and 14, 1839, tests were performed in New York harbor and a Board of Officers reported favorably on India-rubber "air boats," or pontoons.[11] A report dated October 19, 1839 detailed further trials with an India-rubber air-boat, upon which "A 6 pounder without limber was rolled upon it and fired with round shot repeatedly without injury or inconvenience."[12] A patent was issued to the estate of "Colonel John F. Lane, U.S. Army—deceased" in 1840 for an "India Rubber Pontoon Bridge" (U.S. Patent #1,640-See diagram in "Boats and Pontoons").

In the 1840s, trials and experiments were conducted on India-rubber pontoons, wagon floats, floating bridges, and even a horse floating apparatus. Contracts were issued for 86 of Lane's pontoons for service in the Mexican War.[13] These were likely destined to fill a request by Chief Engineer Captain George Cullum for 40 to be sent to General Zachary Taylor's command, 36 to General Winfield Scott's, and the remainder to West Point for training purposes.[14]

Wagon floats, invented by Colonel Henry Stanton and tested on the Schuylkill River in Philadelphia, were also introduced in the Mexican campaign.[15] India-rubber "camels," invented by Captain George W. Taylor for unloading ships in shallow waters, made appearances in the War as well. Although the use of India-rubber pontoons and floats continued on through the Civil War, they never replaced traditional versions constructed of wood, metal and canvas. While easier to transport, their main objections were their vulnerability to firearms projectiles and the tedious process of inflation.[16]

One use for India-rubber cloth was proposed, ahead of its time, to the War Department by James Boyd of Boston on March 6, 1836:

> I forgot to mention to you in conversation that in Washington the idea was suggested of using some of my India Rubber goods as a kind of sheet to spread under the men in their tent. These would effectually prevent the damp of the ground from striking on the men and might be made so that each man would have one or, one might be made large enough for each tent. By the former method, it might be used round the shoulders when on the march in rainy weather and as a sheet at night. If you like the idea, I will send sufficiency for the two experimental tents, which you propose invoicing, the expense for the single ones would be about $1.50 for each man, for this price you could have a piece two yards long and ¾ yd. wide. Please send the dimensions for the tent that is to be made, as soon as convenient.[17]

Boyd's suggestion pre-dates by twenty-five years government approval for India-rubber blankets and ponchos.

At this time, Charles Goodyear was actively promoting his goods to the military but it was evident that India-rubber's problems had not yet been resolved as of July 6, 1839. In a letter to General Irvine on that date, Goodyear wrote, "...the sample of my fabric with Knapsack & Haversack received immediate attention, the Articles were long since complete and I believe perfect with one exception, viz. the objection which is made on account of the smell—this I am working to overcome."[18] This was the year that Goodyear, with direction from Nathaniel Hayward, learned the relationship of sulphur and heat to the gum, leading to the discovery of vulcanization in 1839 and subsequent patent of 1844. Prior to the perfection of the vulcanization process, other items were tested including "India-rubber cylinders" for long guns, which "...left the bore of the gun quite as clean as flannel."[19]

Testimonials after the vulcanization patent indicate that some finished products had indeed overcome the problems that plagued their predecessors. In a letter dated July 3, 1845, Albert Smith, U.S. Commissioner on Boundary Survey, certified:

> Goodyear's heated cloth preferable to any other cloth prepared with India-rubber. It is not affected by heat, cold or water. I have had it exposed to intense heat before campfires, I have packed it away closely when wet; and it never became sticky and peeled off like the other cloth.[20]

In a testimonial given May 16, 1845, Professor of Chemistry Thomas P. Jones noted, "... powder bags made of India-rubber cloth by Mr. Goodyear's process...powder would be secure from deterioration in such bags and would be especially waterproof." Colonel Bankhead, writing from Fort Columbus on September 8, 1845, stated that "...rations of sugar, salt, rice, beans and soft bread

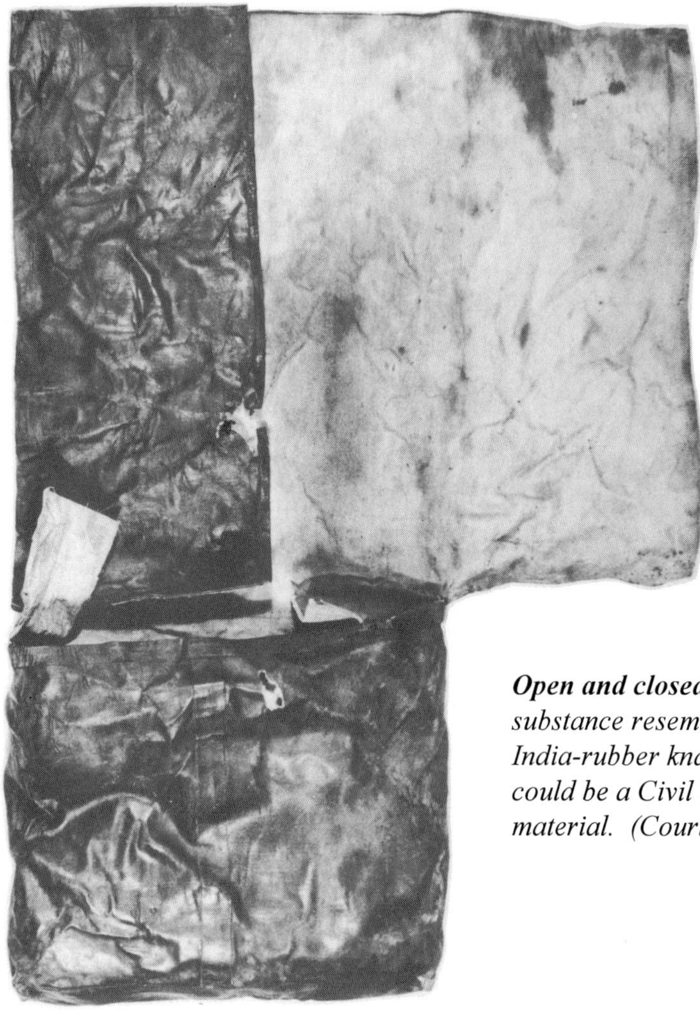

Open and closed view of a "bag" style knapsack *coated with a gummy substance resembling unvulcanized India-rubber. The Army purchased India-rubber knapsacks for the Mexican War. This example, however, could be a Civil War militia item made of a non-rubber "painted" material. (Courtesy Fred Gaede)*

placed in 2 small bags or sacks made of Goodyear's Metallic Rubber and exposed to the weather for 5 weeks were taken out as dry as when they were put in."[21]

On November 24, 1845 Thomas Green, writing to Gamaliel Gay, an agent for The Naugatuck India Rubber Company of New York, said:

> I think that a waterproof cover for each of the cavalry horses to protect them from the sudden cold rains and winds would be, to say nothing more, a decided economy in the government; for I have known animals frequently to suffer great injury from being tied out the night exposed to these changes than they would in months of service in the absence of them.[22]

In the same letter, Green also supported rubber pontoons, provision bags, storage covers, tents, and clothing, and comments that gun covers could extend a musket's life "...for a great number of years." In an unsigned memo that appears to be in Gay's handwriting is a reference to the previously mentioned report of Colonel Bomford regarding India-rubber cases to protect small arms from moisture. The memo detailed annual replacement costs of unprotected arms and was based on 969,165 arms in arsenals and in the hands of troops as of 1845, compared to what could be saved if India-rubber cases were used to protect the arms. The savings is shown to be $5,060,870 over ten years after the cost of $969,165 (1 case for each small arm @ $1 each) for the India-rubber cases is deducted.[23]

Some use of gun covers was made, as evidenced in a letter written at Monterey, Mexico by Captain George D. Ramsay to Colonel George Talcott of the Ordnance Department in Washington. The letter, dated May 10, 1847, included comments on successful battles at Buena Vista, and detailed several suggestions for improvements with artillery equipage. Ramsay also commented, "The Gun Covers of Gum elastic furnished Bragg's battery are highly approved of, as will be, I have no doubt, the accoutrements of the same material for the Artillery now being tried."[24]

On November 20, 1845, Gay made the following proposals for a variety of rubber goods to the Secretary of War, William Marcy:

> Proposals by G. Gay for supplying Army Equipments to be made of "Goodyear's Patent Metallic Rubber," with all the latest improvements, and to be free from sulphurous odor and perfectly water-proof, viz.:

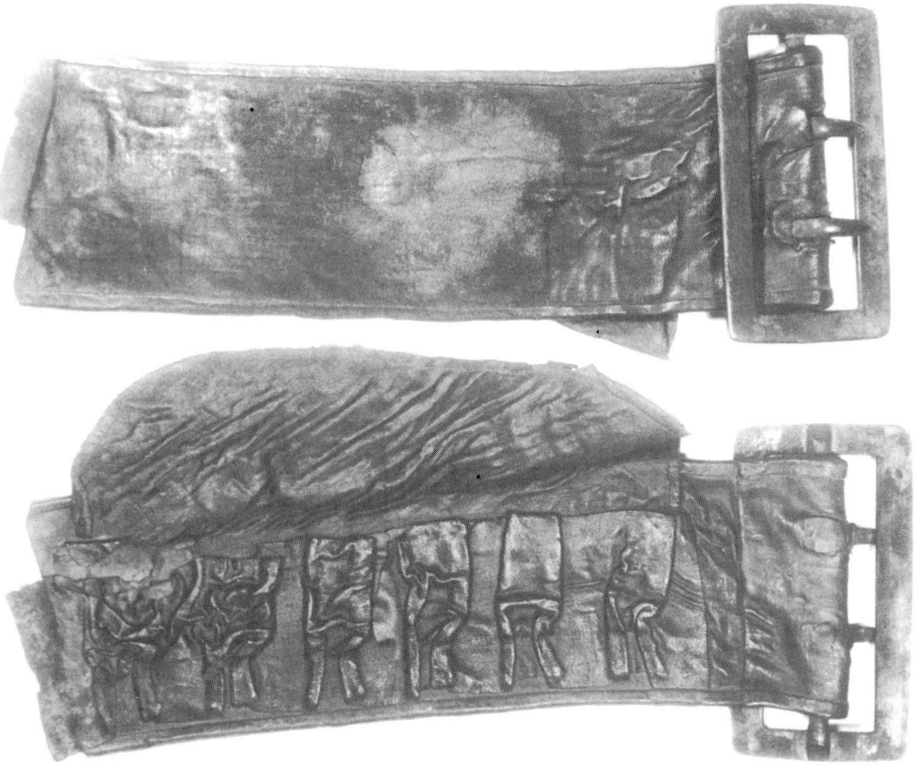

Section of India-rubber cartridge belt. *This section, which was likely part of an experimental cartridge belt of the 1840s or 1850s, has a large brass frame buckle, six rubber pockets for cartridges, and a rubber flap to keep the cartridges dry. Charles Goodyear had suggested Navy Belts in 1855, with "...pockets for cartridges, and appendages for the knife or sabre, and revolver."*

Gun Cases-------------------------1.25 Each
Tents--------------------------------1.00 Per sq. yd.
Overcoats---------------------------6.50 Each
Capes (5/8 Circle & long)---------6.00 "
Ponchos (Black) for Horsemen---5.00 "
Trowsers or Pants (Assorted sizes) 3.50 "
Shoes--------------------------------1.50 Per Pair
Storm Hats or Caps-----------------9.00 Per Doz.
Water Bottles or Canteens(according to size)
.75, 62 ½ & .50 Each
Powder Bags (with Air-tight fastening) 3.00 "
Grain Bags (" " ")----------2.00 "
Provision Bags (" " ")---------2.00 "
Provision Tarpauling or Covers-1.10 Per Sq. Yd.
Provender (" ")---1.10 " " "
Knapsacks--------------------------------2.25 Each
Haversack (water-tight)----------------- .75 "
Cannon Covers-------------------1.10 Per Sq. Yd.
Cannon Swab or Sponge Bag--------- .75 Each
Gunners Haversacks (Stout)----------- 2.25 "
Water Pails (to be folded up for
easy transportation)---- 2.00 "
Water Pails or Fire Buckets
(stiff as leather)------------ 2.00 Each
Horses Feed Bags
(or Buckets to hold water)--------------1.75 "
Covering for Barracks---------- 1.25 Per Sq. Yd.
Waggon Covers------------------1.00 " " "
Leggins----------------------------2.25 Per Pair
Port Fire Cases-------------------2.00 Each
Cannon Cylinders or Cartridges (waterproof)
about the same price of flannel
Pontoons & Pontoon Boats

Should the Department be desirous of having the Army supplied with these articles, or a part of them, and should there be no provision for such expenditure, I would be willing to fill any order or orders and wait for the payment of the same, till an appropriation for the purpose be made by Congress.

G. Gay
Nov. 20th 1845[25]

Five days after Gay's letter, the Secretary of War received a letter from Captain A. R. Hetzel recommending the purchase of some of the articles proposed by Gay. Of tents, knapsacks, and haversacks, Hetzel said:

> ...[they] would cost considerably more than those with which the Army is now supplied but as they would last much longer their introduction would ultimately result in a saving to the Government, to say nothing of the health of the

troops which it is believed would be greatly benefited thereby besides adding greatly to their comfort.

Hetzel went on to recommend the tarpaulins and grain bags to "...reduce heavy losses of forage and subsistence from the weather." He also suggested purchasing a limited number of each article for immediate testing and, if successful, gradual introduction into the service.[26]

As the Mexican War escalated in 1846 and 1847, many orders for India-rubber goods were delivered to the Army including the following:

Haversacks: 17,302
Knapsacks: 19,414
Canteens: 13,016
Grain/Provision Bags: 7,612
Water Bottles: 5,000
Water Tanks: 110
Covers: 13[27]

There were further orders for military India-rubber goods recorded through the late 1840s, when purchases ceased. Knapsacks for Engineers were priced at $3 each while all other knapsacks were $1.50; haversacks and canteens were 50 cents each.[28]

Despite the endorsements of many and the obvious success of the vulcanization process in overcoming the difficulties with India-rubber, it is apparent the manufacturing processes needed some improvement. Some knapsacks and haversacks received were "...not decried equal to the original patterns."[29] Haversacks and canteens were blistered or not sufficiently gummed. In some cases, over half of the knapsacks received on an order were stiff and hard or were missing buttons or buttonholes. In one instance, 86 of 110 water tanks were rejected due to blistering.[30]

But the decade of the 1840s had witnessed the earliest military applications of the improved vulcanized India-rubber. India-rubber had not escaped the trials and tribulations that accompany the introduction of any new products or services. And, in addition to further challenges for the fledgling industry, the 1850s would bring a viable competitor to the military marketplace.

The Trials of the 1850s

Uses of India-rubber continued to be applied to various articles, as evidenced by patents and references in periodicals and military records. Charles Goodyear gives the subject of early military trials honest treatment in his introduction to the "Military" chapter of his book, *Gum-Elastic and Its Varieties*, when he writes in 1855:

> In the early stages of this manufacture, and during the Mexican war, an assortment of articles were made for the government of the United States. They were made to order in great haste, before the gum-elastic fabrics for such purposes were invented, and they were manufactured chiefly from the coated cloths, which were not found to answer the purposes of leather, as the fibrous fabrics have done since that time. It is not surprising, therefore, that mistakes should have been made in the first attempts to introduce some articles for the use of the army and navy. Nor is it surprising that some articles that are now known to be useful, should then have been imperfectly manufactured, and consequently have failed at that time to answer the expectations that were formed of them. Many of the articles were highly approved by the troops and officers in the service; among them which may be noticed camp blankets, tent carpets and water tanks, and provision and powder bags.
>
> In consequence of the introduction of the hard compounds and other new fabrics, a new field is now opened for the manufacture of various military articles that have not been heretofore attempted. Owing, also, to the invention of vulcanizing in moulds, many of these articles that have formerly been made of sole or harness leather, may now be made by this process, and they are a class of articles for which no objection can be raised to the use of gum-elastic. Among these are pistol holsters, sword sheaths, cartridge boxes, etc. some of which are hereafter described.[31]

In his characteristic and optimistic way, Goodyear goes on to describe how the above articles, as well as cannon covers, port-fire cases, cannon sponge covers, military belts, canteens, caps, haversacks, knapsacks, and musket covers of gum-elastic could be adapted to military uses. Goodyear credits the invention of wagon floats to Colonel Henry Stanton and the suggestion of "ambulanches," a mattress of India-rubber supported by elastic springs for transporting the sick and wounded, to "...an officer of the United States' army."[32]

Naval uses detailed included sails, tarpaulins, hammocks, sailors' bags, and various buoys. Inventions by military seamen were submarine armor

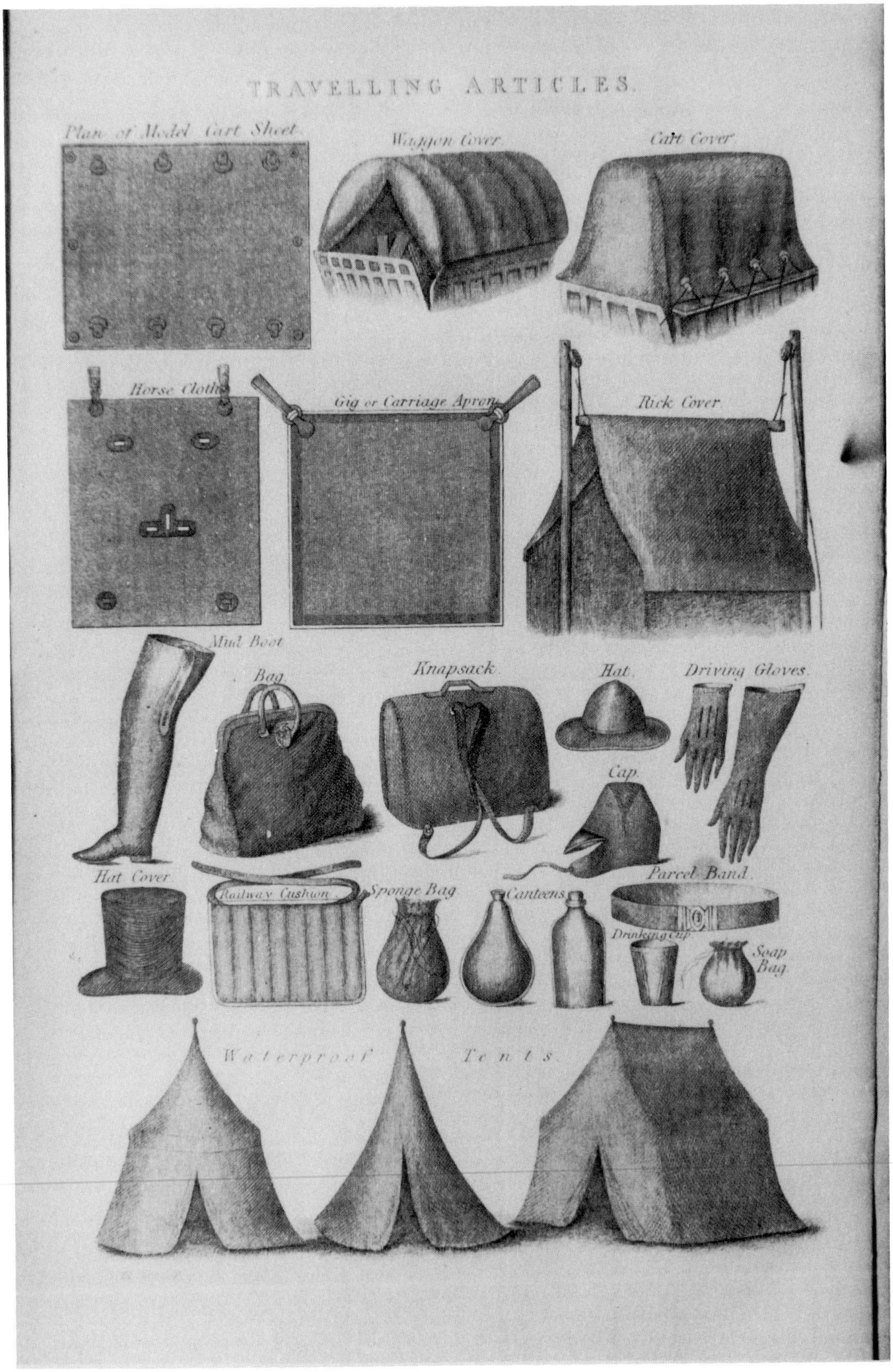

Military products. *Illustrations from Thomas Hancock's* Personal Narrative, *published in England in 1857, including some products tested by the military up to that time.*

for diving, and camels for "...lightering vessels in shoal water," attributed to Captain George W. Taylor. A cannon tompion and a "shot plug" for stopping holes in the sides of ships were invented by Lieutenant Samuel Seely, a former Naval Officer. The latter invention, designed to replace wooden shot plugs that required carpenters to be lowered down the outside of the vessels, contained a self-sealing flap-valve and could be installed from the inside of the ship and away from sharpshooter fire.[33]

Despite the creative suggestions and continued pursuit of contracts by inventive minds and capable manufacturers, there seems to have been no significant increase in India-rubber uses by the military in the 1850s. Some of the goods that were produced, however, may have been held over in arsenals and used in the forthcoming war.

Gutta-Percha in the Military

But the mid-1850s saw an emergence of an entire line of articles manufactured from a rubber-like

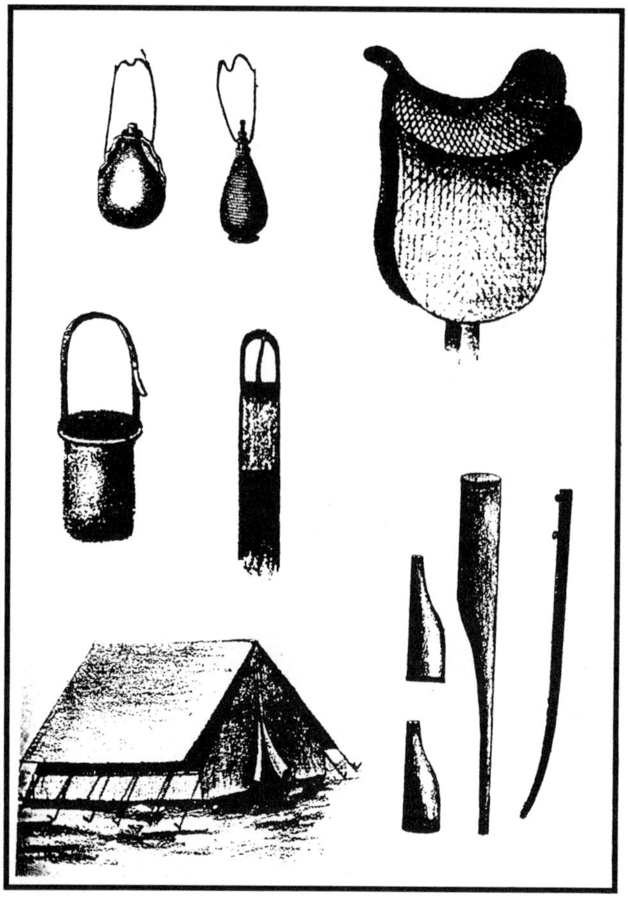

Goodyear's military applications. *Illustrations from* Gum Elastic and Its Varieties, *published by Charles Goodyear in 1853 and 1855. Goodyear considered India-rubber suitable for many military uses.*

substance known as gutta-percha. As with India-rubber, many gutta-percha items were made for the military and put to the test with varying results. A "Catalog of Vulcanized Gutta Percha Fabrics," issued by the North American Gutta Percha Company of New York in 1854, contains a number of military items, including gutta-percha ponchos, tents, blankets, canteens, haversacks, buckets, gun cases, hospital sheeting, etc. Under the heading of "Military and Naval Articles" is the following statement:

> The total failure of India Rubber to answer the requirements of the service, presents an open field for the introduction of Vulcanized Gutta Percha fabrics. Officers in the U. S. service are solicited to give our fabrics a trial, under the guarantee of the Company that they shall answer the purpose.[34]

Several endorsements of gutta-percha, likely solicited by the Riders from military officers, were also printed in this catalog, many referencing the superiority of gutta-percha over India-rubber. Specific items mentioned in letters to the company include gutta-percha tent cloth, canteens, overcoats, caps, leggins, and full suits of protective clothing.

A letter sent on March 4, 1854 from Captain Henry D. Hunter, U.S.R.S. to William Rider, President of the North American Gutta Percha Company, states that:

> Water Proof articles of gutta percha bought at Rider's store are superior to any water proof clothing I've ever used. For various applications for the Army and Navy, your vulcanized fabrics must be invaluable, such as cannon covers, tarpaulins, etc.[35]

On April 12, 1855, a Board of Officers met to examine a number of military articles made of gutta-percha fabrics. They were subjected to various tests to determine their suitability for military applications. A circular printed in the 1860s by D. W. Umberhine, Agent of the Gutta Percha Manufacturing Company, detailed the "Official Report on Vulcanized Gutta Percha Manufactured by the North American Gutta Percha Co. of New York, Under Rider's Patent, Made by Order of the United States War Department." The following extracts are quoted in the circular:

> By Order of the United States War Department. Extract of Proceedings of a Board of Officers, at Philadelphia, April 12, 1855.
>
> The Board proceeded to examine the following articles of Gutta Percha Fabrics, viz.

Knapsacks on duck, Haversacks on duck, Water Tanks on duck-3½ gallons, Military Shawl, Cap Body, Cap Cover and Cape, Pistol Case, Valise, Neck Stocks, Large Sacks for transporting supplies, Canteen on duck.

These were submitted to the following tests: Heat was applied to them at a temperature of 150 deg. Fahrenheit. This did not soften the surface of the articles, or render them adhesive in the least. The canteens were then filled with spirits of wine, vinegar, lard oil and water, all of which were submitted to the degree of heat mentioned above, and the liquors kept in the canteens for twenty-four hours. They were not impaired by the high temperature applied to them, and there was no perceptible action upon the fabrics by either of the liquids. The results of the aforementioned experiments were so satisfactory to the Board, that it was deemed important to ascertain what would be the effect of a long continued high heat upon the fabrics. Arrangements were accordingly made by the Quartermaster for applying this heat and the Board adjourned to meet again on the 20th instant.[36]

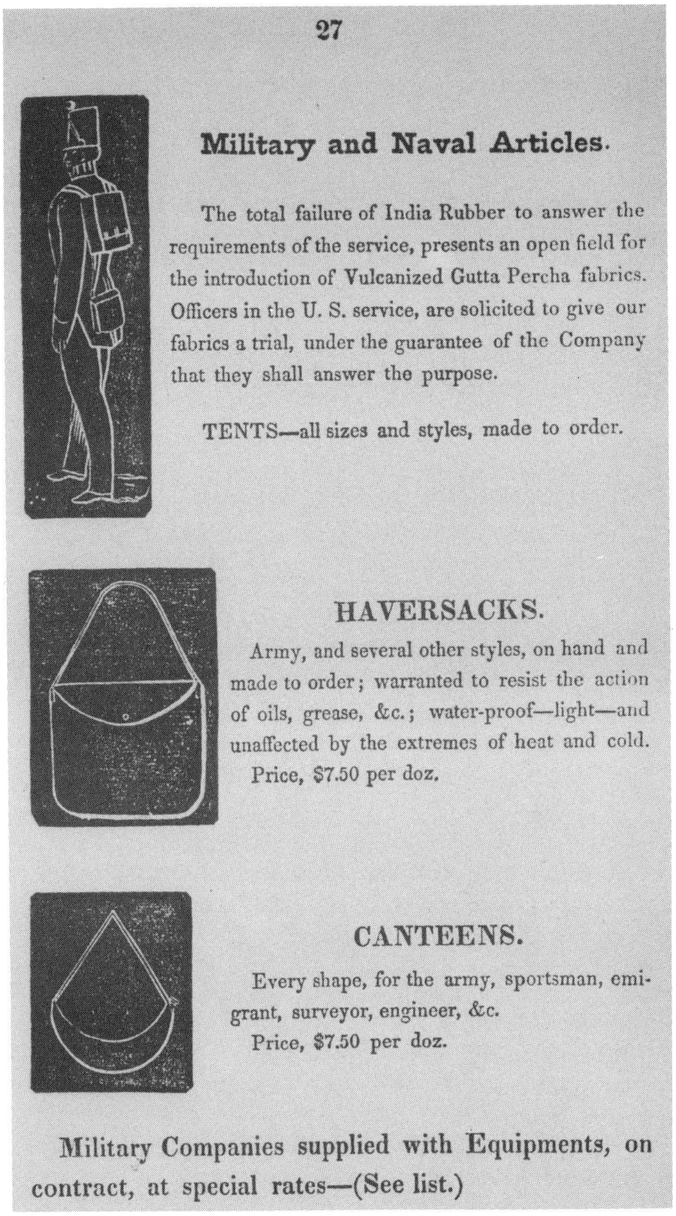

27

Military and Naval Articles.

The total failure of India Rubber to answer the requirements of the service, presents an open field for the introduction of Vulcanized Gutta Percha fabrics. Officers in the U. S. service, are solicited to give our fabrics a trial, under the guarantee of the Company that they shall answer the purpose.

TENTS—all sizes and styles, made to order.

HAVERSACKS.

Army, and several other styles, on hand and made to order; warranted to resist the action of oils, grease, &c.; water-proof—light—and unaffected by the extremes of heat and cold.
Price, $7.50 per doz.

CANTEENS.

Every shape, for the army, sportsman, emigrant, surveyor, engineer, &c.
Price, $7.50 per doz.

Military Companies supplied with Equipments, on contract, at special rates—(See list.)

Military articles. *From the 1854 catalog of the North American Gutta Percha Company. (National Archives)*

The Board met again on April 20th and filed the following report:

An examination was then made of several article which had been folded and rolled up into a compressed package, as also the canteens containing the liquids before mentioned, all of which had been subjected for a week to continuous heat of from 100 to 150 deg. Fahrenheit in a close room.

The articles contained in the package were found to be not impaired by the action of the heat, and were apparently in the same state as before being submitted to the test.

There was a slight taste imparted to the spirits of wine, but no change could be perceived in the oil, vinegar, or water, from the action of the Gutta Percha upon them. Several articles were placed in an ice house, embedded in the ice, and after being kept there for several days, were examined while in a very cold state, and found to be very pliable.

Several articles were submitted to the Board today, which are not enumerated in the list before mentioned. These are a military cap, or hat of a new pattern, a bridle rein, and a sword scabbard, all made of Gutta percha.

All the tests that the Board have been able during the limited time it has been in session, to apply to the fabrics presented, has been perfectly satisfactory to all the members, and they would recommend, for a practical service trial upon the frontiers, the articles herein enumerated, to be placed in the hands of troops, in order to ascertain whether they will prove serviceable in the field.

Several letters from officers in the Army and Navy have been presented to the Board, which give the results of their experience in the use of Gutta Percha, all of which are favorable.

The Board would especially recommend for the use of the enlisted men the

> military shawl, knapsack, haversack, canteen, valise, stock, cap and cover or hat, bridle rein, saddle bags, and sword and bayonet scabbard.
>
> The Board also recommend for the Expedition that has been ordered into the Plains, under General Harney, the large sacks for the transportation of flour, sugar, coffee, medical and ordnance stores, or other articles that are liable to become damaged by exposure to moisture or the weather; also, the tanks for carrying water or other liquids, and the water buckets for the use of trains.
>
> In the economy of transportation, as in the preservation of the contents, these articles are regarded by the Board as very superior to the wooden barrels, etc., or the ordinary cloth grain sacks.
>
> The Board, in conclusion, would unanimously recommend to the consideration of the Department the propriety of adopting the new pattern military hat presented by the Gutta Percha Company. It is believed that this hat, in addition to being water-proof, is much better adapted to frontier service than the cap at present in use. It has a broad rim, which will protect the head from the rain and sun, and can be compressed into any shape without injury, and it is regarded as very serviceable.[37]

The report is signed by Lt. Col. J. E. Johnston—First Cavalry, Major G. H. Crosman—Quartermaster, and Captain R. B. Marcy—Fifth Infantry.

Correspondence in 1855 confirms that some gutta-percha articles were in service early in that year. William Morrine, Commanding the U. S. Naval Forces in the Pacific Ocean, wrote to William Rider from the Flagship Independence on February 5, 1855 that "These goods are much superior from those manufactured from India Rubber...greater pliability and their freedom from the smell & sticking property peculiar to that article."[38] On May 21, the Quartermaster General wrote to Asst. Quartermaster

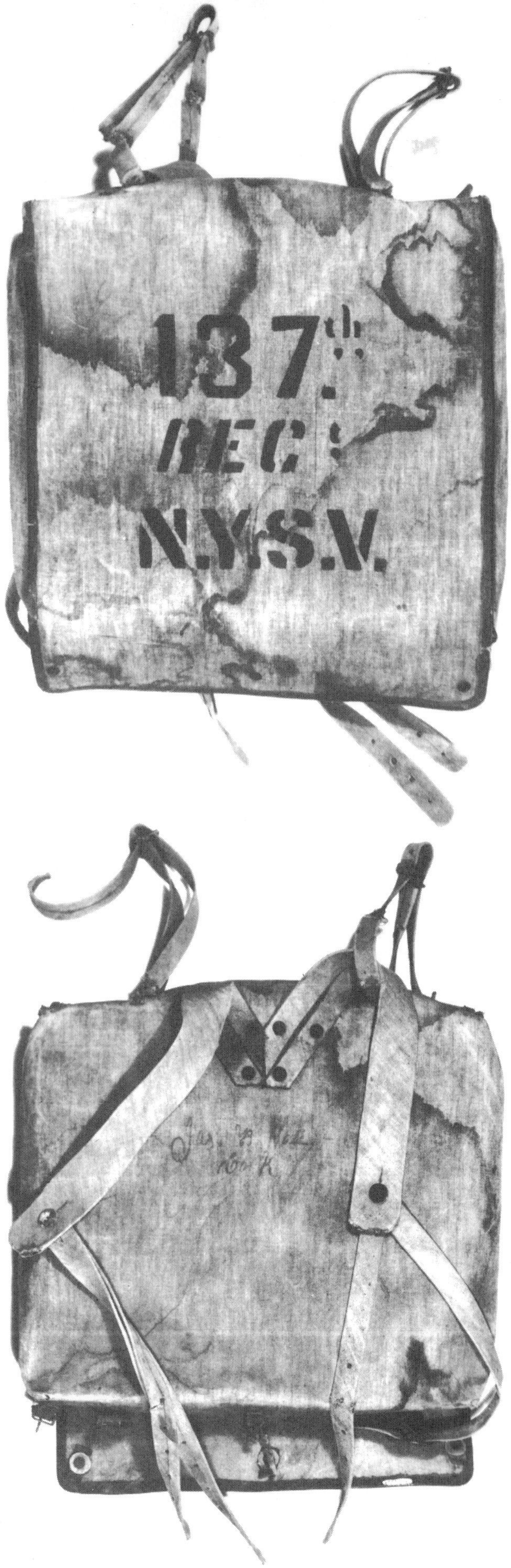

Gutta-percha knapsack, *front and back, worn by James B. Hall of Company K, 137th New York. The gutta-percha is covered with white canvas material on both sides and displays a large faded red "Rider and Murphy" patent marking inside the flap. Although gutta-percha knapsacks were tested in the mid-1850s, the Feb. 18, 1862 patent date on the buckles indicates manufacture during the Civil War. (Courtesy West Point Museum)*

General Major G. H. Crosman telling him "The Secretary of War approved providing Gutta Percha Knapsacks...you will have manufactured a sufficient number for two regiments."[39]

1st Lieutenant Lymun Bissell of the 9th Regiment, U. S. Infantry, reported results of tests on various gutta-percha articles to his commander in a letter from Fort Monroe, Virginia, dated September 13, 1855. His report to Colonel George Wright of the same regiment is more objective than the printed endorsements that were obviously solicited by the Company:

> After having had in use in Company A, 9th Infantry, the following articles of Camp & Garrison Equipage manufactured from gutta percha, viz Haversacks, Canteen, and Uniform Caps, I have the honor to report as follows:
>
> The Cap I consider preferable to the ordinary Uniform Cap, inasmuch as it is not liable to injury from being crushed or broken, or from damp or rainy weather. In weight it is one fourth of a pound heavier than the ordinary cap.
>
> I have kept a Gutta percha Canteen filled with water for twenty days, at the expiration of which time the taste or quality of the water did not appear to have become in the least bit affected.
>
> The Haversack I consider far preferable to the canvass or duck haversack formerly in use in the Army.
>
> There are no Gutta Percha Knapsacks in use in Company A.[40]

In the meantime, the remains of any India-rubber articles were being put up for sale. In a letter to the Military Storekeeper, Captain R. M. Potter, Major Crosman commented on items to be included in a sale to be held November 15, "You may add all the India Rubber goods, except the best of the Canteens and a few Water Tanks, you have on hand." In the same letter, Crosman referenced a question about painted knapsacks and wrote, "I have ordered a year's supply of them from the Gutta Percha factory in New York.[41]

Aside from Crosman's orders regarding India-rubber goods, it is evident that not everyone agreed with the overall endorsements being made, nor was everyone in favor of immediately accepting the articles into general service. On November 22, 1855, Lt. Col. Charles Thomas, Assistant Quartermaster General, wrote the following report to Secretary of War Jefferson Davis:

> I herewith respectfully return the letter General W. S. Harney returned to this Office on the 6th inst. and have the honor to report that, I am unable to state what articles General Harney alluded to in his letter, as he had various Articles of Gutta Percha to experiment upon; and I understand from the Qr. Mr. General that he was as much at a loss when he made the endorsement.
>
> The only articles of Gutta Percha which can be considered as having been in constant use for the past year are Uniform Caps, Cap Covers, Knapsacks, Haversacks and Canteens.
>
> The Cap & Cap Cover are part of the Soldier's allowances of Clothing, and a limited number furnished during the term of service. Should the Gutta Percha Articles be better and more durable, it follows that a less number would suffice.

*:**A kidney-shaped gutta-percha canteen** with a white canvas exterior is seen in this early Civil War image of a Northern militiaman. (Courtesy Rick Carlile)*

service and would recommend its general adaption with out those made of leather or canvas which might answer as well could be furnished more economically.[57]

A few days earlier, Quartermaster General Jesup had written to Swords concerning the worthlessness of India-rubber and gutta-percha canteens, explaining that all those at Philadelphia had been sold and none but tin canteens covered with cloth would be issued hereafter.[58]

In April of 1859, a Board of Officers examined a "Tent Knapsack," patented by Captain William B. Johns of the 3rd Infantry (U.S. Patent No. 23,582) and manufactured by The Gutta Percha Manufacturing Company. The sheet of gutta-percha with straps, buckles, brass studs, and buttonholes was intended to replace the knapsack and horseman's valise. When buttoned together with three other units, the sheets formed a shelter tent. The manufacturer claimed that it "...could be applied to many other useful purposes, as a water-proof cape or talma, as a bed cover, as a pack cover, as a dry shelter for ammunition, and other uses." The Board was "...satisfied with its merits in all uses to which it is proposed to be put, and is of opinion that the Gutta Percha Tent Knapsack may be adopted in the military service with advantage."[59]

At the same time, the Board examined and approved a gutta percha water keg that was two feet long, one foot in diameter, and held ten and a half gallons of water. It could be collapsed when empty and could also be used as a float or pontoon for rafts

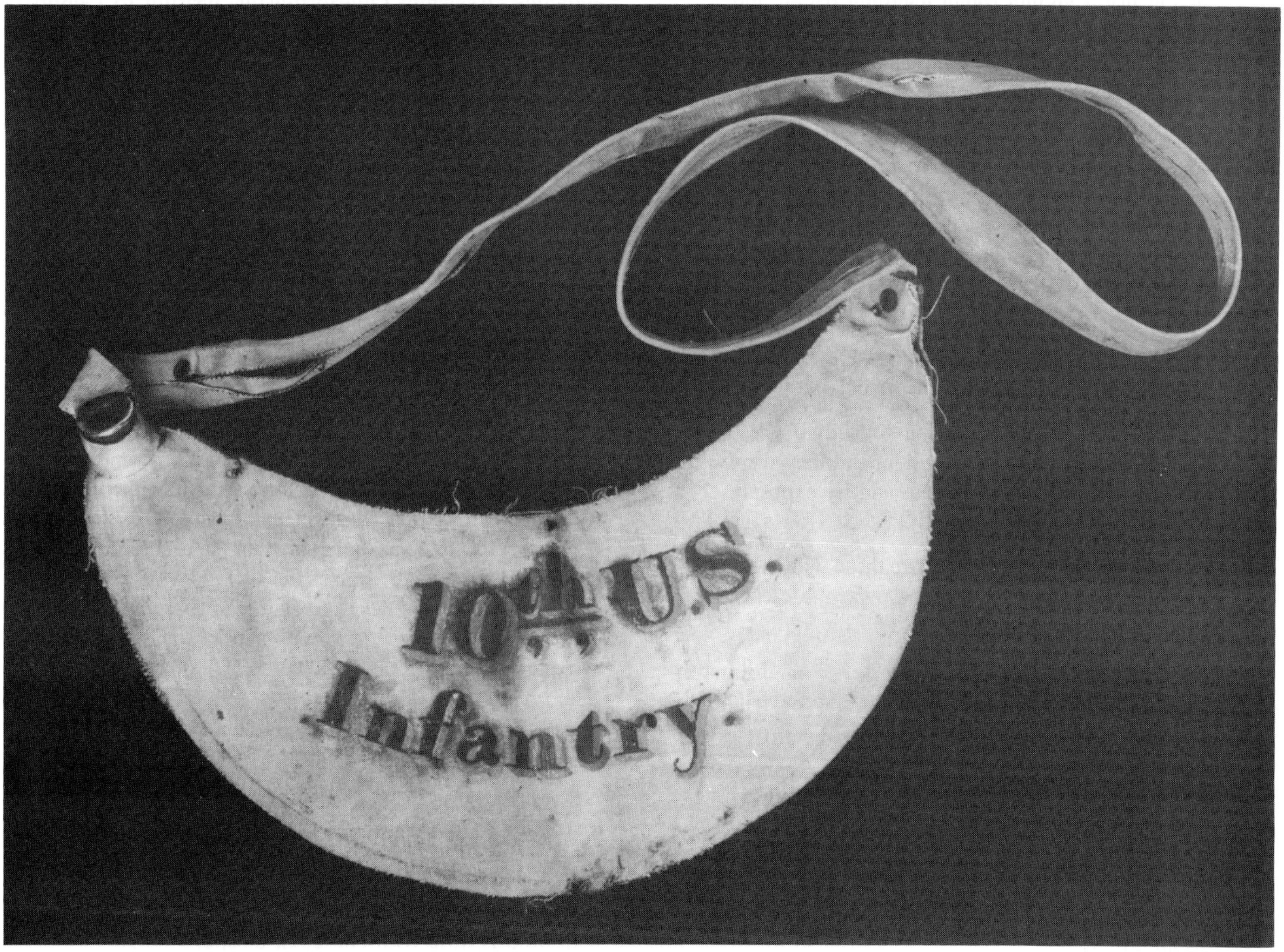

Gutta-percha canteen *that was likely among those tested by the U. S. military in the mid-1850s. Cloth fabric with vulcanized gutta-percha on inside of body and strap. Pewter spout. Marked on front "10th U.S. Infantry." Stenciled on the reverse is "Vulcanized Gutta Percha - Made by North American Gutta Percha Company - Rider's and Murphy's Patent 1852 & 1854." (Courtesy National Museum of American History, Division of Armed Forces History, Smithsonian Institution)*

when crossing rivers. On April 12, Secretary of War John Floyd approved an order of not more than 1,000 tent knapsacks and 300 water kegs for trial, which were ordered from John Rider on April 15.[60] On June 16, 1859, Secretary of War Floyd wrote to the Quartermaster General referencing this order and stated, "The tent knapsacks recently ordered to be procured for trial, will be made with different kinds of side fastenings, both with studs, and with strings as proposed by Captain Johns; for the purpose of testing their relative merits."[61]

On March 25, 1860, the Regiment Quartermaster of the 8th Infantry in Fort Davis, Texas reported to Asst. Quartermaster General Vinton that gutta-percha water kegs were less durable than oak & cedar but were more convenient and when empty were much lighter, took up less room and are were easily packed. They imparted no taste of the gutta-percha to the water. While they were easily rubbed and broken in cold weather, he felt, "In careful hands, the Gutta Percha keg is probably the best."[62] Captain J. H. Carleton of the 1st Dragoons at Fort Tejon, California recommended an extra thickness of heavy twilled cotton be placed outside the gum to prevent chafing and having the gum rub off.[63]

The gutta-percha trials of the 1850s were reminiscent of the India-rubber trials of the 1840s and early 1850s. And, while traditional materials continued to be favored by the military, the troops shared a glimpse of applications for these versatile, waterproof, elastic materials that, when perfected, would be utilized by armies of the future. For many, the experience with some of those future military applications began with the arrival of the 1860s.

NOTES - Section 6

[1] National Archives, Records Group 92, Entry 225, Consolidated Correspondence File, "Rubber."
[2] Bruce S. Bazelon and William F. McGuinn, *Supplement to A Directory of American Military Goods Dealers & Makers 1785-1915* (Manassas, VA: Published by the Authors, 1995) p. 35.
[3] Bruce S. Bazelon and William F. McGuinn, *A Directory of American Military Goods Dealers & Makers 1785-1915* (Manassas, VA: Published by the Authors, 1990) p. 121.
[4] NA, RG 92, Entry 2117, Masterson Entry 35, Commissary General of Purchases, Letters Sent, Book "L," pg. 326.
[5] *The Army and Navy Chronicle*, 22 Sep. 1836, Vol. 3, No. 12, Page 177. Bomford's 1833 report and this citation are mentioned in an unsigned, undated report showing potential savings to the Army by using India-rubber covers for the small arms on hand in 1845. Unsigned report: NA RG 92, Entry 225, CCF, "Goodyear's Metallic Rubber Company."
[6] NA, Office of the Chief of Ordnance, RG 156, Letters Received, 1836.
[7] *Ibid.*, Entry 944, Correspondence relating to Inventions, 1812-1870, IN-7-22.
[8] NA, RG 92, Entry 225, CCF, "India-Rubber." Letter from Captain Mackay to Colonel H. Stanton.
[9] *Ibid.*, "Goodyear's Metallic Rubber Co."
[10] *Ibid.*, "India-Rubber."
[11] *Ibid.*, "Rubber." Letter from S. T. Armstrong to Adjutant General R. Jones, dated September 14, 1839, references the Board of Officers' report.
[12] *Ibid.*, "India Rubber Bags." Report to Colonel George Talcott, Commander—Ordnance, from Captain William H. Bell of Washington Arsenal.
[13] *Ibid.*, "India-rubber."
[14] George W. Cullum, *Papers on Practical Engineering, No. 4, Description of a System of Military Bridges with India-Rubber Pontons* (New York: D. Appleton & Co., 1849) p. 279.
[15] NA, RG 92, Entry 225, CCF, "India-Rubber."
[16] Charles Goodyear, *Gum-Elastic and Its Varieties* (New Haven: Published by the author, 1853 & 1855) Vol. II, p. 137.
[17] NA, RG 92, Entry 225, CCF, "India-Rubber."
[18] *Ibid.*, "Goodyear's Metallic Rubber Co."
[19] *Ibid.* Letter dated January 25, 1844 from Commander William Hudson and Lieutenant Alex Gibson to Captain Stringham, Commander of New York Navy Yard, regarding tests ordered on gun locks, primes and India-rubber cylinders on 32 pounder long guns.

[20] *Ibid.*
[21] *Ibid.*
[22] *Ibid.*
[23] *Ibid.*
[24] NA, RG 156, Office of the Chief of Ordnance, Entry 21, Letters Received, R-132 1847.
[25] NA, RG 92, Entry 225, CCF, "Goodyear's Metallic Rubber Co."
[26] *Ibid.*
[27] NA, RG 92, Entry 25, "Orders for Army Supplies, No. 1, June 17, 1846 - February 15, 1847."
[28] General Orders No. 64, War Department, 29 Nov. 1848; G.O. No. 13, W. D., 12 May 1853; G.O. No. 7, W. D., 12 May 1854, as copied in the Armed Forces Institute of Heraldry, "U.S. Army Equipments, 1834-1870."
[29] NA, RG 92, Entry 25, "Orders for Army Supplies, No. 1, 17 June 1846 - 15 February 1847." Comments of January 5, 1847 relate to the variance in the patterns.
[30] *Ibid.* Rejections dated August 26, 1846 and January 26, 1847
[31] Goodyear, *Gum-Elastic and Its Varieties,* Vol. II, pp. 129-130.
[32] *Ibid.*, pp. 129-139.
[33] *Ibid.*, pp. 141-157.
[34] NA, RG 92, Entry 225, CCF, "North American Gutta Percha Company."
[35] *Ibid.*, "Gutta Percha."
[36] *Ibid.*
[37] *Ibid.*
[38] *Ibid.*
[39] *Ibid.*
[40] *Ibid.*
[41] *Ibid.*, "Crosman."
[42] *Ibid.*
[43] NA, RG 156, Entry 21, Box 148, "1855 M-S."
[44] *Ibid.*
[45] NA, RG 156, Entry 6, Vol. 15, p. 238.
[46] *Ibid.*, Entry 994.
[47] *Ibid.*
[48] NA, RG 92, Entry 332, "Maj. G. H. Crosman, Letters Received, 1835-62."
[49] *Ibid.,* Entry 999, Letters Sent CI, OQMG. Letter from Jesup to Crosman.
[50] *Ibid.,* Entry 1003, File Number C 908, Register of Letters Received Clothing, OQMG. Letter from Crosman to Jesup.
[51] Pise' is defined in unabridged dictionaries as "A building material consisting of stiff earth or clay rammed in between forms."
[52] A drugget is defined as "A course cloth made of felt and printed in various patterns and colors. Not only employed as a carpet but to underlie carpets." Florence M. Montgomery, *Textiles in America—1650-1870,* (New York: W.W. Norton & Co., 1984) p. 226.
[53] "The Tent Knapsack" Booklet, p. 13; NA, RG 92, Entry 225, CCF, Box 518, "Knapsacks."
[54] NA, RG 92, Entry 225, CCF, Subject unknown.
[55] *Ibid.*
[56] *Ibid.*
[57] *Ibid.*
[58] *Ibid.,* Letters Sent, Entry 999, Clothing and Equipment. Letter dated December 1, 1858 from Jesup to Swords.
[59] "The Tent Knapsack" Booklet, p. 10; NA, RG 92, Entry 225, CCF, "Knapsacks."
[60] *Ibid.*
[61] NA, RG 107, Letters Sent, Secretary of War, Vol. 41.
[62] NA, RG 92, Entry 225, CCF, "Gutta Percha."
[63] *Ibid.* Captain Carleton's letter is dated July 26, 1860.

SECTION 7

THE 1860s AND THE CIVIL WAR

"The India Rubber interest is powerful, and of its fabrics some are of a certain value to the soldier. Up to this time the Government has not decided upon any fixed policy as to the use of India Rubber & Gutta Percha."

Quartermaster General M. C. Meigs to Secretary of War Simon Cameron, September 25, 1861

At the close of the decade of the 1850s, with the nation becoming increasingly polarized on key political issues, the following comments made by the writers of *The Scientific American* at the conclusion of their tour of the New York Rubber Belting Company seem optimistic and, at the same time, ominous:

> As we journeyed homeward, we thought that this was the true age of Iron and Gold, the age of the Steamship and the Electric telegraph, of the Railroad and the Rubber Factory; the age which finds its wonders, its miracles, and its romance in the peaceful achievements of science and the mechanical arts, not in the strife of men or the battlefields of nations.[1]

Nineteen months later, Americans were using the products of science and invention to outfit their armies for war against one another. Military applications for India-rubber and gutta-percha were minimal as peace continued on into the 1860s. But now the War, and the need to outfit thousands of troops with the best equipment available, provided opportunities for the evolving rubber and gutta-percha industries.

Early in the War, many volunteers were issued equipment by their individual states or even specific units. While some of this equipment—including most India-rubber and gutta-percha

CDV shows Federal sergeant *with popular waterproof kepi cover. (Courtesy of Mike McAfee)*

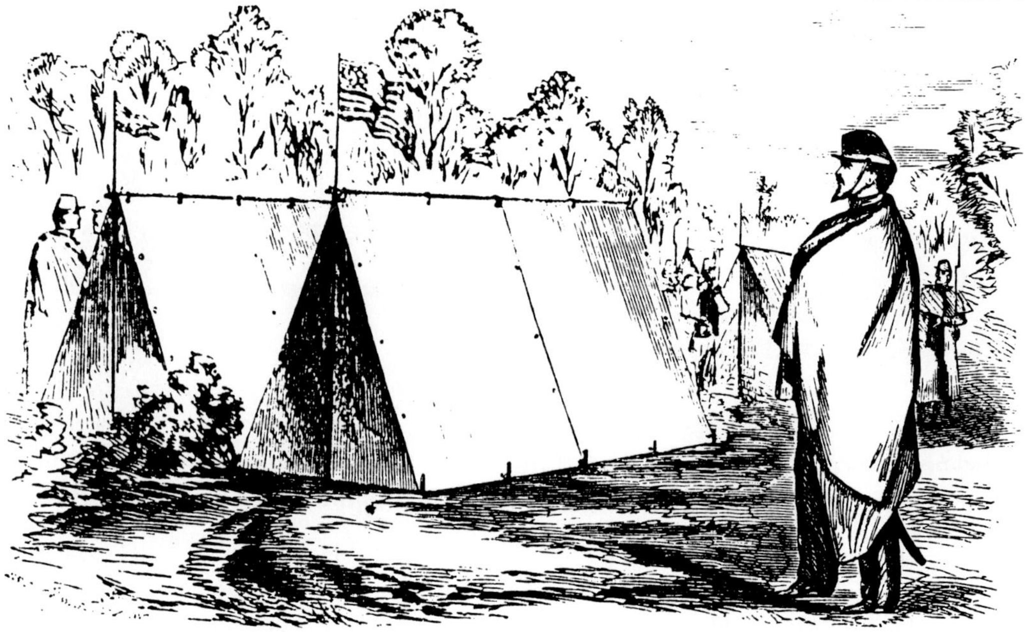

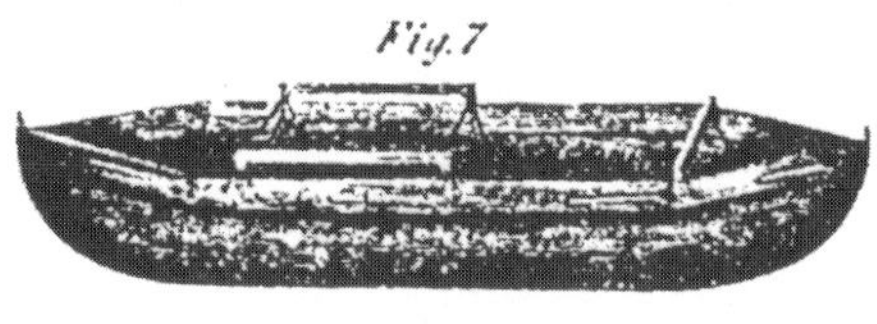

PORTABLE BOAT

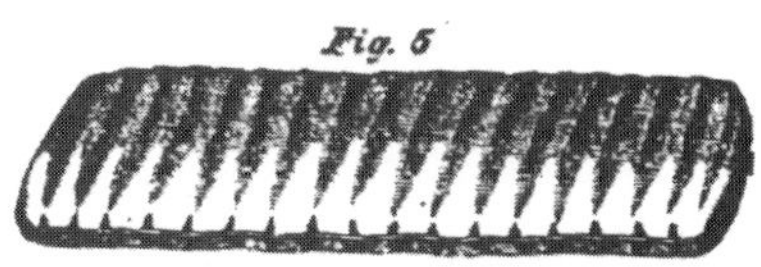

AIR MATRESS

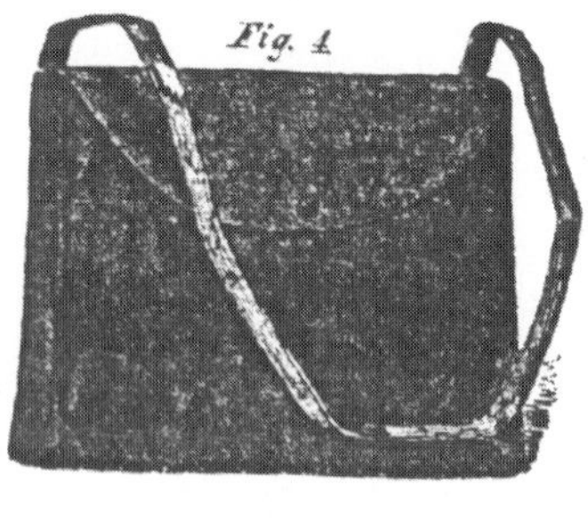

HAVERSACK

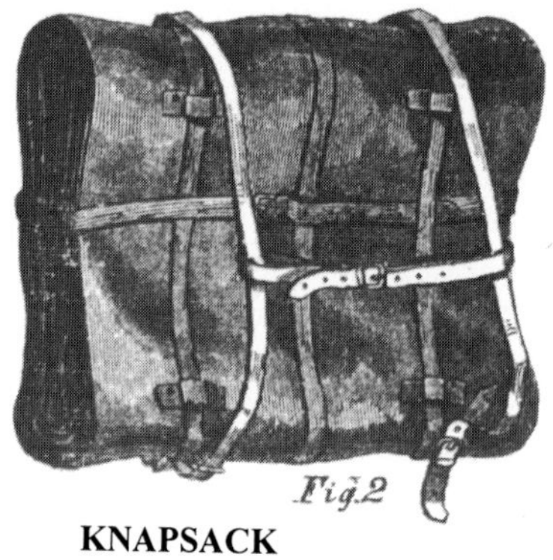

KNAPSACK

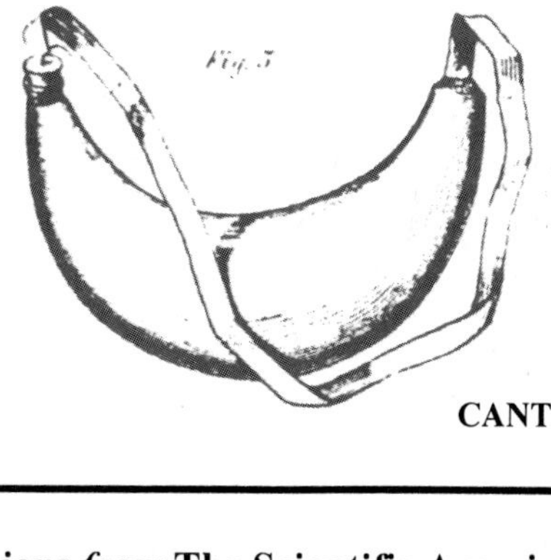

CANTEEN

LEGGINGS

Illustrations from* The Scientific American *of June 22, 1861*, showing various military items made by the Rubber Clothing Co. The article on the company mentions an order for 40,000 yards of enameled India-rubber cloth placed shortly before the start of the War for shipment to Europe. When it was determined that the goods were actually to be diverted to the secession states, the company refused to ship them.*

articles—had not been formally tested and accepted by the Federal Government, their use by the volunteers opened the door to aggressive sales efforts by the India-rubber and gutta-percha companies.

John Rider of the Gutta Percha Manufacturing Company had sent a sample of a gutta-percha knapsack directly to the Secretary of War on June 21, 1861. His letter of the same date referenced a sample army cot sent earlier in the year.[2] He sent the following letter to Quartermaster General Meigs on July 22, 1861:

> When in your Office before I left Washington you informed me you wanted soon two thousand Gutta-Percha Army Camp Blankets. Did you expect me to go and make them; or will you send me the order. I can make them in about ten days, at $2.00 two doll's each, all complete; same as Sample Exhibited to you in the War Office.[3]

Controversy and Acceptance

Neither India-rubber or gutta-percha blankets had been approved as standard issue, but comments in official correspondence indicate that sales were made directly to some units of the service and debates ensued. The following letter from the Quartermaster General to the Secretary of War summarizes his feelings:

> Quarter-Master General's Office
> Washington City, Sept. 25th 1861
>
> Hon. Simon Cameron,
> Secretary of War
>
> Sir—The India Rubber interest is powerful, and of its fabrics some are of a certain value to the soldier. Up to this time the Government has not decided upon any fixed policy as to the use of India Rubber & Gutta Percha.
>
> Some of the States have furnished their volunteers with India Rubber blankets. These seem to be much liked by the men, and doubtless as long as the men are not moved much—not compelled to make long marches—they are of great use to them. But as the soldier must carry every article in his possession upon his back, it becomes a serious question whether, in giving him these water proof blankets in addition to the heavy load which he is now obliged to carry under the regulations, we are not doing him more harm than good.
>
> I have, at the earnest solicitation of the Governor of West Virginia, ordered a large quantity of water proof blankets and talmas for the troops of that state. I have received a requisition from Gen. Fremont for 24,000 water proof blankets and 8000 talmas for his army. This will involve an expenditure of some $80,000. If charged to the soldier upon the muster roles the expenditure will be ultimately in some measure refunded; but many of them will be lost, many of them will be thrown away by order upon the battle-field, and for these the government must ultimately pay.
>
> I think that the Department should no longer leave this question undetermined. Some quartermasters buy them; others, not finding them in the regulations, refuse.
>
> To provide them for the whole army will involve an expenditure, not estimated for, of probably a million of dollars. I do not think that

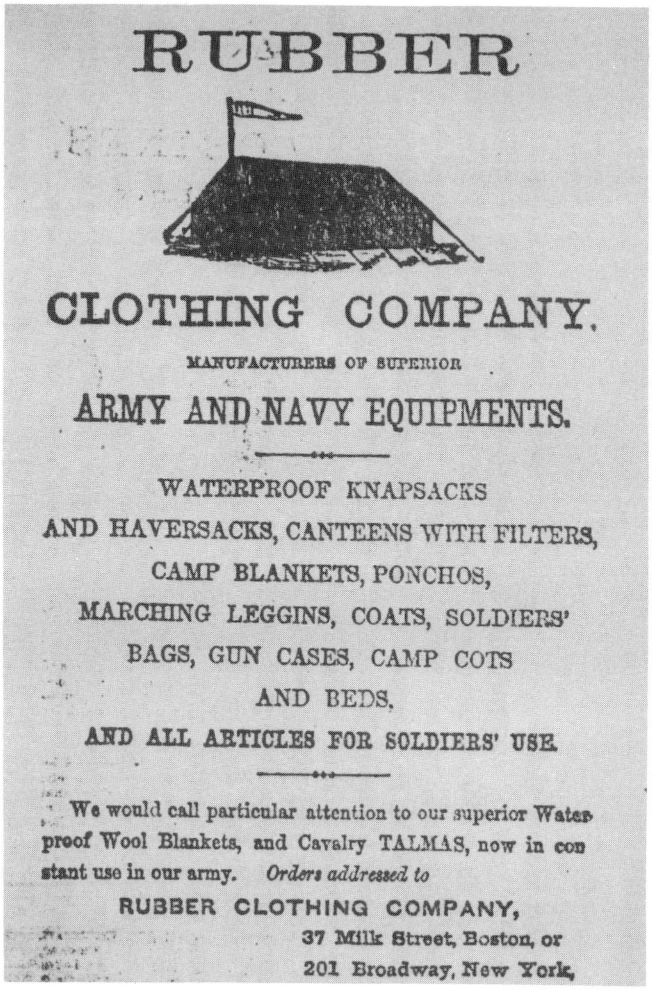

***Rubber Clothing Company advertisement** from* Patten's Infantry Tactics, Bayonet Drill, and Small Arms Exercise, *published in 1863, showing a variety of military goods. (Courtesy of Denis E. Reen)*

any other articles but the blankets should be purchased, and I am doubtful for the reasons above given whether these should be?

If it is determined to make them articles of supply, they should be provided in sufficient quantity to be issued to all soldiers who desire them, and should be charged to the muster roles as issued.

To provide so large a quantity large contracts must be made on proper advertisements and proposals.

I await the instructions of the Secretary.

M. C. Meigs
Qr. Mr. Gen'l.[4]

On September 27, 1861, Meigs sent Lt. Col. G. H. Crosman, Assistant Quartermaster General, in charge of the U.S. Army Office of Clothing and Equipage, Philadelphia, the following letter:

> Your letter of the 25th inst. relating to the purchase of India Rubber or Gutta Percha Blankets is at hand. In reply I have the honor to inform you that in special cases and on application of Governors of States some purchases and issues have been made. The whole matter is now before the Secretary of War for his decision whether purchase shall continue to be made or no. Until decided, do not buy.[5]

Even as waterproof blankets were being considered and subsequently approved as standard Union Army issue on November 1, 1861,[6] India-rubber and gutta-percha companies were continuing their pursuit for more business. On October 7, 1861, Rider referenced the report on the experiments of 1855 in the following note to General Meigs:

> I have taken the liberty to enclose to you some copies of Official reports on Gutta Percha Equipments made by order of Sec'y of War & Navy. If you will be good enough to look them over you will much oblige.[7]

Rider continued to press for considerations of his gutta-percha products by the Army as a letter dated December 26, 1861 indicates:

> Brig. Genl. Meigs
> Qr Mr Genl.
> US Army
>
> Sir—I would respectfully request you allow me to have a certified copy of Capt. Hancock's Report on Vulcanized Gutta Percha Army Buckets dated Benecia California Dec 11th 1858. Also Col. A. S. Johnston Report on Cavalry Talma dated San Antonio Jan 19th 1857.
>
> I would respectfully request you call for a report on Gutta Percha French Gray Knapsacks issued to the 5th Maine Regiment & Blanket-Knapsacks issued to the Garibaldi Guards.
>
> John Rider
> 139 Broadway New York[8]

The following letter from the Gutta Percha Manufacturing Company's brochure indicates that Rider went directly to the "Garibaldi Guards" for his report on the blankets and knapsacks:

> Quartermaster's Department
> Garibaldi Guard, 39th N.Y.S.V.
> January 30, 1862
>
> Emory Rider, President
> Gutta Percha Manufacturing Co, New York
>
> Sir—In reply to your letter of the 25th instant, I have the pleasure in reporting to you that the Vulcanized Gutta Percha Blankets and Knapsacks made by you, and with which I have been fortunate enough to supply my Regiment, are doing most excellent service; and the Blankets, especially, are nearly, if not absolutely indispensable, during this wet weather which we have had for a time.
>
> All of our officers and men speak in the highest terms of them, and I believe them to be the best article of the kind ever made and the only water-proof blankets I have ever seen.
>
> Lieut. E. D. Lazell
> Quartermaster, 39th N.Y.S.V.[9]

Rider was ultimately awarded contracts for 27,040 gutta-percha knapsacks. The Gutta Percha Manufacturing Company provided 63,000 gutta-percha blankets and 35,000 ponchos under contracts with the United States Quartermaster's Department.[10]

India-rubber Patent and License Infringement

The right to manufacture India-rubber goods for the military became a legal issue in late 1861. On November 6, 1861, H. G. Hadden, President of the Union India Rubber Company, wrote to Quartermaster General Meigs stating that his company "...proposes to furnish the Government with Vulcanized India Rubber Blankets, at a reasonable price, and with an article of the best quality." Hadden proceeded to give a detailed history of Goodyear's vulcanization patent, and emphasized the fact that the Union India Rubber Company had been assigned full

Cape or Cloak, (no sleeves.)

	EACH.
Vulcanized White Rubber, ...	$14 37
do. Dull finish, ...	12 12
Cured—fine light, ...	8 50
Macintosh, light, ...	8 50
Unvulcanized Black or White, ...	
Ladies' Capes, Flocked, Blue or Brown, ...	15 00

***Above, cloak illustrated** in The Rubber Clothing Company's catalog dated October 1, 1864. Note button hole design similar to Kellogg's, at right.*

***India-rubber cloak** worn during the Civil War by Capt. Henry Kellogg of the 33rd Illinois. Kellogg's diary entry for May 19, 1863 notes, "Passed a sleepless night with only my rubber cloak for bed—was very cold commence our advance at 5 o'clock...." The unfortunate captain was the only member of his unit killed during the ensuing assault on Vicksburg. (Courtesy Mike Cunningham)*

***Close-up, below,** shows rubber-coated fabric and distinctive design around button hole.*

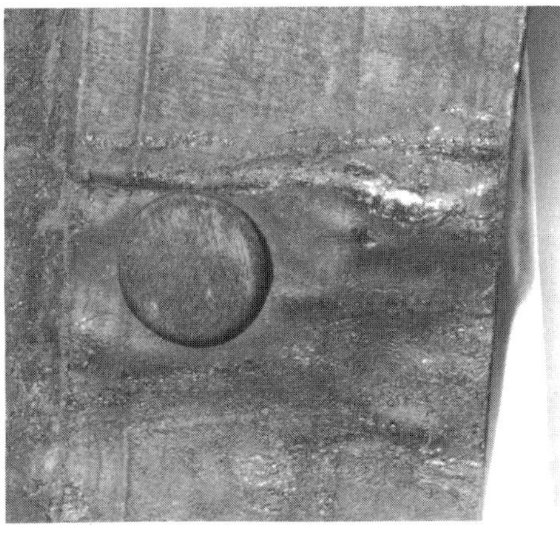

rights to making "...all Army and Navy Equipments (Embracing Blankets)...on July 1, 1848." He further stated, "There are many counterfeits which are worthless, and will prove so on use," and admonishes the Government to not contract with them in conflict with the patent they had issued.[11]

On May 1, 1862, the Union India Rubber Company assigned a license to Warren Ackerman & Company to "...make and sell rubber goods in the same manner as the Union India Rubber Company." On July 31, 1862 that license was transferred to the Phoenix Rubber Company, which manufactured many of the Union's rubber blankets.[12] Blankets were also manufactured by the Boston Rubber Shoe Company, under license from the Union Company.[13]

A bid letter was written by George Cornwall of the Union India Rubber Company on June 18, 1862, stating that his Company:

> Will contract top deliver pure vulcanized rubber blankets or ponchos...have the facilities for manufacturing from 3,000 to 5,000 of these Blankets or Ponchos daily and we warrant our goods in all cases to give full satisfaction—being thoroughly waterproof—not affected by any extreme of heat or cold—never softening or stiffening.[14]

During this time, the Government continued to solicit proposals and bids for India-rubber goods.[15] On October 30, 1862, a court notice was issued by the Rhode Island District Circuit Court of the United States on a injunction by the Union India Rubber Company and the Phoenix Rubber Company against the Providence Rubber Company. The Providence Rubber Company notified General Meigs that they had only one contract for 20,000 blankets. Judge Advocate General I. Holt informed Secretary of War Stanton that payment on the blankets should be withheld, although the attorney for Providence claimed that Holt did not intend to include the blankets already delivered in his notice. The N. Hayward and Company was also asked by Judge Holt to respond in this case but hadn't done so as of November 21.[16] Interestingly, the Providence Rubber Company received government contracts for 115,000 vulcanized India-rubber blankets and 50,000 ponchos during 1863 and 1864.[17]

Painted and Saturated Alternatives

Other companies manufactured waterproof goods for the military without infringing on the Goodyear or Rider patents. These products were accepted by some and considered inferior by others. A February 9, 1865 letter from Quartermaster General Meigs to Colonel Olcutt, a special commissioner investigating fraud, stated that painted blankets were lighter and as durable, strong, and impermeable as India-rubber or gutta-percha blankets. Meigs commented that it was his understanding that the Navy had adopted painted cloth instead of the more expensive and less durable rubber cloth. He admitted that some alternatives to Goodyear blankets may be inferior but would provide good service. Mention was made of India-rubber blankets made by the "sulphide of carbon process" that had been used with success in Europe.[18]

In addressing quality problems on 50,000 India-rubber blankets manufactured under government contract by Samuel Myers Hurlbert of Boston using the sulphide of carbon process, Colonel Perry wrote to Colonel W. W. McKim of the Cincinnati depot on February 13, 1865, saying the

C D V of unidentified soldier wearing a waterproof cape and kepi cover. *Waterproof clothing made of vulcanized India-rubber and gutta-percha became popular during the Civil War. (Courtesy of Ronn Palm)*

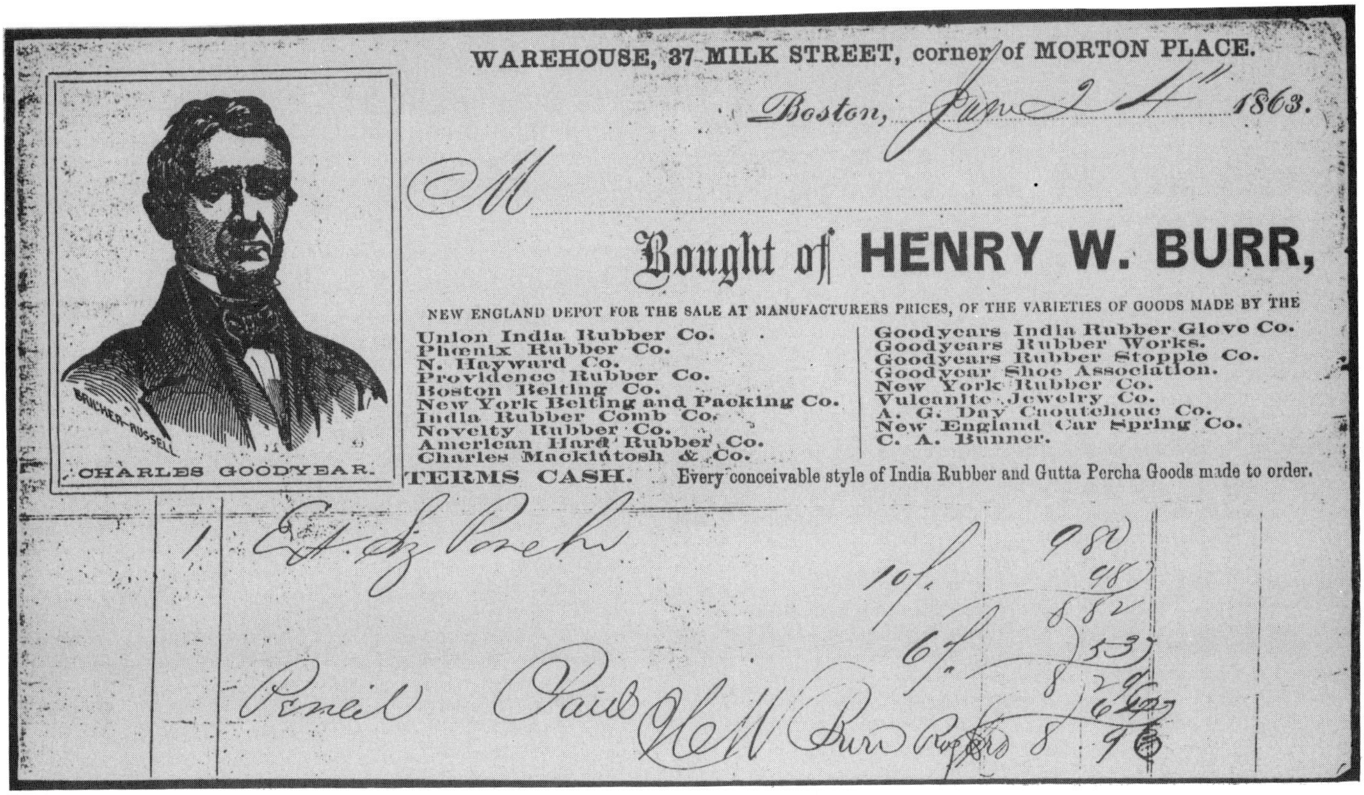
WAREHOUSE, 37 MILK STREET, corner of MORTON PLACE.

Boston, Jan 24th 1863.

M

Bought of HENRY W. BURR,

NEW ENGLAND DEPOT FOR THE SALE AT MANUFACTURERS PRICES, OF THE VARIETIES OF GOODS MADE BY THE

Union India Rubber Co.	Goodyears India Rubber Glove Co.
Phœnix Rubber Co.	Goodyears Rubber Works.
N. Hayward Co.	Goodyears Rubber Stopple Co.
Providence Rubber Co.	Goodyear Shoe Association.
Boston Belting Co.	New York Rubber Co.
New York Belting and Packing Co.	Vulcanite Jewelry Co.
India Rubber Comb Co.	A. G. Day Caoutchouc Co.
Novelty Rubber Co.	New England Car Spring Co.
American Hard Rubber Co.	C. A. Bunner.
Charles Macklntosh & Co.	

CHARLES GOODYEAR.

TERMS CASH. Every conceivable style of India Rubber and Gutta Percha Goods made to order.

1 Ext Siz Poncho 9 80

10/. 98

8 82

6/. 53

8 29

Pencil 67

Paid H W Burr 8 96

1863 Receipt. *The interesting receipt for "1 Ext Siz Poncho" (Extra Size?) and a pencil, likely made of hard rubber, bears the image of Charles Goodyear and lists a number of the India-rubber and hard rubber manufacturers in business at the time.*

blankets were of poor quality and were "...more or less a fraud upon the United States."[19] But, in a letter sent to Colonel Moulton on March 20, 1865, Colonel Perry supported Meigs' position on the use of alternatives to vulcanized India-rubber:

> Quartermaster General has always declined to adopt the Goodyear Patent Rubber Blanket to the exclusion of all others to avoid giving a monopoly to any parties in supply of waterproof blankets and has authorized purchase of large quantities of painted blankets containing india rubber.

He goes on to say that these products have given good service and saved money, and that inferior products were likely due to errors in manufacture.[20] Quartermaster's Department records show contracts for 230,000 painted or saturated blankets and 144,733 ponchos.[21]

Charles Goodyear's nemesis, Horace Day, also surfaces again in the wartime records as a manufacturer of India-rubber and other waterproof goods. The New York Tribune of April 26, 1861 advertised 1,000 of Day's rejected government knapsacks with the statement, "...this article will answer for the emergency."[22]

General Stewart Van Vleit of the Quartermaster's Office of the Army of the Potomac, was a strong advocate of using Day's "Poncho Tents," a combination waterproof blanket and poncho which could be buttoned together with two others to form a tent. The India-rubber poncho tents retailed for $10 each, while those of painted cloth sold for $6.50 each.[23] On January 2, 1862 a Board of officers, including Generals Van Vleit, Porter, Stoneman, and Sykes, convened to examine the merits for military purposes certain leggins, ponchos, and water proof blankets. According to Day, the Board concluded:

> The Shelter Tent Poncho submitted by Horace Day of New York is the best submitted to its inspection and recommends its use with the addition—that button holes and buttons be placed one three sides instead of one and that eyelet holes in the corner be made of metal.[24]

On January 6, 1862, General Van Vleit requested General Meigs order "20,000 ponchos of the kind adopted by the Board for the Army of the Potomac." On January 17, Van Vleit wrote to General Meigs stating, "I have the honor to report that it is the General's intention to furnish them to footmen, as well as Cavalry. They will be furnished in lieu of Water proof Blankets, and be used also, as Shelter tents."[25]

The quality of some of the Day ponchos delivered to the troops was so poor that, on January 18, 1862, General Van Vleit had been requested to personally examine them.[26] In a letter to General Meigs dated the same day, Lt. Colonel Crosman stated, "I have received your telegram and have telegraphed Horace Day that no more ponchos or blankets would be received from him."[27] However, a January 21, 1862 letter from Crosman to Meigs reads:

> No order has been issued from this office to "H H Day" for painted or oil cloth ponchos—I have a sample...said to be like those furnished the Department at Washington but I declined ordering them. ...The only order extended to "H H Day" is for 7000 India rubber tent blankets, 46 x 71 inches, not yet furnished. When supplied they will be rigidly inspected and also tested and reported by Prof. Booth before they are accepted.[28]

From this time through mid-1862, Day was at odds with the Government, admitting that, while there were problems with his early shipments due to the fact that the goods were "painted" as opposed to India-rubber, the later goods were vulcanized India-rubber and of good quality. The Government not only suspended orders but also withheld payments on some shipments due to the poor quality of the goods. General Meigs wanted to suspend the entire contract, but Van Vleit wanted to accept the good ponchos, adding "Unless Mr. Day delivers a good article, I would not, of course, permit them to be received."[29]

Printed testimonials praising "Day's Water-Proof Poncho Tent" were circulated. One letter, signed by twenty officers of the 8th PA Cavalry, was sent to "Horace Day or General Van Vliet [sic]" on March 22, 1862. It read:

> This Regiment has had in use the linen painted Poncho Tents about one month and finds every reason to speak highly of them. We cannot imagine anything in their place that could give more general satisfaction. We have used them as tents, and we also find them an excellent shelter from storm while on march. In short, we would not wish to be without them.[30]

On April 17, 1862, Day wrote to Meigs asking him to reconsider the use of his poncho tents and, on the same day, appealed to Secretary of War Stanton, saying the Government was buying "...many other substitutes of his perfect invention," while he, "...urged by my patriotic solicitude for the welfare of my countrymen, battling for my country's cause, my professional pride, and over $40,000.00 in invested capital, have patiently been waiting the action of the Department." Apparently receiving no satisfactory response, Day wrote directly to President Lincoln on May 22, 1862, claiming that "...one-half of the sickness of the past 3 months in our Armies would have been avoided had each Soldier one of my India Rubber Ponchos, 3 of which also form a Tent." Day also advised Lincoln that he "...has not made a dollar of profit out of anything I have done for either State or Federal Government."[31]

An 1865 Quartermaster General's synopsis of

Pair of India-rubber leggings with leg and foot straps. *One legging is marked: "Rubber Clothing Co., Beverly, Mass.-Patents January 19, 1858, June 29, 1858, October 19, 1858 and June 14, 1859." U. S. Patent #19,172 was issued to Hiram Hall on January 18, 1858 for a process of restoring waste rubber and assigned to the Beverly Rubber Company.*

the "Case of Horace H. Day - Ponchos," recalls:

> On January 18, 1862 General Van Vleit was requested to call and see some of the Day ponchos that had been furnished to the troops in the Army of the Potomac which, after one day's use in the sleet were brought to the office and found upon examination to be absolutely worthless, the varnish being cracked and peeled in the short time it had been used, so as to permit the water to pass through it freely.[32]

In 1867, attorneys for Day resubmitted bills to the Quartermaster General for 2,723 ponchos at a total cost of $4,765.25. On April 27, 1869, a further claim to the Secretary of War for re-examination is noted, "Payment was withheld on account of the worthlessness of the Poncho tents delivered."[33]

Other Sources of Military Goods

A full-length article in *The Scientific American* of June 22, 1861 featured India-rubber blankets, knapsacks, haversacks, leggings, canteens, and other military products made by the Rubber Clothing Company, which "...operated business places in New York, Boston, Philadelphia, Cincinnati, and Chicago."[34] They provided 26,382 India-rubber blankets, 29,950 knapsacks, 35,736 haversacks, and 12,120 canteen straps to the Quartermaster General of Massachusetts as of January 31, 1862.[35] It does not appear that this company had a Goodyear license, but they marketed India-rubber goods under the name "black enamel rubber."[36] Several patents for restoring waste India-rubber and gutta-percha were assigned to the Beverly Rubber Company in 1858 and 1859 (U.S. Patents #19,172, 20,242, 20,678, 22,217, 22,265, 23,740, 23,805, 25,160). Dates of some of these patents are found on Rubber Clothing Company artifacts, so the company may have been licensed by or associated with the Beverly Rubber Company. Examination of a surviving pair of leggings made by the Rubber Clothing Company in Beverly, Massachusetts reveals no noticeable difference between vulcanized India-rubber and their article. The company's catalog of October 1, 1864 listed many India-rubber articles and also offered "unvulcanized," "white," and "flocked" variations.[37] These articles may have been produced by some other process, such as solarization. The white and flocked varieties were likely made with a cloth material or fur on the outside of the rubber article.[38]

In addition to India-rubber and gutta-percha goods contracted by the state and federal governments, many military articles were no doubt sold directly to individuals. Advertisements for India-rubber canteens, knapsacks, tents, and other military articles frequently appeared in newspapers and periodicals.

While manufacture of India-rubber goods in the South was limited, there were at least three recorded manufacturers in the late 1850s and 1860—D. B. Camp of Savannah, Georgia and Brands & Korner[39] and Eagle Manufacturing Company of Columbus, Georgia.[40] It is not certain if these companies had been licensed prior to the War.

In one case, weeks of experiments and dogged persistence resulted in the successful production of a small quantity of India-rubber in the South. On May 25, 1863, Confederate Assistant Engineer C. G. DeLisle sent a report to General Beauregard on the progress of "...a line of electric

Unidentified soldier sporting a waterproof kepi cover. *Various styles of waterproof caps and cap covers were used in the 1850s and 1860s. (Courtesy of Ronn Palm)*

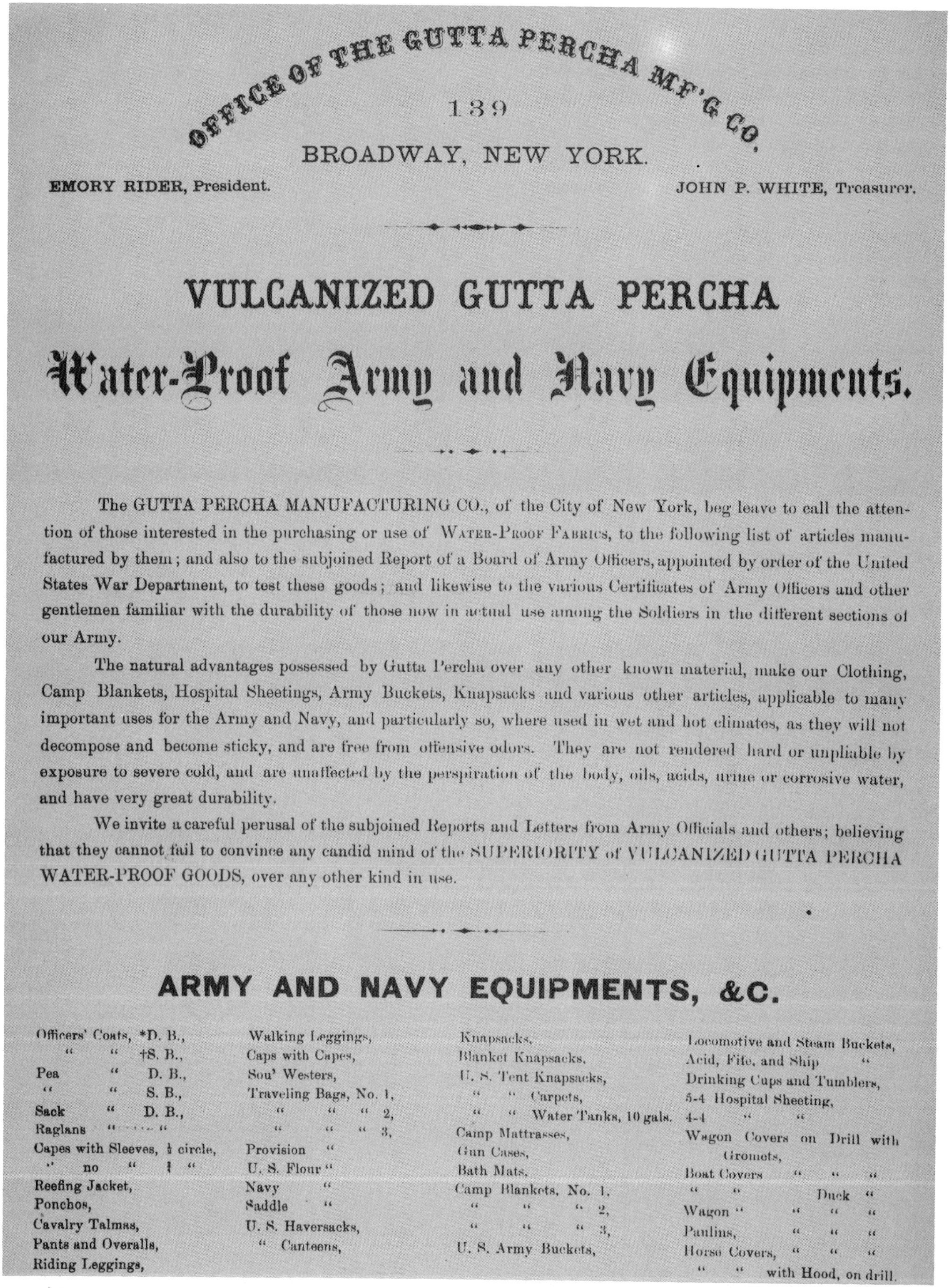

OFFICE OF THE GUTTA PERCHA MF'G CO.

139

BROADWAY, NEW YORK.

EMORY RIDER, President.

JOHN P. WHITE, Treasurer.

VULCANIZED GUTTA PERCHA

Water-Proof Army and Navy Equipments.

The GUTTA PERCHA MANUFACTURING CO., of the City of New York, beg leave to call the attention of those interested in the purchasing or use of WATER-PROOF FABRICS, to the following list of articles manufactured by them; and also to the subjoined Report of a Board of Army Officers, appointed by order of the United States War Department, to test these goods; and likewise to the various Certificates of Army Officers and other gentlemen familiar with the durability of those now in actual use among the Soldiers in the different sections of our Army.

The natural advantages possessed by Gutta Percha over any other known material, make our Clothing, Camp Blankets, Hospital Sheetings, Army Buckets, Knapsacks and various other articles, applicable to many important uses for the Army and Navy, and particularly so, where used in wet and hot climates, as they will not decompose and become sticky, and are free from offensive odors. They are not rendered hard or unpliable by exposure to severe cold, and are unaffected by the perspiration of the body, oils, acids, urine or corrosive water, and have very great durability.

We invite a careful perusal of the subjoined Reports and Letters from Army Officials and others; believing that they cannot fail to convince any candid mind of the SUPERIORITY of VULCANIZED GUTTA PERCHA WATER-PROOF GOODS, over any other kind in use.

ARMY AND NAVY EQUIPMENTS, &C.

Officers' Coats, *D. B.,
" " †S. B.,
Pea " D. B.,
" " S. B.,
Sack " D. B.,
Raglans " "
Capes with Sleeves, ½ circle,
" no " ¾ "
Reefing Jacket,
Ponchos,
Cavalry Talmas,
Pants and Overalls,
Riding Leggings,

Walking Leggings,
Caps with Capes,
Sou' Westers,
Traveling Bags, No. 1,
" " " 2,
" " " 3,
Provision "
U. S. Flour "
Navy "
Saddle "
U. S. Haversacks,
" Canteens,

Knapsacks,
Blanket Knapsacks,
U. S. Tent Knapsacks,
" " Carpets,
" " Water Tanks, 10 gals.
Camp Mattrasses,
Gun Cases,
Bath Mats,
Camp Blankets, No. 1,
" " " 2,
" " " 3,
U. S. Army Buckets,

Locomotive and Steam Buckets,
Acid, Fife, and Ship "
Drinking Cups and Tumblers,
5-4 Hospital Sheeting,
4-4 " "
Wagon Covers on Drill with Gromets,
Boat Covers " " "
" " Duck "
Wagon " " " "
Paulins, " " "
Horse Covers, " " "
" " with Hood, on drill.

Cover page from an advertising booklet for gutta-percha army and navy equipment. *The booklet was printed in 1862 and included testimonials from officers on the use of vulcanized gutta-percha blankets, knapsacks, and talmas in the first year of the Civil War. (National Archives)*

torpedoes intended to obstruct the channel between Forts Moultrie and Sumter."[41] The main portion of the copper cable was covered with cloth and saturated in a simple solution of melted India-rubber (possibly worn-out vulcanized goods). DeLisle then presented the challenge and the solution:

> Each branch (fifty in number) had to receive a special contrivance to secure the current within the line after the shooting of one or several torpedoes; ...An elastic India-rubber tube of a certain diameter was absolutely necessary to complete the line. This article existing not in the limits of the Confederacy, Mr. Waldron applied to General Ripley for a lieutenant who had some experience in the India-rubber manufacture. After fifteen days of trials and experiments, resulting in very imperfect samples, the lieutenant abandoned the work. I then undertook the same, and succeeded at last in making the needed article. I had also to manufacture some brass wire to make the necessary springs for the circuit-closer of the branches.[42]

Virtues Extolled

While the value of gutta-percha and India-rubber to the military was questioned at the beginning of the War, there seemed to be no doubt at the end of the War—especially where India-rubber was concerned. The following excerpts from "A Medical Report upon the Uniforms and Clothing of the Soldiers of the U. S. Army," submitted by the Surgeon General's Office in 1868, provide insights on objections and benefits of India-rubber garments:

> The waterproof covering now allowed is, each year, one blanket for foot troops and one poncho for mounted men.
>
> Their value is best stated in the words of Dr. Parkes, of the British Army, the eminent writer on military hygiene. "Like leather articles, the india-rubber is an exceedingly hot dress, owing to the same causes, viz: impermeability to wind, and condensation and retention of perspiration. It is objected to on these grounds, and especially the latter; If, however, woolen underthings are worn, the perspiration is sufficiently absorbed by these during the comparatively short time waterproof clothing is worn, and the objection is probably not valid, unless the waterproof is continually worn.
>
> The great use of waterproof is, of course, its protection against rain, and in this respect it is invaluable to the soldier, and should be largely used. By the side of this great use, all its defects appear to me to be minor evils."
>
> It is recommended that the present allowance be continued.
>
> (The Territory of Alaska is so exceptional as regards rain that a special allowance should be made for that district. Assistant Surgeon A. H. Hoff, who accompanied the Expedition of Occupation, considers that "outside india-rubber covering is indispensable." He recommends that "a poncho, india-rubber hat with a stiff brim at least three inches wide, thick double-soled kip-skin boots, with the leg reaching to the knee, and a pair of rubber gloves, should be furnished each man to enable him with safety to his health to perform the duties required of him as a soldier.")[43]

Rubber heel stirrups *which buttoned to the bottom of an identified Union officer's pants legs to secure them within his boots. The nonregulation items were purchased privately in February, 1865, following his release from a Rebel prison. Courtesy Kim Hughes and Marianne Moe.*

The most eloquent summation of the use of India-rubber in the Civil War is written by Charles Goodyear biographer, Reverend Bradford K. Pierce, in 1866. While the widespread use of all of the applications he mentions may be exaggerated in dramatic fashion, all were utilized to some extent in the course of the four-year war:

> Some of our readers have been out on the picket line during the war. They know what it is to stand motionless in a wet and miry rifle-pit, in the chilling rain of a southern winter's night. Protected by India rubber boots, blanket, and cap, the picket man performs in comparative

comfort a duty which, without that protection, would make him a cowering and shivering wretch, and plant in his bones a latent rheumatism to be the torment of his old age. Goodyear's India rubber enables him to come in from his pit as dry as when he went into it, and he comes in to lie down with an India rubber blanket between him and the damp earth. If he is wounded, it is an India rubber stretcher, or an ambulance with India rubber springs, that gives him least pain on his way to the hospital, where, if his wound is serious, a water-bed of India rubber gives ease to his mangled frame, and enables him to endure the wearing tedium of an unchanged posture. Bandages and supporters of India rubber avail him much when first he begins to hobble about his ward. A piece of India rubber at the end of his crutch lessens the jar and the noise of his motions, and a cushion of India rubber is comfortable to his arm-pit. The springs which close the hospital door, the bands which exclude the drafts from doors and windows, his pocket-comb and cup and thimble, are of the same material. From jars hermetically closed with India rubber he receives the fresh fruit that is so exquisitely delicious to a fevered mouth. The instrument case of his surgeon, and the store-room of his matron, contain many articles whose utility is increased by the use of it, and some that could be made of nothing else. In a small rubber case the physician carries with him and preserves his lunar caustic, which would corrode any metallic surface. His shirts and sheets pass through an India-rubber clothes-wringer, which saves the strength of the washerwoman and the fiber of the fabric. When the government presents him with an artificial leg, a thick heel and elastic sole of India rubber give him comfort every time he puts it to the ground. In the field, this material is not less strikingly useful. During the late war armies have marched through ten days of rain, and slept through as many rainy nights, and come out dry into the returning sunshine, with their artillery uninjured, because men and munitions were all under India rubber. When Goodyear's ideas are carried out, it will be by pontoons of inflated India rubber that rivers will be crossed. A pontoon train will then consist of one wagon drawn by two mules; and if the march is through a country that furnishes the wooden part of the bridge, a man may carry a pontoon on his back, in addition to his knapsack and blanket.[44]

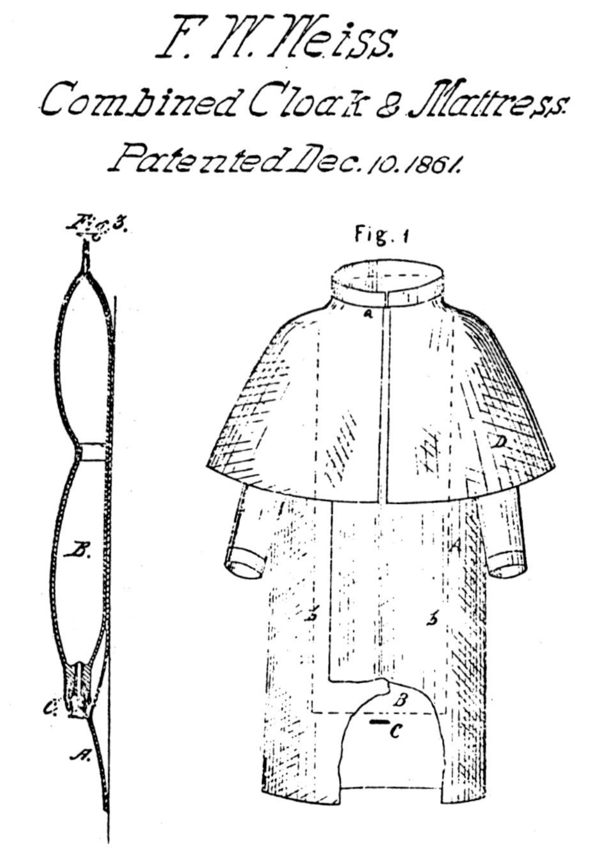

Weiss' Combined Cloak & Mattress. *The patent drawing shows another multi-purpose India-rubber article that combined waterproof clothing with inflatable cushions or pillows.*

In a later Goodyear biography, P. W. Barker summarizes the use of India-rubber goods by the Union army as follows:

> The troops of the North used many rubber blankets, ponchos, knapsacks, shoes, and other equipment.
>
> Prices of this equipment varied considerably: blankets, \$2.55 to \$5; tent blankets, \$3.56 to \$5.25; poncho tent blankets, \$2.72 to \$5.60; infantry tent blankets, \$4.81 to \$5.70; ponchos, \$4.65 to \$6.40; cavalry ponchos, \$7.60. Northern army purchases in the period from May 1, 1861 to June 30, 1865 included 1,893,007 rubber blankets and 1,596,559 rubber ponchos, roughly estimated as costing \$17,000,000. Blankets and ponchos of gutta-percha also were sold to the army, as were a few knapsacks of the same material.
>
> Other rubber goods for the 'boys in blue' included a large number of pontoons, manufactured by the Union India Rubber Company, the number and price of which are not known. Purchases of rubber boots and shoes amounted to more than \$10,000,000. It will be

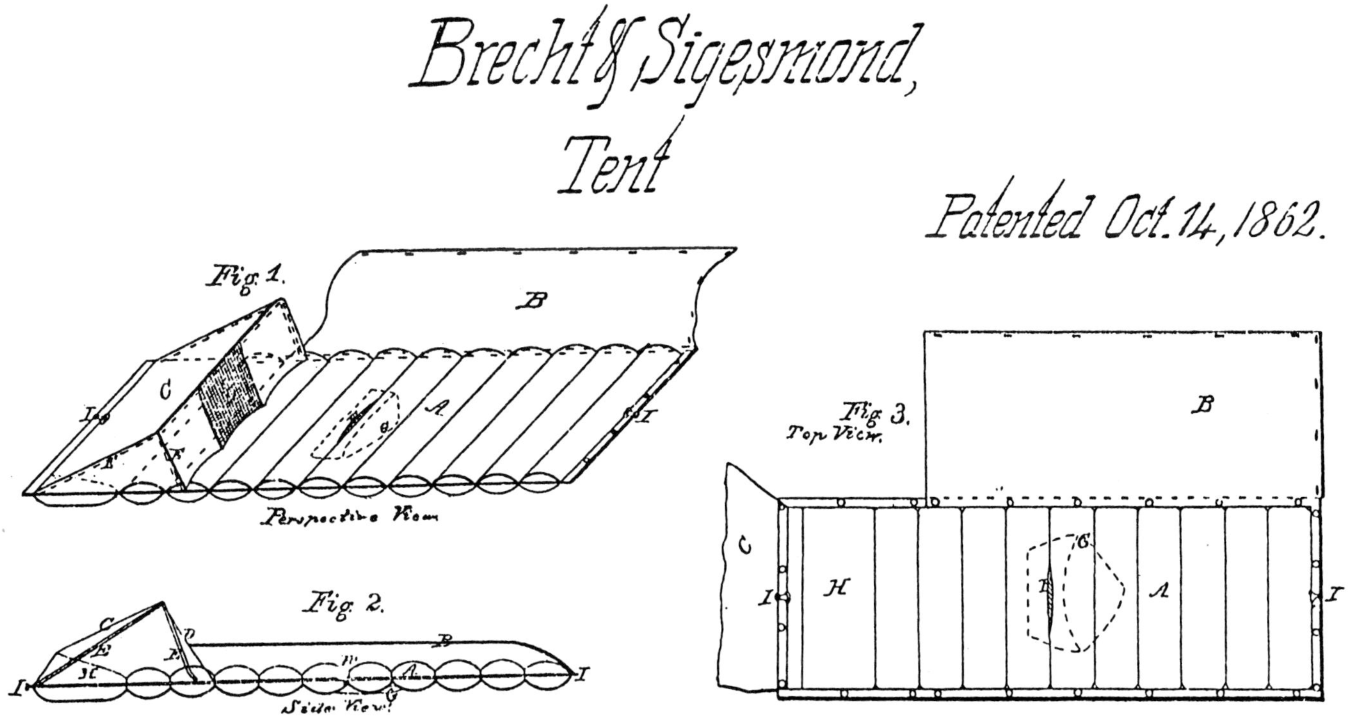

Patent drawing of Brecht and Sigesmond's Combined Cloak, Bed, etc. *showing details of the components. This could be one of many inventions that never made it to production or actual use.*

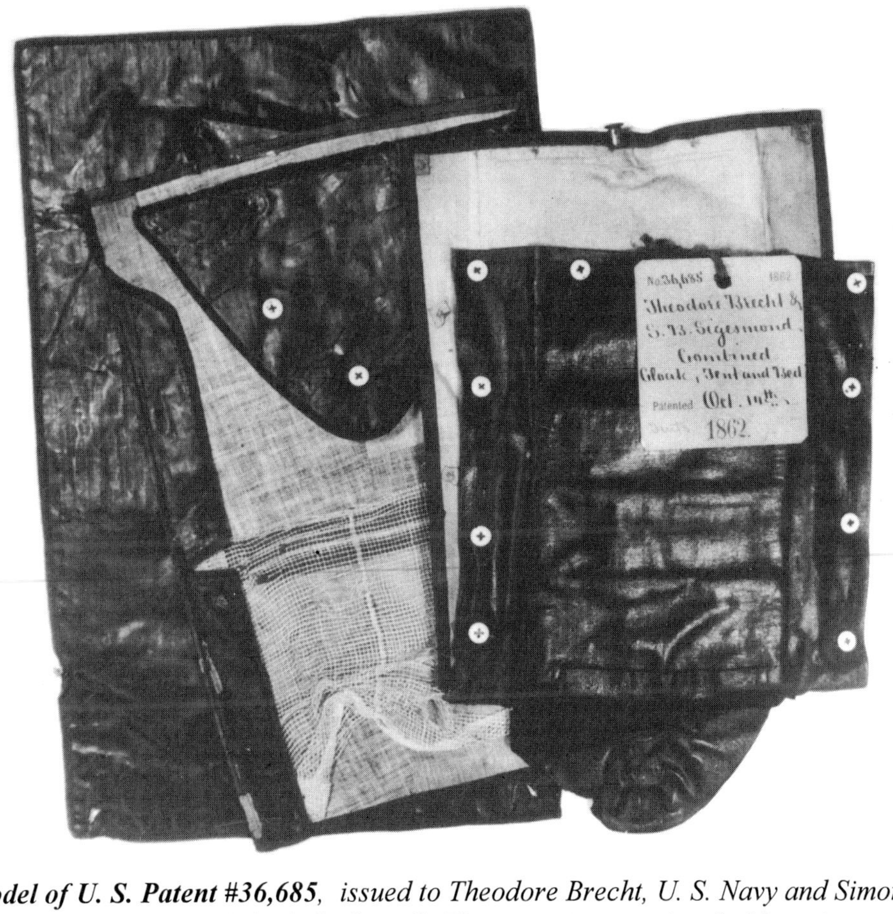

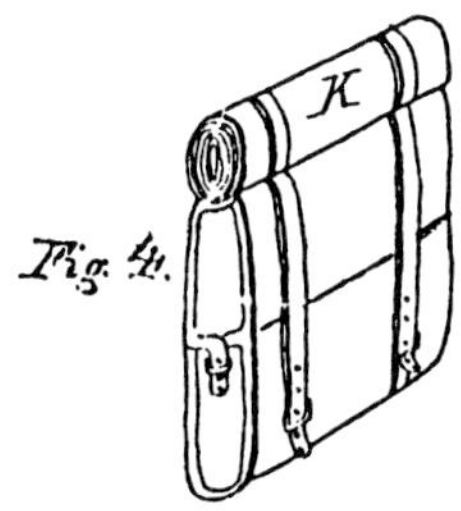

Model of U. S. Patent #36,685*, issued to Theodore Brecht, U. S. Navy and Simon Sigesmond on October 14, 1862, for a "Combined cloak, bed, etc." This invention consisted of five pieces of rubber-coated cloth, some with buttons, buttonholes, flaps, and valves for inflating the mattress and pillow. The pieces could be configured in various combinations to form a cloak, tent, air-mattress, hammock, life-preserver, etc.*

seen, therefore, that the Northern army spent more than $27,000,000 for rubber goods.

The Northern navy used over $500,000 worth of mechanical rubber goods, some of the prices being: 4-ply hose, $1.62 a foot; 3-ply hose, .90; packing, $1.58 a pound; gaskets and valves, $1.58; and tubing, .35 a foot. Also, $130,000 was spent for submarine telegraph cable.[45]

The author's source for the number of blankets and ponchos purchased was likely *The 1865 Report of the Quartermaster General.* However, this report uses the number 1,893,007 to represent the total number of "Blankets, rubber and painted" as well as 1,596,559 "Ponchos, rubber and painted."[46] An analysis of Quartermaster Department contracts detailing 2,863,534 blankets and ponchos shows that approximately 82.5% of the total items were India-rubber, 13% painted or saturated, and 4.5% gutta-percha.[47] (See Section 8, Blankets and Ponchos)

Other articles listed on government contracts for 1864 include 34,710 India-rubber talmas,[48] 7,640 gutta-percha knapsacks, 500 vulcanized rubber linen canvas wagon covers, and 20,000 India-rubber pouches for horsemen.[49] Additional India-rubber, gutta-percha, and painted articles were purchased by individual states during the War.

The 1865 Report of the Quartermaster General contains a "Statement showing the highest and lowest prices paid by the department for articles of clothing and equipage during the past four years," reflecting the following:

Blankets, rubber	$2.00 to $5.00
Blankets, painted	$1.99 to $2.35
Ponchos, rubber	$1.87 to $5.60
Ponchos, painted	$1.83 to $2.35[50]

Hard Rubber and Gutta-percha

Hard rubber and hard gutta-percha articles were also utilized extensively throughout the War. The use of hard rubber uniform buttons by the Union Army and Navy[51] and the inclusion of hard rubber and gutta-percha syringes in the Medical Corps are well documented. Two known cavalry sabers used in the War have handles of hard rubber.[52] Various personal items sold in stores and by camp sutlers were made specifically for individuals in the military.

Throughout his years of research, the author has observed many hard rubber artifacts associated with Civil War military personnel. Tokens, army and navy shaving soap boxes, various combs, and other

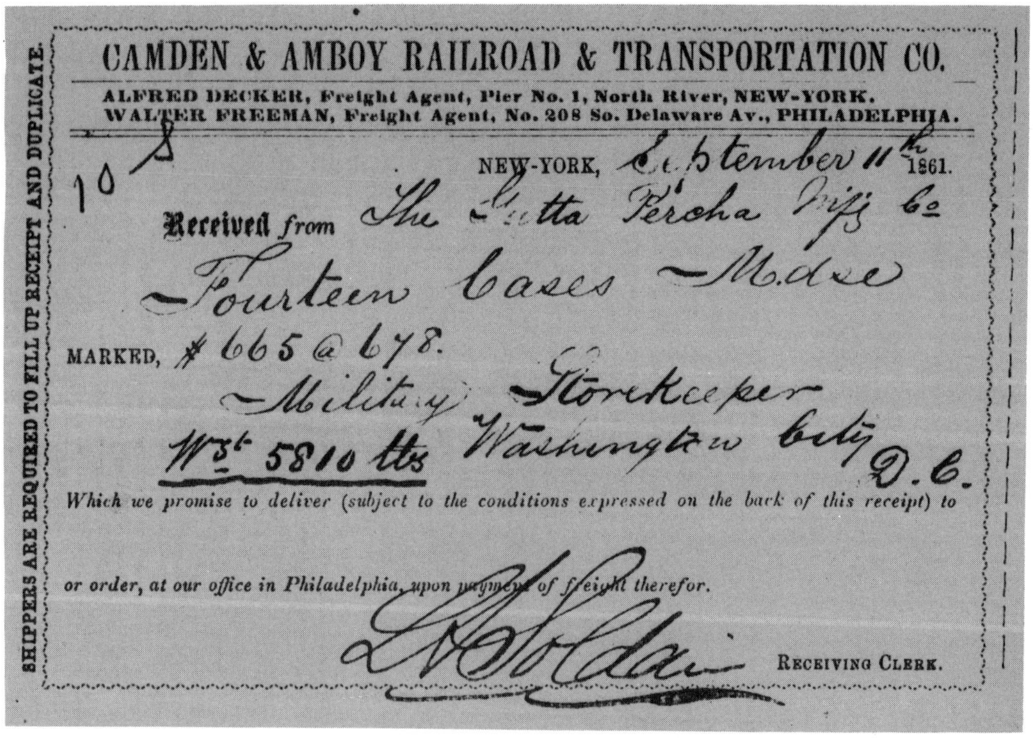

SHIPPERS ARE REQUIRED TO FILL UP RECEIPT AND DUPLICATE.

CAMDEN & AMBOY RAILROAD & TRANSPORTATION CO.

ALFRED DECKER, Freight Agent, Pier No. 1, North River, NEW-YORK.
WALTER FREEMAN, Freight Agent, No. 208 So. Delaware Av., PHILADELPHIA.

10 S NEW-YORK, September 11th 1861.

Received from The Gutta Percha Mfg Co

—Fourteen Cases —Mdse

MARKED, # 665 @ 678.

—Military Storekeeper

Wt 5810 lbs Washington City D.C.

Which we promise to deliver (subject to the conditions expressed on the back of this receipt) to

or order, at our office in Philadelphia, upon payment of freight therefor.

L. Holden RECEIVING CLERK.

Shipping label for blanket knapsacks *sent to Quartermaster General Meigs by The Gutta Percha Manufacturing Co. Note total shipping weight of 5,810 pounds. (National Archives)*

items with military motifs or markings have been found with the personal effects of soldiers and sailors. In addition, many varieties of non-military hard rubber combs, buttons, penholders, mechanical pencils, thimbles, whiskey and powder flasks, pipes, and even toys have been passed down through descendants or excavated from Union and Confederate campsites. Based on surviving examples of personal handiwork, many soldiers passed the time carving hard rubber buttons into rings and other creative articles.

Confederate cotton haversack with rubber cover *possibly cut from a blanket or poncho. "Battle of Olustee" on old tag with "Colonel Hawley's Charge Feb. 20th" inked on the material. February 20, 1864, was the date of the Federal defeat at Olustee, Florida. Colonel Joseph R. Hawley and his 7th Connecticut Regiment made no charges on that occasion. However, extensive souvenir hunting was reported days earlier in captured Rebel camps as the U.S. Army marched inland. (Courtesy Richard Ferry)*

A Multitude of Applications

A number of military applications are included in the following sections, "Products and Applications" and "Patents." But the chapter on military use of India-rubber certainly did not end with the Civil War. What started as novel applications of India-rubber, hard rubber, and gutta-percha evolved to the use of other materials and products designed to outfit and protect the military forces of the United States through the century and for a century to come. Rubber ponchos and blankets of the Indian and Spanish-American Wars, plastic canteens of Vietnam and chemical-resistant rubber suits of the World Wars and Desert Storm have shown that the dreams and innovations of Charles and Nelson Goodyear, John Rider, Thomas Hancock, and all the early rubber pioneers have become realities time and time again.

NOTES - Section 7

[1] *The Scientific American,* 10 Sep. 1859.
[2] National Archives, Records Group 92, Entry 225, Consolidated Correspondence File, "Gutta Percha."
[3] *Ibid.*, "Blankets."
[4] *Ibid.*, "Rubber."
[5] NA, RG 92, OQMG-Letters Sent, Clothing & Equipage, v. 18, p. 475.
[6] *Official Records of the War of the Rebellion* (Washington: Government Printing office, 1880) Series III, Vol. 1, p. 615. Letter from Secretary of War Cameron.
[7] NA, RG 92, Entry 225, CCF, Subject unknown.
[8] *Ibid.*
[9] *Ibid.*
[10] E. J. Coates and F. C. Gaede, "U.S. Army Quartermaster Contracts, 1861-1865," (Lutherville, MD: Production Draft No. 4, December, 1994).
[11] NA, RG 92, Entry 225, CCF, "Goodyear."
[12] *Ibid.*, "Union India Rubber Company."
[13] G. Leander Bishop, *A History of American Manufactures from 1608 to 1860* (Philadelphia: Edward J. Young & Co., 1868). p. 309.

Early India-rubber items and manufacturers. *This advertisement appeared in the* Historical Picture Gallery, *published by John R. Chapin in Boston in 1856.*

SECTION 8

PRODUCTS AND APPLICATIONS

"A large proportion of the fabrics, and a great number of the applications and newly invented articles that appear have never, until now, been presented to the public."

Charles Goodyear, writing in *Gum-Elastic and Its Varieties*, 1855

From the earliest experiments with India-rubber and gutta-percha, on through the discovery of vulcanization processes and into the manufacturing of soft and hard compounds, numerous uses for these materials were generated and marketed. Numerous ideas that never actually made it to the production stage are described and illustrated to show the creative uses that were invented and patented. While many of these applications are documented in previous sections, the following pages provide more detail on specific items and areas of interest.

Though military applications are prevalent within these alphabetical listings, many general uses are also detailed. References to specific patents and other sections of this work are made where appropriate.

ARTILLERY

Cannon covers...are designed to...protect guns when mounted, and their carriages, from the effects of the sun and weather."

Charles Goodyear, Writing in *Gum-Elastic and Its Varieties,* 1855

Several patents relating to the use of India-rubber and gutta-percha in artillery and ordnance were issued during the Civil War years. Three patents issued in 1862 are best summarized by the following article, which appeared in *The Scientific American* of January 10, 1863:

ELASTIC BREECH CANNON AND SABOT

Numerous patents have recently been taken, both in this country and Europe, for devices to lessen the strain and liability of explosion in ordnance by the use of india-rubber or gutta-percha applied in the breech to confine the air, against which the exploded powder will act, whereby the sides of the bore are relieved from the immense strain of the ignited charge. The objects of these inventions are to lessen the danger of explosion and enable the gun to give a greatly increased velocity to the shot by using a

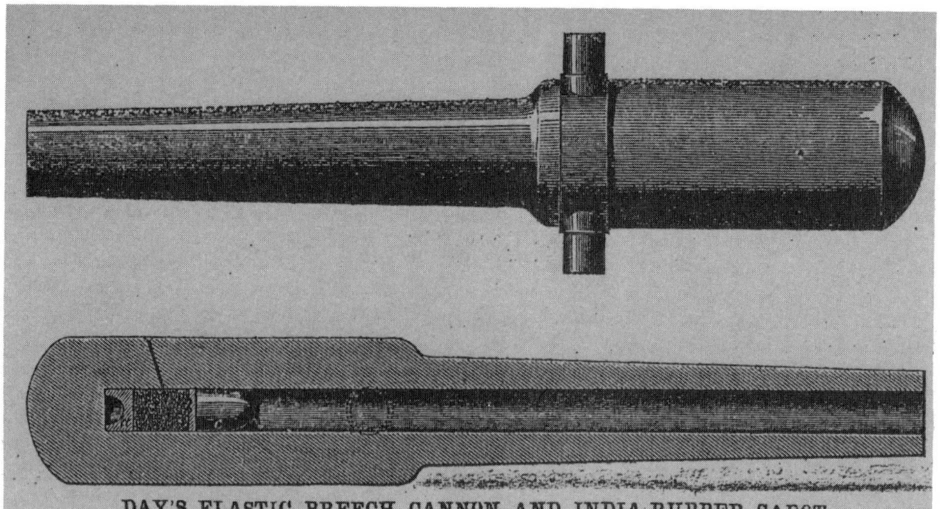

DAY'S ELASTIC BREECH CANNON AND INDIA-RUBBER SABOT.

***Horace Day's elastic breech cannon and India-rubber sabot**, as pictured in* The Scientific American *of January 10, 1863. Day claimed that the invention was readily adapted to any gun, and could be used in connection with all kinds of shot.*

larger charge of powder than is allowed or deemed safe on the old kind of guns. One of these inventions is that patented by Horace H. Day, the celebrated india-rubber manufacturer, of this city; another is that patented by Mitchell [sic] Ritner of Vincennes, Ind., and a third is that patented by Col. J. W. Bird of Trenton, N.J.

Numerous experiments have been made with these devices, and we are informed that they were very satisfactory. Col. Bird's invention is a solid wad, made of india-rubber combined with fibrous substances; it is cut in disks to fit the bore of the gun, and is placed between the powder and the shot, and serves to rotate the latter. The elastic property of the rubber acting against the shot, it is claimed, gives it a perfect rotation, and at the same time it fills the space in the gun and prevents "windage." Much of the strain due to all the wedge-shaped leaden rings is avoided in Col. Bird's wad and projectile, and the inventor claims an increased power of at least one-tenth.

The invention of Mr. Ritner is to confine air in sabots of vulcanized rubber, to be placed between the powder and the shot, so that the sudden concussion against the projectile is lessened by the yielding property of rubber and confined air, and enabling the artillerist to use a larger charge of powder, as previously stated. The object claimed to be gained is that which has long baffled the world of inventors-the use of a larger charge of powder without increasing the strain upon the gun.

We here illustrate the last patent—that of Mr. Day. The tendency of the sudden blow imparted to the base of the missile, by the explosion of the powder, was to upset the base of the projectile by expansion, rendering it, if not immovable, at least dangerous to the gun. To

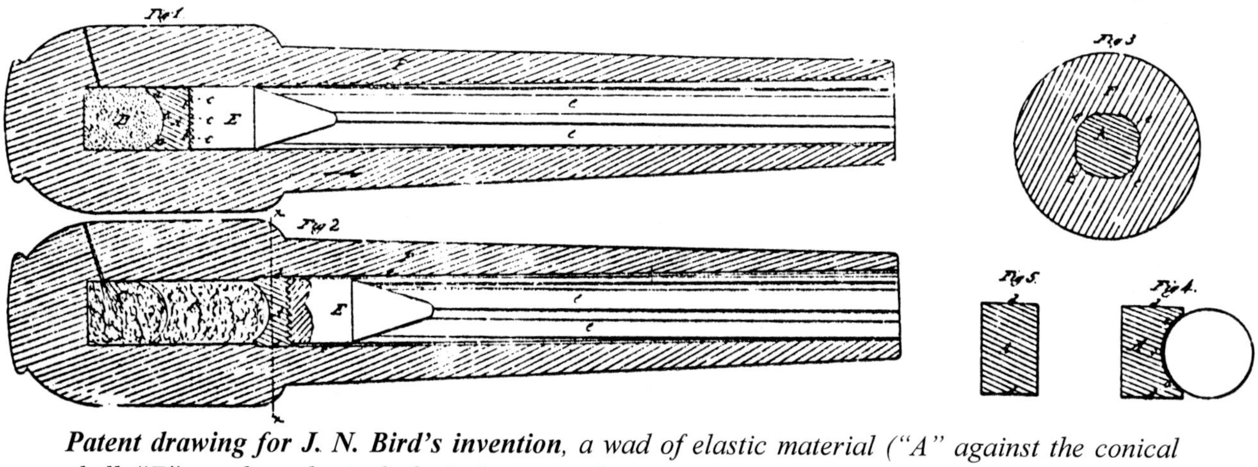

***Patent drawing for J. N. Bird's invention**, a wad of elastic material ("A" against the conical shell "E" or the spherical shot) that served to assist in rotating the projectile, prevent all windage, and lessen recoil, without fouling the bore or sticking to the projectile after firing.*

overcome this, the inventor placed the hollow sabot between the powder and the shot. Mr. Day uses the india-rubber cushion, *a*, having a conical recess at its base. The charge is inserted upon the top of this, and another thinner disk, *b*, of the same substance, is placed on top of it; the projectile, *c*, is then put in, and the weapon is fired in the usual manner. The cushion, *a*, is not destroyed by the explosion; and it is said one of them will last nearly as long as the piece itself; the thin wad, which obviates all liability of "windage," is of course fired without the charge. The effect of the elastic cushion is to impart a gradual movement at the moment of explosion, which starts the bolt gently from its seat; the gases then follow it up and expel it with as much force as the powder is capable of exerting. The sabot has been tried (the inventor says) in a 100-pounder gun, and the strain at the breech was ascertained to be 48,000 lbs., while with the same gun and the same charge, namely, 10 lbs., without the use of this invention, it was found to be 69,000 lbs. The cushion has always been worked in a rifled 50-pounder with a charge of 8 lbs., from which the charge has been increased to 12 lbs., without any apparent evil effect on the gun; using shot of various sizes and weight, from 40 to 50 lbs. Two of the holes made in the target so long laying in Wall Street, this city, were made by shot in connection with this elastic breech, fired with a 10 lb. charge from a rifled 50 lb. gun.[1]

Michael Ritner's patent was U.S. Patent #35,544, Colonel Bird's U.S. Patent #36,879, and Horace Day's U.S. Patent #37,035.

Other patents issued during the War include a "Removable Chamber for Explosive Shells," consisting of a tube or pouch of India-rubber (U.S. Patent # 39,267) and elastic or India-rubber "Breeches for Ordnance" (U.S. Patents #44,905 and 45,586). India-rubber was included as an ingredient for filling projectiles and constructing sabots in Patrick Devlan's patent of 1863 (U.S. Patent #39,897).

12 Pound spherical case shot with Bormann time fuse and India-rubber washer. *Recovered in 1972 at the site of Allegheny Arsenal near Pittsburgh, PA. A letter dated June 9, 1864 from Captain Peter Simonson, Chief of Artillery, 1st Division, 4th Corps stated, "Many of the cartridges had been wet; the powder in many of the shell and spherical case was so hard that we were unable to get it out. No attention appears to have been paid to putting rubber gaskets under the Bormann fuse, and many of them exploded prematurely."*[2]

(Also see **Buckets** and **Firearms and Accessories**.)

NOTES - Artillery

---

[1] *The Scientific American*, 10 Jan. 1863.

[2] *Official Records of the War of the Rebellion; The Civil War CD-ROM* (Washington: Government Printing Office, 1880; Carmel, Indiana: Guild Press of Indiana, 1996) Series I, Vol. XXXVIII/1 [S# 72] p. 488.

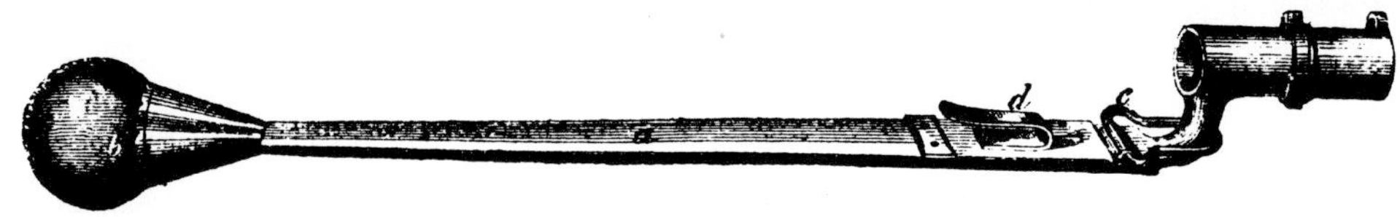

***Ernst's Bayonet Guard**, as depicted in* The Scientific American *of September 27, 1862. The guard was designed to be placed directly on the bayonet and scabbard, so the recruits and soldiers could practice with the actual muskets and bayonets instead of lighter practice weapons.*

BAYONET GUARDS & FENCING BAYONETS

"At West Point the cadets are taught to fence with india-rubber bayonets at the ends of their muskets..."

From *The Scientific American,*
September 27, 1862

Issue of a fencing bayonet by the U.S. Ordnance Department was well established prior to the Civil War. It was both described and illustrated in McClellan's *Bayonet Exercises for the Army* in 1852, and included the prescription of a leather or India rubber ball.[1] The fencing bayonet was also pictured in Kelton's 1864 *A New Manual of the Bayonet.* While the government would supply the sockets and blades, Kelton directed the buyer to the "Gutta-percha Co., 139 Broadway, N.Y." to acquire the "bayonet points."[2]

Another example of the use of India-rubber in bayonet training was described in the following excerpt from a *Scientific American* report on a novel invention:

IMPROVED BAYONET GUARD

The skillful handling of the bayonet...is one of the most important arts for the soldier to learn, and in military schools a great deal of attention is devoted to it. At West Point the cadets are taught to fence with india-rubber bayonets at the ends of their muskets, and the rapid and furious manner in which they thrust the points into each other's faces and against all parts of their bodies is perfectly terrific.

John G. Ernst, of York, PA., has invented a guard to be attached to the end of an ordinary bayonet scabbard, so that the teaching of the bayonet exercise may be continued, as a part of the regular drill while the persons are in camp, without any danger to their persons. It consists simply of an india-rubber ball attached to the end of the scabbard and of a device for fastening the scabbard securely to the bayonet, so that there will be no danger of the scabbard being thrown off in the exercise.

In fencing with india-rubber bayonets the arm is lighter than in service, and the muscles are, consequently, trained and developed to a less degree of strength than is required in battle, but with this guard the weapon is a few ounces heavier than it is with the bayonet naked, and the soldier trained to its use will handle his arm in battle with greater ease.[3]

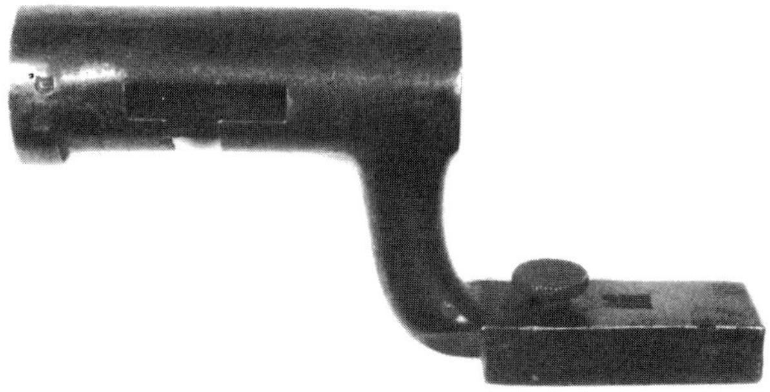

John Ernst's original bayonet guard, patented August 5, 1862 (U.S. Patent #36,081), consisted of an elastic ball secured to the bayonet by an arrangement of straps or bands. Ernst's improved bayonet guard, featured in the above report, was defined in U.S. Patent #37,222 dated December 23, 1862. It was manufactured by Welch & Company of York, Pa.

__Fencing bayonet with a whalebone blade and rubber ball__ on the tip. Close-up shows the socket consisting of a cut-down Model 1816 bayonet with a steel box added including a screw to secure the whalebone blade. Known by many collectors as the Model 1858 fencing bayonet, this style was used until around the turn of the century.[4] (Courtesy George Kegerreis)

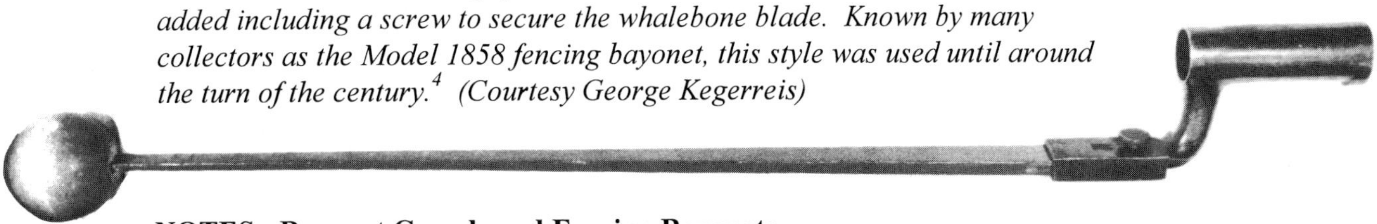

NOTES - Bayonet Guards and Fencing Bayonets

---

[1] George B. McClellan, *Bayonet Exercises for the Army* (Washington, D. C., undated), p. 4.
[2] J. C. Kelton, *A New Manual of the Bayonet* (New York, 1864), p. 50.
[3] *The Scientific American*, 27 Sep. 1862.
[4] George T. Wheeler, "The Model 1858 Fencing Bayonet," *Military Collector & Historian*, Vol. XXXIX, No. 2, (Summer 1987) p. 85.

BLANKETS AND PONCHOS

"Will contract to deliver <u>pure</u> <u>vulcanized</u> rubber blankets or ponchos...and we warrant our goods in all cases to give full satisfaction--being thoroughly waterproof--not affected by any extreme of heat or cold-never softening or stiffening."

Bid letter from George Cornwall of the Union India Rubber Company, June 18, 1862

The concept of a rubber or gum blanket, as a singular piece of equipment or combined as a poncho or tent, was devised as early as March 6, 1836, when James Boyd of Boston wrote to the War Department:

> I forgot to mention to you in conversation that in Washington the idea was suggested of using some of my India Rubber goods as a kind of sheet to spread under the men in their tent. These would effectually prevent the damp of the

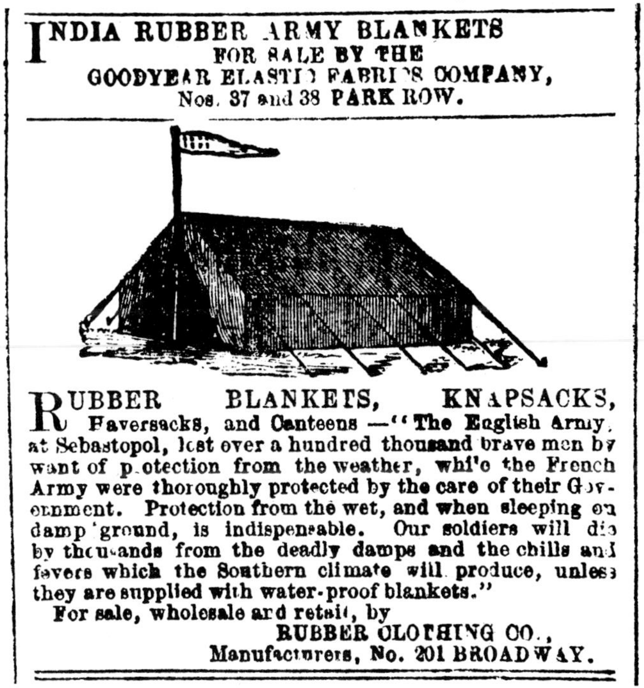

INDIA RUBBER ARMY BLANKETS
FOR SALE BY THE
GOODYEAR ELASTIC FABRICS COMPANY,
Nos. 37 and 38 PARK ROW.

RUBBER BLANKETS, KNAPSACKS, Haversacks, and Canteens —"The English Army, at Sebastopol, lost over a hundred thousand brave men by want of protection from the weather, while the French Army were thoroughly protected by the care of their Government. Protection from the wet, and when sleeping on damp ground, is indispensable. Our soldiers will die by thousands from the deadly damps and the chills and fevers which the Southern climate will produce, unless they are supplied with water-proof blankets."
For sale, wholesale and retail, by
RUBBER CLOTHING CO.,
Manufacturers, No. 201 BROADWAY.

***Advertisements published by two rubber suppliers** in a May, 1861 New York City newspaper listing rubber blankets and other waterproof items for the military.*

> ground from striking on the men and might be made so that each man would have one or, one might be made large enough for each tent. By the former method, it might be used round the shoulders when on the march in rainy weather and as a sheet at night.[1]

Charles Goodyear acknowledged the value of this application, when he recommended in *Gum-Elastic and Its Varieties*, "...camp blankets...as an invaluable article for the soldier, or others who are obliged to sleep on the ground, or are in any way exposed to storms."[2] He touted his "tent carpet... constructed with eyelet holes, that they may be used upon occasion as tents or awnings...must be a great comfort, and a great security to health."[3] And, in 1858, Major Henry C. Wayne wrote, "The [gutta-percha] floor cloths were excellently adapted to the purposes intended, and I found much comfort in their use in camp, as a protection from the damp ground."[4]

Early in the Civil War there was some stir caused over units ordering India-rubber blankets without approval. Perhaps the memories of some of the unsuccessful trials of the 1840s and 1850s remained in some military minds. Quartermaster General Meigs had written to Secretary of War Simon Cameron in September of 1861, asking for a firm policy on the purchase of India-rubber or gutta-percha goods, particularly blankets.[5] On November 1, 1861, the "waterproof blanket" was approved as standard issue, as Secretary of War Simon Cameron wrote, "The Quartermaster General is directed to make arrangements to procure for issue to the troops, to be charged to those who issue them as other articles of clothing are, waterproof blankets for use in camp."[6]

The Soldier's Guide - A Complete Manual and Drill Book for the use of Volunteers, Militia and the Home Guard advised all volunteers to "Buy a small India rubber blanket to lay on the ground or to throw over your shoulders when on guard duty during a rain storm," and noted that "Most of the Eastern troops are provided with these."[7] Veteran soldiers were likely to discard knapsacks and instead make a "blanket roll" by rolling all of their possessions into the rubber blanket and carrying the waterproof case diagonally across their backs.[8]

Rubber blankets could frequently be found with a checkerboard drawn or painted on the canvas back.[9] Other creative uses included using pieces of a rubber blanket to make ration bags or to cover holes in tents.[10] In his book, *Hardtack and Coffee*, Union veteran John D. Billings described India-rubber blankets and ponchos, and detailed their use in the construction of winter huts:

Post-war (1866) photograph of a War Department model wearing an India-rubber poncho. *While India-rubber blankets were standard issue for the Union Army after November of 1861, they were prized by the Confederate troops, who frequently picked them up after battles or obtained them in trades with Northern prisoners. (Courtesy Smithsonian Institution)*

> When four men occupied a stockade, it was built accordingly and covered by four half-shelters. In each case these were stretched over a framework of light rafters raised on the walls of the stockade. Sometimes the gables were built up to the ridge poles with smaller logs, but just as often they were filled by an extra half-shelter, a rubber blanket, or an old poncho. An Army poncho, I must say here, is specified as made of

unbleached muslin coated with vulcanized India rubber, sixty inches wide and seventy-one inches long, having an opening in the center lengthwise of the poncho, through which the head passes, with a lap three inches wide and sixteen inches long. This garment is derived of the woolen poncho worn by the Spanish-Americans, but of different proportions, these being four feet by seven. The Army poncho was used in lieu of the gum blanket.[11]

In his diary, George D. Carrington of the 11th Illinois Infantry, detailed three types of waterproof blankets, "1st Like oil cloth, 2nd Poncho hole in the center to put head through, 3rd Rubber blanket."[12]

Some soldiers converted the rubber blanket to a poncho by merely cutting a slit towards the middle and slipping it down over their head onto their shoulders. Converting to a poncho or tent was also made possible by the "Improved Shawl Pin," patented by H. M. Hall (U.S. Patent #36,003) on July 29, 1862. This was a hook-shaped pin with a straight shank and a button on the head. An improved "Spring Eyelet Hook" was patented by Abel Putnam, Jr. (U.S. Patent #36,933) on November 11, 1862 for the same purpose. The patent specification stated, "This invention relates to an improved spring hook or

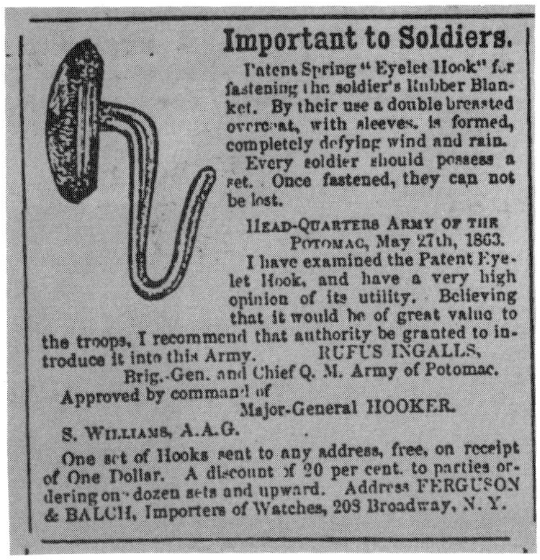

Important to Soldiers.

Patent Spring "Eyelet Hook" for fastening the soldier's Rubber Blanket. By their use a double breasted overcoat, with sleeves, is formed, completely defying wind and rain.

Every soldier should possess a set. Once fastened, they can not be lost.

HEAD-QUARTERS ARMY OF THE POTOMAC, May 27th, 1863.

I have examined the Patent Eyelet Hook, and have a very high opinion of its utility. Believing that it would be of great value to the troops, I recommend that authority be granted to introduce it into this Army. RUFUS INGALLS, Brig.-Gen. and Chief Q. M. Army of Potomac.

Approved by command of Major-General HOOKER.

S. WILLIAMS, A.A.G.

One set of Hooks sent to any address, free, on receipt of One Dollar. A discount of 20 per cent. to parties ordering one dozen sets and upward. Address FERGUSON & BALCH, Importers of Watches, 208 Broadway, N. Y.

Advertisement for the "Spring Eyelet Hook," *patented in 1862.*

fastening, more especially designed for soldier's india-rubber wrappers or blankets together to form tents or coverings for a plurality of men." A newspaper advertisement with testimonials from Generals Hooker and Ingalls further claimed, "...a double-breasted overcoat, with sleeves, is formed, completely defying wind and rain." Examples of both of these inventions have been dug at Civil War campsites, the latter being the type most commonly seen. At least one homemade version, with a brass wire double hook soldered to the back of a standard Massachusetts button, has been noted.[13]

Excavated and non-excavated examples of the Spring Eyelet Hook, *patented November 11, 1862. This hook or a similar patented "Shawl Pin" could be used to form a poncho or combined with other blankets to form a tent. Other hooks, including one with an Eagle "I" button and a variant made from a Massachusetts button, have been noted.*

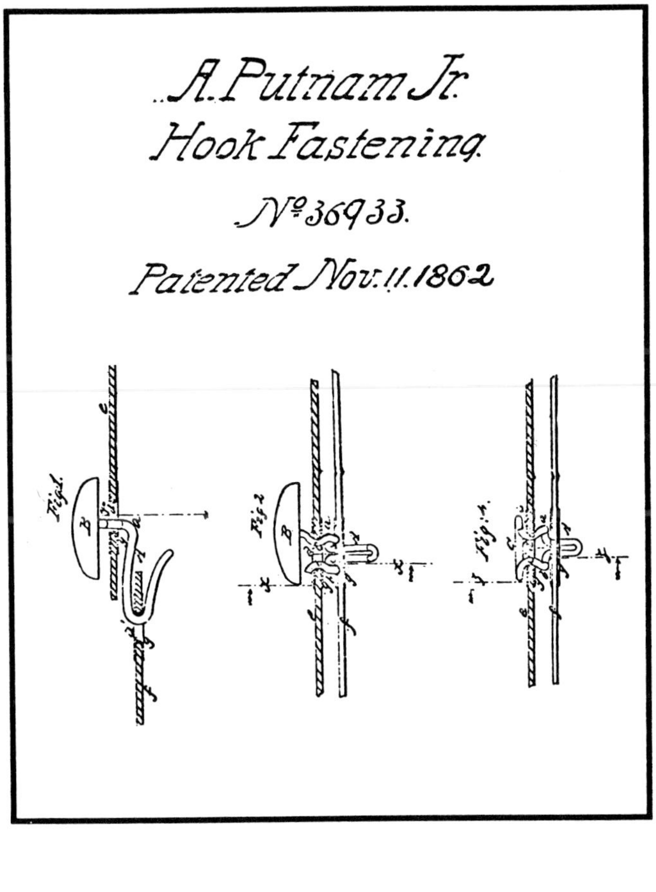

Drawing of the Spring Eyelet Hook *from the original patent specification.*

The value of the India-rubber blanket was supported in a letter sent by Acting General Nathan A. M. Dudley of the 30th Massachusetts on May 19, 1863. In an operation near Merrit's Plantation, Louisiana, Dudley ordered a battery and at least one regiment of infantry to "...start at once. Let them march with only rubber blankets and two days' uncooked rations. Camp kettles can be brought out in wagons."[14]

The versatile India-rubber blanket added to the comfort of the wounded in some cases. Writing from Cold Harbor on June 6, 1864, Surgeon Thomas A. McParlin summarized challenges in treatment of wounded soldiers:

> The hospitals of the Sixth Corps had to be pitched upon wet and muddy ground, but by the use of boards, obtained from an adjacent saw-mill, India-rubber blankets, and evergreen boughs, the condition of the wounded in them was rendered very tolerable.[15]

In a report issued from Mobile, Alabama on May 24, 1865, Surgeon Charles B. White detailed similar circumstances while treating the wounded at Spanish Fort:

> The wounded did not endure any exposure to wet, cold, or heat that was of long duration or sufficient to cause suffering. Rain fell heavily three days, but those wounded on the skirmish line, though remaining on the field until night, were protected by the rubber blankets of their comrades.[16]

Confederate soldiers often carried rubber blankets "obtained" from their Union counterparts. In testimony detailed by the U.S. Sanitary Commission in 1864, Sgt. Alfred Jones, Co. C, 1st Mass. Cavalry said, "I sold my India rubber cover to a guard to buy bread with. A good many who went to the prison when I did had their blankets taken from them." The Commission also reported that, in some case, all items—including rubber blankets—were taken from the prisoners.[17]

Lt. Colonel H. C. Hobart of the 21st Regiment of Wisconsin Volunteers was taken prisoner at Chickamauga. Hobart noted that, while being transported with 1,800 other prisoners,

> ...At Tunnel Hill, Ga., all the non-commissioned officers and privates of the above prisoners were ordered by the Confederate commander of the post to stand in line and give up their rubber blankets, which was done.[18]

One of Stonewall Jacksons's Foot Cavalry is quoted as saying:

> I have a splendid rubber cloth, a blanket, a pair of jeans drawers and a pair of woolen socks, every article captured from the enemy. The socks and drawers were placed in the blanket, the blanket was rolled up with the rubber cloth to the outside, the ends drawn together and fastened with a short strap. To carry this we put it over the head and let it hang over the shoulder.[19]

Unidentified soldier wearing an India-rubber blanket as a poncho. *Virtually all of the legally manufactured vulcanized India-rubber blankets and ponchos used by the Union army were made by the Union India Rubber Company or an affiliate. (Courtesy of Ronn Palm)*

On March 18, 1865, General Ulysses Grant wrote to the Honorable C. A. Dana, Assistant Secretary of War, in Washington:

> I understand General Ord is trying to buy some tobacco that is now at Yorktown to issue to his colored troops. They are selling their rubber blankets to rebels daily to procure the article, not having pay or credit to get it.[20]

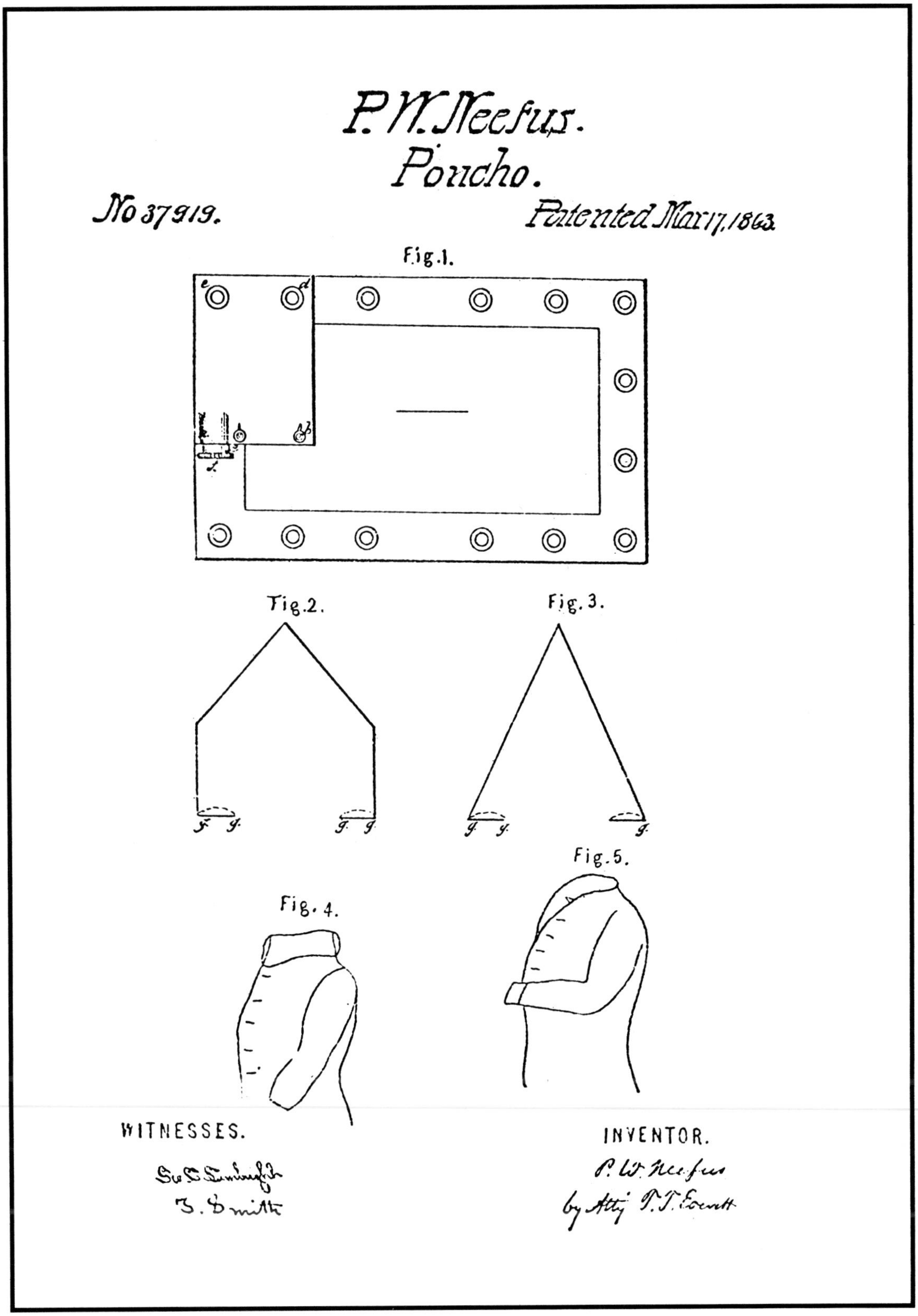

Patent drawing of an improved poncho *patented by Peter Neefus on March 17, 1863. The patent specified an inflatable pillow built into a corner of a standard poncho, rubber blanket, tent, or article of clothing. This improvement was deemed by the inventor, "...indispensable for tents or camp life."*

Contracts and Variations

By late 1861, contracts for India-rubber, gutta-percha, and other waterproof blankets were being negotiated by the United States War Department. In one case, on February 26, 1862, Brigadier General Meigs directed Assistant Quartermasters Captain William McKim in Boston and Lt. Colonel D. H. Vinton in New York to advertise for India-rubber blanket proposals. The responses that follow provide an excellent overview of typical bid responses, and the lack of uniformity in the types of blankets at that point.[21]

6

PONCHO.

	EACH.
Vulcanized—White Rubber, ...	$14 57
do. Dull finish, ...	11 62
do. do. large, ...	12 75
do. Lustre, ...	9 41
do. do. Common, ...	7 50
Cured, fine light, ...	8 00
Macintosh, light, ...	8 00
Unvulcanized, Black or White, ...	
Flocked Ponchos, Made to Order, ...	

***Poncho illustration** published in the October 1, 1864 catalogue of New York City's Rubber Clothing Company - "Manufacturers, Importers and Jobbers of India Rubber Clothing, etc."*

Alfred Hale & Company of Boston proposed "India Rubber Blankets like the sample sent 6 ft. long and 52 inches wide, made of the best quality of imported India Rubber, for Two dollars and 25/100 each ($2.25)," to be manufactured at the rate of 400 per week.

Howes Hyatt & Company of New York City proposed to furnish India Rubber Blankets made by the Union India Rubber Company at a cost of $2.25 each. The offer indicated they could provide 3,000 immediately, 5,000 more in ten days, and 10,000 more in twenty days.

George H. Hood quoted several different styles and sizes, ranging from an unbound blanket on sheeting, 1 yard by 2 yards, with 2 grommets and tie strings, at 90 cents each to an unbound blanket on drill, 50 inches by 2 yards, with 2 grommets and tie strings, at $1.67 each. Hood promised 3,000 to 4,000 blankets per day.

N. Hayward and Company of Boston quoted 6,000 to 10,000 blankets per week, at prices and styles similar to George Hood's.

William Judson of Philadelphia submitted the following:

> I will furnish the Government Army Blankets, made out of Elastic and India rubber Cloth, according to the regulation standard dimensions being 46 x 71 inches, bound with rubber binding and having 18 metal grommets, inserted in stays-possessing the very best qualities and properties that is capable of being produced, for the following prices, viz.
>
> (All on Elastic cloth and Vulcanized)
>
> 1st quality $ 2.25
> 2nd quality 2.00
> 3rd quality 1.80
>
> Or I will furnish Blankets of the above respective qualities and properties, at the above prices less ten cents on each Blanket when the eyelets only are inserted in the stays for Shawls instead of 18 grommets for tents-
>
> 10,000 by the 20th of March
> Between March 20th and April 20th - 75,000

The Rubber Clothing Company of Boston proposed the following rubber blankets:

> # 1 Made on drill 2 yds. long & 1 ¼ yds wide-bound all around, 2 grommets & tie string to be used as shawl or cape - $ 1.75 each
>
> # 2 Made on drill 2 yds. long & 1 ¼ yds wide-bound all around, 18 grommets & tie string to be used as shawl or cape and also two or more can be laced together and used as tent - $ 2.00 each
>
> These have been used by the government and have been several times ordered by Col. Crosman of Philadelphia.

Additional bids for India rubber blankets and poncho blankets were received from The Union India Rubber Company, The New York Belting and Packing Company, Goodyear's India Rubber Manufacturing Company, D. M. Whitney, P. D. Davis, Daniel G. Burnet, F. C. Wagner, Horace Day—all of New York, R. S. Walker of Alexandria, Virginia, and W. B. Tobey of Providence, Rhode Island. The proposal from Tobey offered an interesting variety that appears to be a deluxe India-rubber blanket:

> Blanket 46 x 71 inches - 2 ¼ to 2 ½ lbs. with 2 grommets; strong cotton cloth base with which the india rubber is thoroughly combined and the india rubber surface is wholly covered with flocks of fur. The fabric is dyed an indigo blue. The india rubber is perfectly vulcanized and will stand all the changes of temperature without cracking from cold or becoming sticky from heat. The blanket is soft, flexible and not disagreeable to the touch, like the rubber surface. Price $2.75 each.

Tobey's deluxe blankets, 60 by 71 inches with full grommets, were offered at the rate of 150 per day at $4.25 each. Lt. Colonel Vinton noted that, "This article comes higher than the ordinary Rubber blanket and I have no knowledge of its merits or that it has ever been properly tested."

In a summary of bids received, Vinton suggested to General Meigs that he felt the New York Belting and Packing Company's proposal should be considered first. Among other styles, they had proposed 11,000 blankets, 45 inches by 72 inches with 18 metallic grommets, at $2 each, to be delivered in one week. Vinton next recommended the Union India Rubber Company, which had proposed blankets 46 inches by 71 or 72, with 18 grommets, at a cost of $1.75 each. The Union Company, which also offered the same blanket in a lower grade, promised 10,000 blankets in ten days and 3,000 daily thereafter.

At about the same time this contract was being considered, the Beverly Rubber Company of

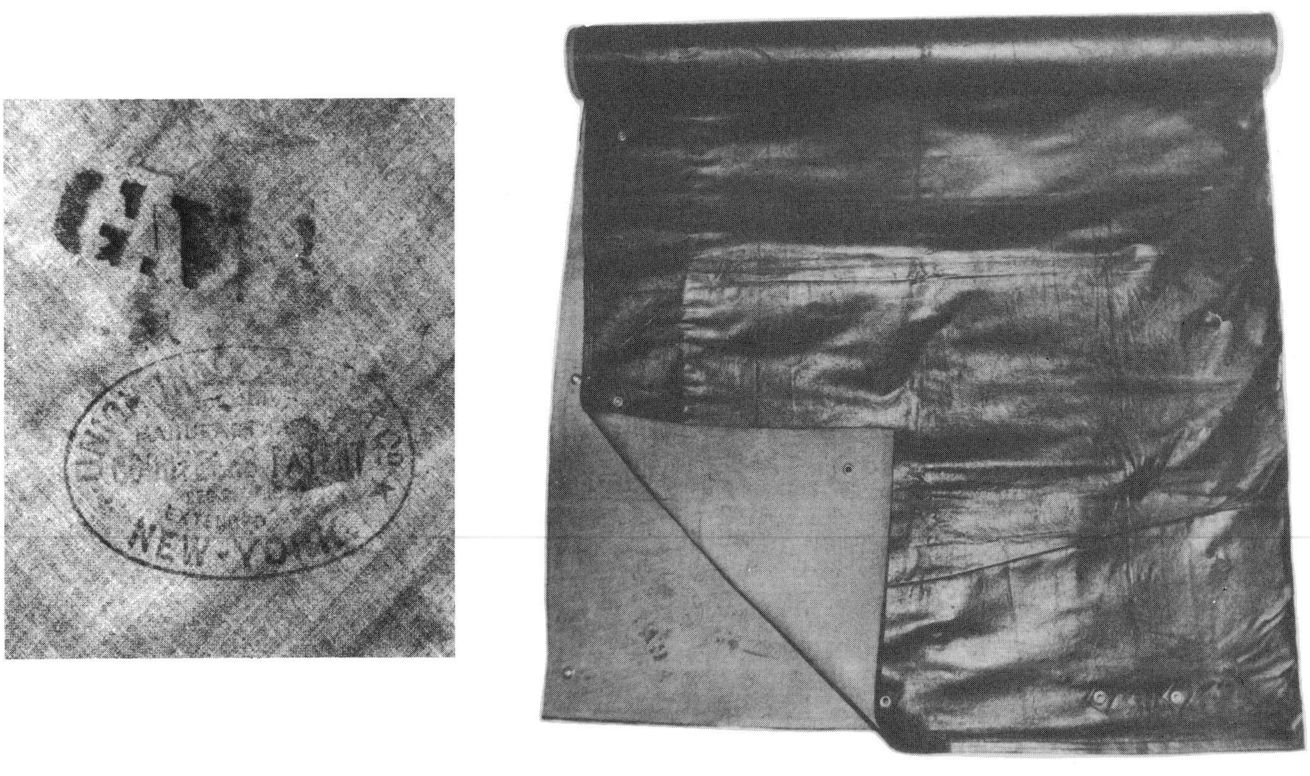

Soldier's India-rubber blanket *carried in the War by Corporal James T. Cowan, Co. E, 34th Mass. Vol. Infantry, who was wounded at 3rd Winchester. The minie passed through the blanket roll, leaving eight symmetrical holes in the blanket. It is marked on the duck or canvas underside: "Union India Rubber Company - New York - [Exclusive] Manufacturer's [Under] Goodyear's Patent Dec 25, 1849 - Extended 1858." Unknown "G. W." marking.*

Beverly, Massachusetts was awarded a contract for 70,000 rubber blankets. But the Beverly factory was destroyed by fire the night of March 5, 1862, the same night that the business agent, Mr. Green, returned from Washington with the contract.[22]

The manufacture of India-rubber goods under government contract for the War Department created some legal problems. The Union India Rubber Company, formed in 1848, held several exclusive Goodyear licenses, including those for wearing apparel and army and navy equipment. Numerous contracts for rubber blankets were awarded to the Union Company, some of which were met by its subsidiary, the Phoenix Rubber Company. Some other companies had contracted with the government for blankets and other items, but found that the Union Company had quietly effected the stoppage of all payments on these articles unless the manufacturer could produce a written order from them. The Boston Rubber Shoe Company had asked the Union Company to consider their services if contracts became more than they themselves could handle. *A History of American Manufacturers from 1608 to 1860* notes:

> This honorable procedure was met in a similar spirit by the managers of the Union Company and thousands of blankets were manufactured at the works of the Boston Company, few of which failed to pass the rigid inspection of the United States officials.[23]

While the Union India Rubber Company and its affiliates manufactured a large number of the India-rubber blankets and ponchos during the War, other companies manufactured them as well, either illegally or out of different materials. Various names were also applied, as India-rubber and gutta-percha blankets were referred to or ordered as "camp blankets," "tent blankets" or "poncho blankets," some being designed to be used for multiple purposes. On January 3, 1863, Lt. Colonel Crosman contracted with the Gutta Percha Manufacturing Company in New York for "Twenty thousand Vulcanized, Water-Proof, Gutta Percha Tent-Blankets on Drills, to be full grommetted...at Two dollars and seventy-five cents" each.[24] On October 2, 1863, a contract was issued to Ellis D. Lazell of New York for "Twenty-Five thousand Rubber Poncho Tent Blankets...at Two Dollars/Fifty-Six Cents for each.[25]

Other waterproof blankets of unknown materials were produced for the Union Army and states' forces. *The Scientific American* of November

***India-rubber blanket** carried in early 1865 by Moses E. Conyers, Co. B, 69th Indiana. A corner of the blanket was cut by Conyers and made into a "housewife" for carrying personal items. The blanket bears the markings of the Union India Rubber Co. Closeup shows reinforced grommet. (Courtesy Jonathan Kearns Cooper-Wiele)*

29, 1862 recorded, "A firm on Philadelphia are making 60,000 water proof blankets, in the form of Spanish cloaks, for the Pennsylvania troops."[26] Several contracts were awarded, specifying only "waterproof blankets," "waterproof infantry blankets," or "waterproof tent blankets."[27] These blankets may have been India-rubber, gutta-percha, or some other waterproof material. Blankets and ponchos that were saturated or "painted" with India-rubber or gutta-percha compounds or other substances were also ordered. The following letter was sent by Lt. Col. Crosman to James E. Bowen of Philadelphia on September 10, 1862:

> You are authorized to make and deliver at the Schuylkill Arsenal twenty thousand Ponchos, painted, on cotton sheeting, like sample in this office, by which they are to be inspected, and which sample is approved, and recommended by Genl R. C. Dale, (QMG of the state of PA) and to be marked "PA. Vols.," at two 10/100 dollars ($2.10) each; to be 60 X 71 inches in size, and provided with flaps of cotton, 3 inches wide, like sample in all respects.
>
> Deliveries to commence in ten days, and to be delivered at the rate of 1500 to 300 per day, after that time, and subject to inspection: your name is required, by law, to be attached to every Poncho. Bond and contract will be made out in a few days, and sent to you for execution.[28]

In some cases, it is possible that the term "painted" referred to vulcanized articles. Ongoing military correspondence, however, did reference both India-rubber and painted products. In response to an inquiry from Quartermaster General Meigs regarding the durability of "India-rubber or painted blankets," Assistant Quartermaster General Thomas Swords at Louisville, Kentucky wrote on May 2, 1863 that "...no painted blankets as far as I know, have been used in this Dept.-those that were sent to this depot, we sent to the Department of the Cumberland for issue."[29] In September of 1863, Lt. John Matthews, Asst. Quartermaster of the 3rd Regiment at Ellicott Mills, wrote to Meigs asking whether the Regiment is chargeable with "gum or painted blankets," saying that "...the men are disposed to believe that they are not chargeable with the same, that the blankets are a part of their Camp and G. Equipage."[30]

Contracts and surviving examples with manufacturer's markings indicate that the vast majority of waterproof blankets and ponchos were indeed vulcanized India-rubber. The following summaries are taken from a compilation of Union Army contracts:

Blankets:

India-rubber:	1,410,401
Gutta-percha:	63,000
Painted/Saturated:	230,000
Total:	1,703,401

Ponchos:

India-rubber:	955,400
Gutta-percha:	60,000
Painted/Saturated:	144,733
Total:	1,160,133[31]

As a single article of issue, the rubber, gum, or waterproof blanket and poncho were certainly the largest contributing item towards the personal comfort of the Civil War soldier.[32] It may well have preserved the health and lives of many, just as James Boyd and Charles Goodyear had predicted.

(Also see **Section 6**, **Section 7**, **Knapsacks**, and **Medical Applications**.)

NOTES - Blankets and Ponchos

---

[1] National Archives, RG 92, Entry 225, Consolidated Correspondence File, "India-Rubber."
[2] Charles Goodyear, *Gum-Elastic and Its Varieties* (New Haven, CT: Goodyear, 1853 and 1855) Vol. II, p. 133.
[3] *Ibid.*, pp. 47-48.
[4] "The Tent Knapsack" Booklet, p. 13; NA, RG 92, Entry 225, CCF, "Knapsacks."
[5] NA, RG 92, Entry 225, CCF, "Rubber."
[6] *Official Records of the War of the Rebellion* (Washington: Government Printing Office, 1880) Series III, Vol. 1, p. 615.
[7] An Officer in the U.S. Army, *The Soldier's Guide - A Complete Manual and Drill Book for the Use of Volunteers, Militia, and The Home Guard* (Philadelphia: T. B. Peterson and Brothers, N.D.) p. 11.
[8] Philip Van Doren Stern, *Soldier Life in the Union and Confederate Armies* (New York: Bonanza Books, 1961) pp. 198, 236.
[9] Francis A. Lord, *The Civil War Collector's Encyclopedia* (Harrisburg: The Stackpole Co., 1965) p. 196.
[10] Van Doren Stern, *Soldier Life in the Union and Confederate Armies*, pp. 49, 82.
[11] *Ibid.,* p. 30.
[12] *Diary of George D. Carrington of Company B, 11th Illinois Infantry*, Translated and edited by Ken Bauman; Original in Chicago Historical Society.
[13] Jack Magune, Civil War List of "The House of Billy Yank" (Worcester, MA, August, 1992.)
[14] *Official Records of the War of the Rebellion; The Civil War CD-ROM* (Washington: Government Printing Office, 1880; Carmel, Indiana: Guild Press of Indiana, 1996) Series I, Vol. XXVI/1 [S# 41], p. 38. Letter to Asst. Adjutant General, Captain Halsted.
[15] *Ibid.,* Vol. XXXVI/1, [S# 67], p. 231.
[16] *Ibid.,* Vol. XLIX/1, [S# 103], p. 153.
[17] U.S. Sanitary Commission, *Narration of Privations and Sufferings of U.S. Officers and Soldiers While Prisoners of War in the Hands of the Rebel Authorities* (1864) pp. 134, 155.
[18] *Official Records of the War of the Rebellion,* Series II, Vol. VII, [S# 120], p. 60. Report to Major Southard Hoffman.
[19] William A. Albaugh and Edward N. Simmons, *Confederate Arms* (New York: Bonanza Books, 1957) p. 183.
[20] *Official Records of the War of the Rebellion,* Series I, Vol. XLVI/3, [S# 97], p. 28.
[21] N.A., Record Group 92, Entry 225, CCF, "Rubber."
[22] *The Scientific American*, 22 Mar. 1862.
[23] G. Leander Bishop, *A History of American Manufactures from 1608 to 1860* (Philadelphia: Edward J. Young & Co., 1868) p. 309.
[24] National Archives, Records of the Treasury Department, Hereinafter referred to as Records Group 217, 2nd Controller's Office, QM Contracts, 1863, Box 22 "L."
[25] *Ibid.*
[26] *The Scientific American*, 29 Nov. 1862.
[27] Bruce S. Bazelon and William F. McGuinn, *A Directory of American Military Goods Dealers and Makers 1785-1915* (Manassas, VA: Published by the Authors, 1995) Supplement, pp. 42, 44, 46, 64, 76, 99.
[28] N.A., RG 92, Entry 2195, Book 5, Page 256.
[29] N.A., RG 92, Entry 225, CCF, "Blankets."
[30] *Ibid.*
[31] E. J. Coates and F. C. Gaede, "U.S. Army Quartermaster Contracts, 1861-1865" (Lutherville, MD: Production Draft No. 4, December, 1994).
[32] Author's opinion.

BOATS AND PONTOONS

"...the bridge was kept two hours to pass over our wounded men and stragglers. It was nearly worn out (India-rubber floats,) having been in use over two years. ...Without it my whole command would, in all probability, have been lost."

From General Steele's report on the retreat from the Battle of Jenkin's Ferry, April 30, 1864

According to Charles Goodyear, "Boats were among the first things that were attempted to be made of India rubber, not only in the United States, but also in Europe."[1] In 1834, Gulielmus B. Millard and James Brown constructed a fishing boat consisting of two cylinders lashed together. The cylinders were made of duck coated with India-rubber dissolved in spirits of turpentine, a common process prior to vulcanization. Over the next three decades, various other boats, pontoons, and floats were made utilizing air-filled cylinders. Some early boats had a wooden frame covered with India-rubber cloth or canvas saturated in liquid India-rubber. At least one employed cork covered with India-rubber canvas.[2]

In 1841, a one-hundred foot boat with India-rubber canvas sides, "...was loaded with nearly one-hundred tons of wood and wine, which it safely conveyed from Auxerre to Paris, down a shallow and much interrupted stream. It was taken to pieces in three or four minutes, and all the material packed in two carts, which took it to Auxerre for another cargo."[3] While the accuracy of the details and logistics of this event are questionable, it does appear the boat's capabilities were impressive.

Colonel John Charles Fremont used rubber boats made by Horace Day while surveying the River Platte in 1842 and on the Great Salt Lake in Utah in 1843.[4] Horace Day also designed and patented an India-rubber life boat in 1845 (U.S. Patent #4,356) that closely resembled the rubber rafts used today. This boat had four cylinders of inflatable India-rubber cloth with 45 degree beveled ends to fit together and form the outer frame. The bottom of the boat consisted of a sheet of India-rubber cloth wrapped around and over the cylinders and secured by straps and buckles. Loops were provided in the cloth above the cylinders which would receive rods of wood to add stiffness if desired. When not in use the cylinders could be folded into a small package or simply transported individually by unbuckling and disconnecting them.

In the late 1840s, Day was advertising a "Patent India-Rubber Boat" that could be converted in a few minutes into a "most convenient tent" Prices for the boat, made to carry 6 passengers and weighing 75 pounds with accessories such as mattresses, ranged from $40 to $75.[5]

Other India-rubber life boats were patented through the years, including a design by William and Thomas Schnebly in 1849 (U.S. Patent #6,053). A "Bed-boat or Life Preserver," patented by Joseph

Early model for U. S. Patent #1,634, *issued June 17, 1840 to Colonel J. F. Lane for a "Pontoon Bridge." The model consists of three rubber and canvas pontoon cylinders (now flattened) with brass inflating nozzles. Each is stenciled "Roxbury I. R. Mfg. Company," the only surviving rubber company of the late 1830s. The stiff, inflexible India-rubber precedes Goodyear's vulcanization patent, so the sulphur smell used in that process is not present.*

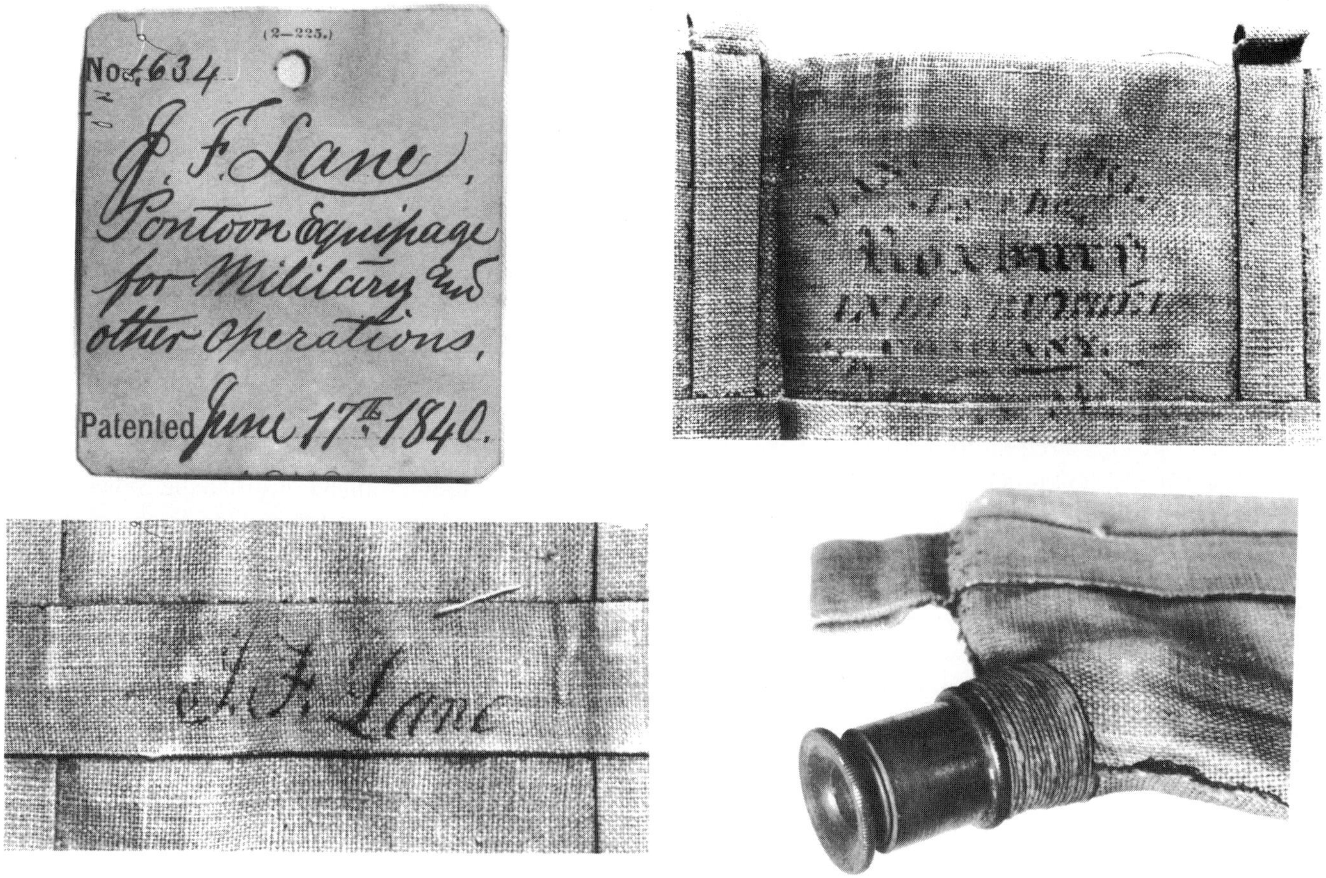

Detail of the patent office tag, *Roxbury markings, and the inflation valve on Colonel Lane's pontoon bridge.*

Stevenson in 1855 (U.S. Patent #12,450), was one of several combination boats that were invented. Stevenson's design was an India-rubber mattress that was stored in a berth and used as a bed, but could be converted to a boat. In 1861, F. W. Weiss patented a military cloak of India-rubber with an inflatable compartment that converted it to a bed or a float, "…that will sustain or buoy up the wearer with fifty pounds surplus weight quite easily" (U.S. Patent #33,910). U.S. Engineer Herman Haupt developed simple "Blanket-boats" made of a rubber blanket over a frame of round wooden sticks, used for carrying relatively light loads.[6]

Documented use of India-rubber boats during the Civil War is included in a letter from R. J. Kimball, United States Consular Agent at Toronto, to Secretary of State Seward on January 3, 1865. Kimball detailed activity of Confederate agents who traveled from Toronto to pass information to the Confederate Government. The agents used British passports to get to Baltimore. Then, as Kimball noted:

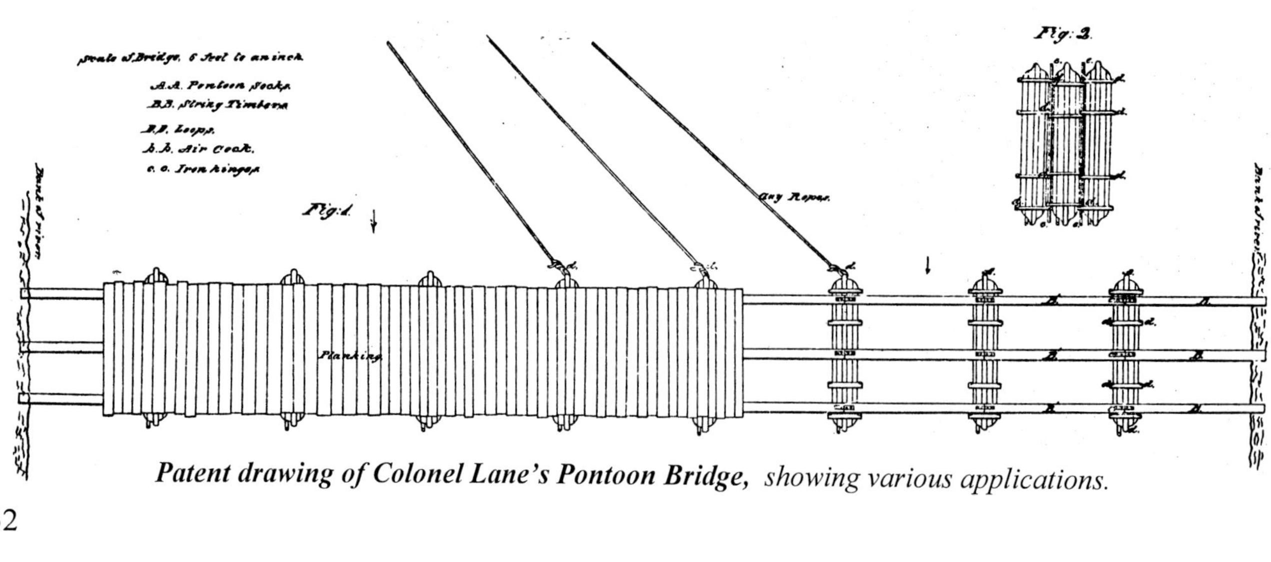

Patent drawing of Colonel Lane's Pontoon Bridge, *showing various applications.*

S. S. Broad's "Life Saving and Troop Landing Raft," *as pictured in an 1862 advertising booklet. According to the patent issued May 8, 1860, the inflatable rafts could be made of India-rubber or gutta-percha. (National Archives)*

> The messenger arriving at Baltimore receives additional instructions from "B------," and proceeds to Washington. Here he undergoes a thorough examination, is searched and permitted to pass; he takes a southeasterly direction to Port Tobacco, where he is sheltered by a widow, "Mrs. F------," and at dead of night crosses in an India rubber boat to the south side of the Potomac; thence he goes to Bowling Green, where his rebel passport is used, to Guerrilla "B------," who hastens him on to Richmond.[7]

Various gutta-percha life boats and accessories, including folding buoys and settees with air-tight compartments, were displayed at the Great Exhibition in 1851.[8] Gutta-percha boats were reviewed by *The Scientific American* on November 8, 1851. The article introduced gutta-percha sledge boats for use in icy regions:

> Fitted with a skate, the boat served as a sledge; floated it would carry five or six persons, with ample provisions; at other times it might be folded up, or converted into a wrapper or bed-

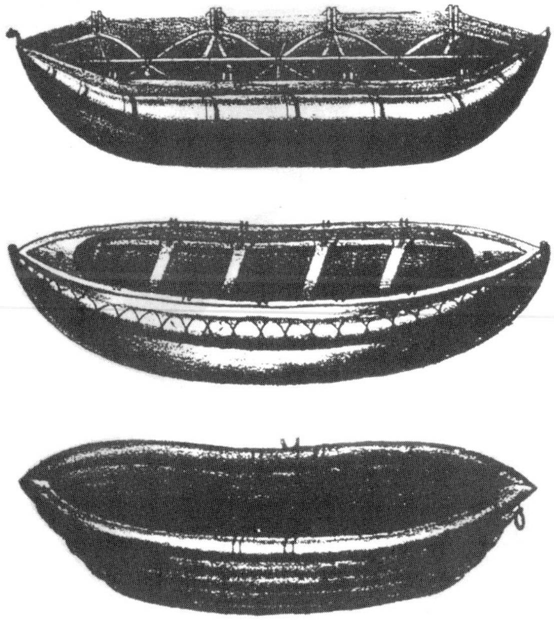

PORTABLE BOATS.

India-rubber boats, *as pictured by Charles Goodyear in* Gum-Elastic and Its Varieties.

tent, safe against the cold, that three or four men might sleep under. Its weight was only eighteen pounds. Moreover, after undergoing all the rough work of the voyage, it returned to England not in the least bit damaged, and in almost as good a condition as went it left.[9]

On October 24, 1857, the same periodical printed a plea for inventors to develop a suitable lifeboat for saving human lives lost at sea. A gutta-percha lifeboat was made at the time by the Lorcha Gutta Percha Life-boat Company of New York but, according to the article, was too small to be practical. Their choice of materials at that time, however, was clear: "In our opinion, as regards the durability, strength, and capabilities of working gutta percha, it is the very substance for a life-boat."[10]

The September 4, 1858 issue of *The Scientific American* records the results of tests on a gutta-percha lifeboat patented by E. B. Larcher in 1855 (U.S. Patent #13,315). The lifeboat, which was 16 feet long, 5 feet wide, and 3½ feet deep, was capable of holding 100 people. The boat was cast and molded from one solid piece of gutta-percha, so it contained no joints or seams. The article stated:

> The spot chosen for the trial was the beach at Long Branch, N. J., opposite the National Hotel, and as there was a southwest wind, the "rollers" or surf was pretty high. The surf-boat men took her and tried hard to capsize her, but could not succeed, and then filling her with water, sent her adrift among the rollers, which brought her safe and uninjured to shore. Altogether the experiments proved that gutta percha was an excellent material for a life-boat.[11]

Pontoons And Wagon Floats

Pontoons and wagon floats were traditionally made of wood, metal, and canvas, and consisted of a series of air-tubes, or cylinders, much like modern pontoons. "Camels" employed similar cylinders and were used for "...lightering [unloading] vessels in shoal water."[12] Pontoons is spelled "pontons" in most period references and oftentimes used synonymously with boats, wagon floats, rafts, floating bridges, etc.

Military trials of India-rubber pontoons began in August of 1835, when Lieutenant J. F. Lane, with the support of the Quartermaster General, Major General T. Jesup, was placed on special duty by the Secretary of War to experiment with the use of India-rubber cloth for military purposes.[13] In May of 1836, the Office of Ordnance informed Lieutenant Lane that a Major Craig had been instructed to "...afford assistance required by Lieutenant Lane in making

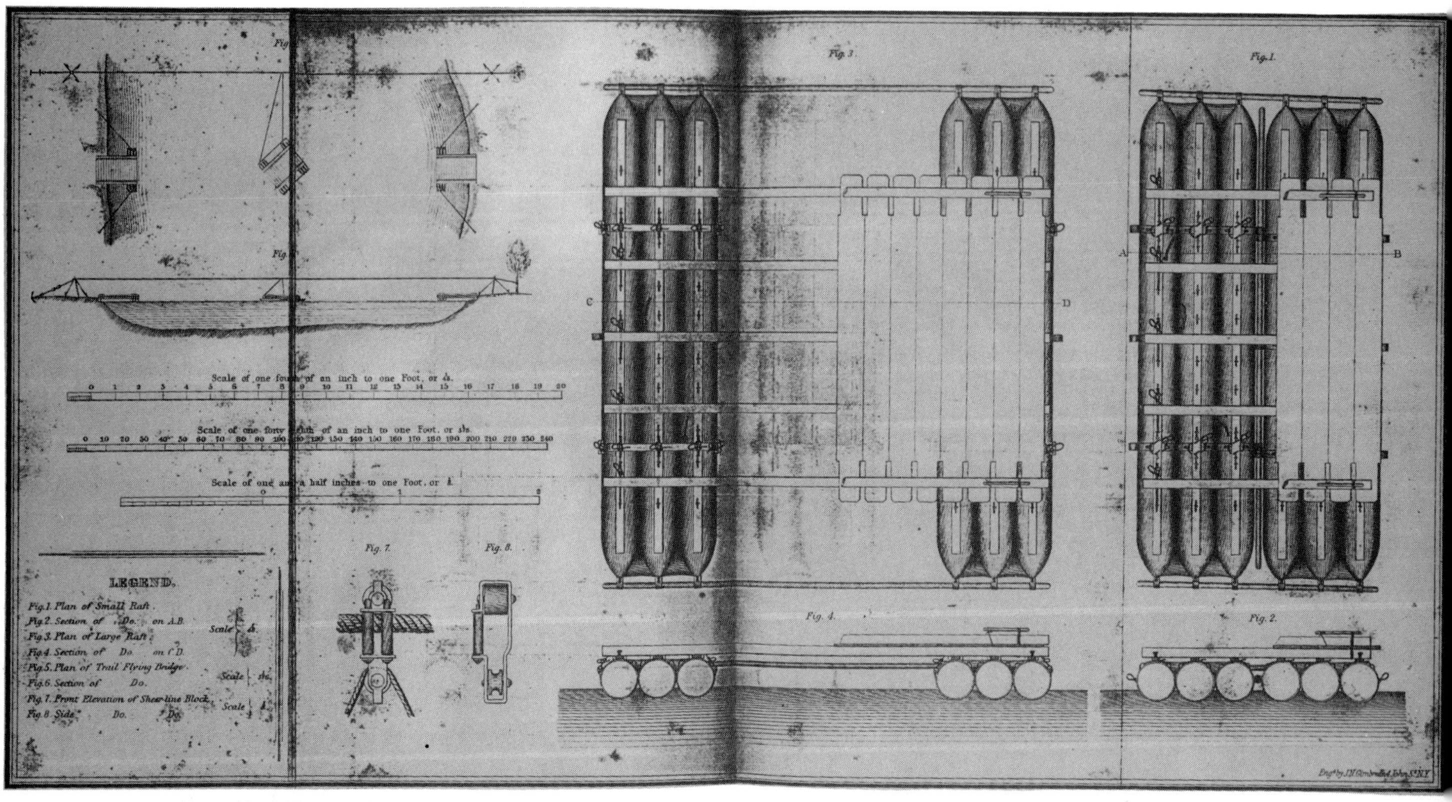

Detailed illustrations from **"Description of a System of Military Bridges with India-Rubber Pontons,"** *by George W. Cullum, published in 1849.*

experiments on India rubber cloth provided the same can be performed by the enlisted men under his command without deranging the operation now conducting at the post."[14]

In 1836, Lieutenant Lane (later promoted to Captain in the Regular Army and then Colonel of the Creek Volunteers before his death later that year) constructed an experimental 350 foot bridge, supported by sixteen India-rubber pontoons across the Tallapoosa River in Alabama.[15] He also built a 375 foot, 40-pontoon bridge over the Chatahooche River near Fort Mitchell in Alabama.[16] In the same year William H. Harlow of Roxbury, Massachusetts, constructed the first pontoon boat of three cylinders united together.[17] In 1837, Captain C. A. Waite of Boston submitted a plan for an India-rubber pontoon with a sketch titled "Ponton Wagon Body."[18] Patent #1,634 was granted on June 17, 1840 to the estate of John F. Lane for pontoon equipages similar to William Harlow's design.

In June of 1838, Colonel Kearney of the 1st Dragoons informally asked the Quartermaster General if his Office would approve a requisition for an India-rubber boat to carry the baggage of the Dragoons during their expeditions over Western waters. On August 29, 1838, a formal request was made by Kearney to Colonel Henry Stanton, Acting Quartermaster General, asking that an order for such a boat be placed with Mr. S. T. Armstrong, agent for the Roxbury India Rubber Company. The boat was to be forwarded to the Office of the Assistant Quartermaster at Fort Leavenworth.[19] The requisition was authorized and the order placed on September 5. The boat was to be manufactured per Colonel Kearney's specifications and furnished within a week to ten days at a cost of about $150.[20]

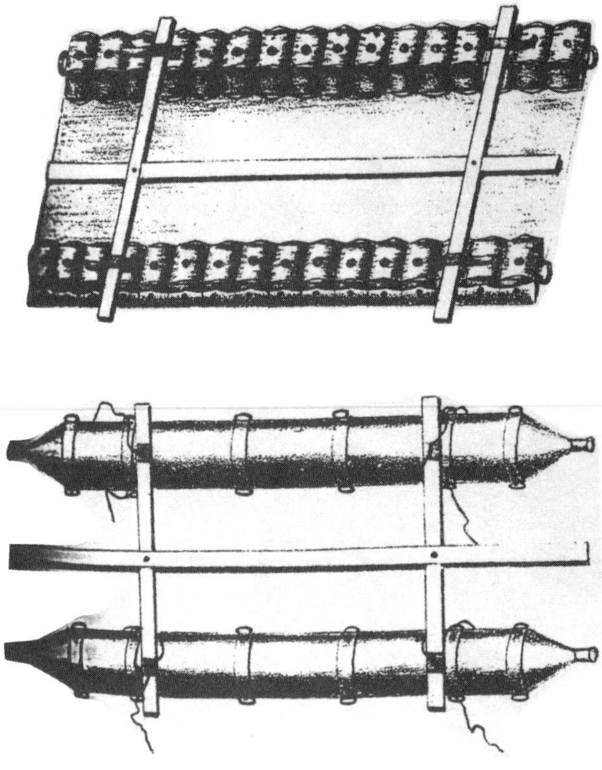

India-rubber pontoons - *from* Gum-Elastic and Its Varieties.

On September 14, 1839, Armstrong, again representing Roxbury India Rubber Company, sent U.S. Army Adjutant General R. Jones a letter saying he had witnessed experiments made with the India Rubber air boats (likely pontoons) in New York harbor, and that Major General Macomb "...speaks very favourable [British spelling] of the boats." He offered to make some under contract, "...should the department think favourable of them."[21] On August 2, 1839, a proceeding of a Board of officers who had witnessed experiments on India-rubber air boats made by the Roxbury India Rubber Company unanimously stated, "...these boats claim a decided preference for military purposes over any other means within the knowledge of the Board."[22]

Captain William H. Bell of Washington Arsenal also reported on experimental India-rubber air boats to Colonel George Talcott, Commander of the Ordnance Department. In a letter dated October 19, 1839, Captain Bell described the boats as having "...three touching cylinders, strongly attached along their lines of contact....made of strong canvas of a double thickness saturated with gum." Each cylinder was 18 feet long and 18 inches wide making the total size of the pontoon 18 feet by 4½ feet. The total weight was 117 pounds with a buoyancy of 4,800 pounds and capacity of 13 men. Bell noted:

> A 6 pounder without limber was rolled upon it and fired with round shot repeatedly without injury or inconvenience. It is thought that the air boat might be advantageously applied in the construction of floating batteries of heavy artillery for the defense of harbors.[23]

Years later, Lieutenant J. Henry Carleton, documenting the Dragoon campaigns to the Pawnee villages in 1844 and the Rocky Mountains in 1845, tells of an "...India-rubber boat which had been brought along by the quarter-master...," and was used in the crossing of the Blue River. At one point, in the

NATICK

BOOT AND SHOE

STORE.

THE Subscriber would respectfully announce to the inhabitants of NATICK and Vicinity, that he has taken the Store on Main St., nearly opposite MR. JOSEPH T. HOWE'S Drug Store, where may be found a general assortment of

Ladies', Gent's, Misses', Boys' and Children's

Boots, Shoes and Rubbers,

which have all been bought for cash, and mostly of the Manufacturers, with whom he has made arrangements for fresh supplies, which will be sold as low for cash, as can be bought in the City or elsewhere.

Having been engaged in the SHOE BUSINESS for the last twenty years he flatters himself that he shall be able to give satisfaction to those who may favor him with a call, and hopes to merit and receive by strict attention to business a liberal share of patronage.

☞ N. B.—Having made arrangements with an experienced Custom Workman he is now prepared to make and repair Ladies' and Gent's Boots and Shoes.---Also, Boots and Shoes soled with rubber, and Rubbers repaired.

HENRY COGGIN.

NATICK, JUNE 19, 1854.

J. C. French, Book, Card and Job Printer, 15, Washington-St., Boston.

***Broadside of the Natick Boot & Shoe Store**, advertising "Rubbers," one of the most popular India-rubber products of the 1850s.*

Child's Civil war era rubber boots. *Marked on sole: "Goodyear's Patent 1858 - Boston R. S. Co." Marked on heel: "Boston Rubber Shoe Co. - Hayward's Patent Heel." One boot appears to have a manufacturing defect at the tip. The 1858 date refers to the extension of the Goodyear vulcanization patent that expired in 1865. In addition to a variety of boots and shoes, the Boston Rubber Shoe Company manufactured rubber blankets for the Union army under a sub-contract from the Union India Rubber Company.*

Misses' and children's gaiter boots..	0.75
Misses' and children's long boots, felt lined.....	1.75
Misses' and children's long boots, net lined......	1.625[7]

Improvements were made in basic designs, such as the "Ventilated India Rubber Overshoe," patented by Julius Pease on November 14, 1854 (U.S. Patent #11,947). The inner surface of Pease's overshoe was ribbed or corrugated to allow circulation of air between it and the boot or shoe over which it was worn.

References are made to the fact that many impoverished individuals, unable to afford leather boots and shoes, could buy the rubber footwear for considerably less. Moreover, the invention of several types of rubber soles and heels provided economical extension of the life of any shoe. In 1853, a rubber sole with a leather edge for attachment was produced and, around the same time, a removable rubber sole with elastic straps was invented by Dr. Charles Stearns of Springfield, Massachusetts.[8]

Christopher Meyer of New Brunswick, New Jersey invented the "Rough-soling Machine," which added "diamond" designs to the bottoms of the soles to prevent slipping and stamped them in relief with the device of the manufacturer. While Goodyear suggested attaching the soles with cement, later innovations of the 1850s and 1860s included rubber heels and soles with nails semi-embedded for attaching to the leather.

In the mid-1860s, Charles Goodyear, Jr. became interested in a machine for stitching light "turned" soles patented by Auguste Destouy. After improving the process and inventing complementary machinery, the entire process and resulting product became known as the "Goodyear Welt," a term still utilized in the trade.

In 1855, the Newark India Rubber Company was turning out 2,000 to 2,500 pairs of shoes per day. A contemporary article noted, "At the present time 200 to 230 hands are employed, who receive per year wages to the amount of $50,000.00; An expert girl will make from 30 to 40 pairs a day, for which she is paid by the piece."

Each year of the early 1850s saw the production of rubber footwear double from the previous year. By 1860, the entire industry output was estimated at 1,200,000 pairs valued at $795,000, and by the end of the decade the annual output in the boot and shoe industry had increased to $8,000,000.[9]

In *A History of American Manufacturers from 1608 to 1860,* published in 1868, the Boston Rubber Shoe Company is cited for its growth and the premium quality of its products. Mention is also made of the company's involvement in manufacturing rubber blankets for the Union Army.[10]

On April 12, 1864, Frederick Vanderpael of New York received a contract from the U.S. Office of Army Clothing and Equipage for "Ten thousand pairs of Bootees with India Rubber Heels and Soles...@ $1.99 per pair" to be delivered by November 1, 1864. However, the contract specified "...thirty percent to be retained from each payment until official report of their service in actual use prove that they are better for Army use than the standard sewed bootees now used."[11]

The Richmond Daily Dispatch of October 20, 1863 contained an advertisement for "a large assortment of men's and misses' rubber overshoes" available at C. Dunham, 6th and Broad Streets. These likely came through the blockade from England.

George Washington Rains, the master-mind and driving force behind the famed Augusta Powder works of the Confederacy, implemented several safety features into the works to reduce the possibility of disastrous explosions. In addition to natural timber walls between buildings and a crude sprinkling system, Rains insisted that workers wear rubber-soled shoes to prevent sparks that could ignite an explosion.[12]

(Also see **Gaiters and Leggings**.)

NOTES - Boots and Shoes

---

[1] Institution of the Rubber Industry, *History of the Rubber Industry* (Cambridge, England: W. Heffer, 1952) p. 240
[2] *The Illustrated London News,* 9 Aug. 1851.
[3] P. W. Barker, *Charles Goodyear* (Boston: Godfrey L. Cabot, Inc., 1940) p. 56.
[4] *Ibid.*, Pg. 53.
[5] Institution of the Rubber Industry, *History of the Rubber Industry*, p. 241
[6] *Eighth Census of the United States,* p. lxxx.
[7] Barker, *Charles Goodyear*, p. 52.
[8] Charles Goodyear, *Gum-Elastic and Its Varieties* (New Haven: Charles Goodyear, 1853 & 1855) Vol. II, p. 323.
[9] Bishop, G. Leander, *A History of American Manufacturers from 1608 to 1860* (Philadelphia: Edward Young & Company, 1868).
[10] Chauncy M. Depew, *One Hundred Years of American Commerce* (New York: D. O. Haynes & Co., 1895), p. 501.
[11] National Archives, Records Group 217, Entry 236, 2nd Controller's Office, Quartermaster Contracts, Box 44, "1864."
[12] Theodore P. Savas, "Bulwark of the Beleaguered Confederacy - The Augusta Powder Works," *Civil War,* (Sep.-Oct. 1991): p. 15.

BUCKETS

"The Board also recommend...[gutta-percha] water buckets for the use of the train." [1]

From "Extract of Proceedings of a Board of Officers," April 12, 1855

India-rubber or gutta-percha water buckets and fire buckets are seen in collections of Civil War period military, fire, and railroad artifacts. Fire buckets, taller and slimmer than regular water buckets, were filled with sand and kept in buildings, warehouses, plants, and on the railroads to extinguish fires.[2] Fire buckets and standard water buckets are seen made entirely of India-rubber or gutta-percha plated fabrics and reinforced inside the fabric of the rim with metal or other material. Heavy-duty water buckets are reinforced with a heavy iron or steel band around the outside of the rim.

India-rubber and gutta-percha buckets are often referenced in early military documents. In an 1845 proposal to the U.S. Army, the following articles made of "Goodyear's Patent Metallic Rubber with all the latest improvements" are detailed:

> Water Pails (to be folded up for easy transportation) $2.00 each
>
> Water Pails or Fire Buckets (stiff as leather) $2.00 each
>
> Horse Feed Bags (or Buckets to hold water) $1.75 each[3]

Captain Winfield S. Hancock, writing to Lt. Colonel Thomas Swords, Deputy Quartermaster General, from Benecia Arsenal in California on December 1, 1858, commented:

> I had a few Gutta Percha Water Buckets issued to me when I left Fort Leavenworth last Spring. The band on them was as follows—Vulcanized Gutta Percha made by North American Gutta Percha Company, New York, Rider and Murphys Patents 1853 & 1854.
>
> I had many varieties of Water Buckets with my train and among the wooden buckets those of the best construction of Oak and Cedar but none stood the knocks of rocks in the road or the heat and cold so well as the Vulcanized Gutta Percha Bucket in question.[4]

Early battlefield relic rubber artillery water bucket *with painted words, "Found On Bull Run Battlefield July 1864 Jos. Miller." Embossed into rubber on bottom is "D. Hayward's Patent 1854 -Manufactured By The N. Hayward Co. Stoneham, Mass." Hayward's patent (U.S. Patent #11,608) was for an early recycling process that utilized worn-out rubber goods in the manufacture of buckets, shoe heels, etc. This larger size and standard bucket shape differentiates it from fire buckets.*

This endorsement was referred to when the Union Army was given the suggestion to use gutta-percha buckets in a letter from John Rider to General Meigs dated December 26, 1861.[5]

Wiard's System of Field Artillery lists among implements and equipments for one six gun field battery, "12 Gutta Percha Water Buckets" for the gun carriages and "12 Gutta Percha Water Buckets" for the caissons.[6]

(Also see **Section 6**.)

Early rubber fire bucket *with "G. I. F. Co." stenciled on the outside. Since this bucket was purchased in Gettysburg, the initials could stand for "Gettysburg Independent Fire Company." While the bucket lacks patent and manufacturer's markings, the construction appears to be pre-Civil War. The markings may have worn off or the bucket may have been illegally manufactured without a license.*

Artillery water bucket. *Stenciled on front: "1st Battery" and underneath "Goodyear's India Rubber Glove Mfg. Co. New York." A paper label underneath reads "India Rubber Goods—W. W. Wilder, Dealer In Rubber Goods Of Every Description. A great variety of fancy goods always on hand. No. 193 Washington Street, Boston." Lack of a patent marking on an article that has other clear markings indicates that this bucket was likely manufactured after Goodyear's patent vulcanization patent expired on June 15, 1865.*

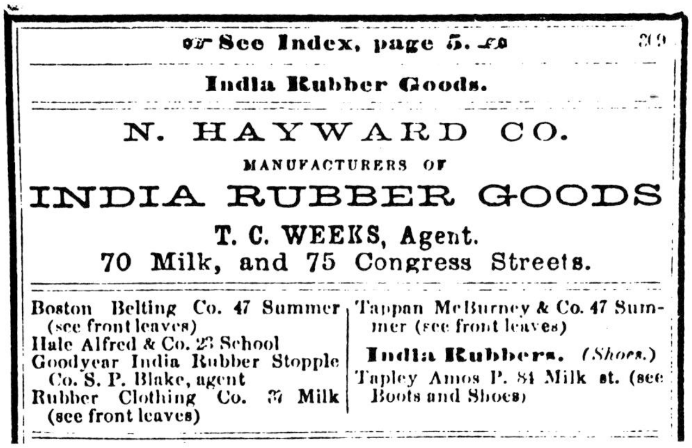

☞ See Index, page 5. ☜

India Rubber Goods.

N. HAYWARD CO.

MANUFACTURERS OF

INDIA RUBBER GOODS

T. C. WEEKS, Agent.

70 Milk, and 75 Congress Streets.

Boston Belting Co. 47 Summer (see front leaves)
Hale Alfred & Co. 23 School
Goodyear India Rubber Stopple Co. S. P. Blake, agent
Rubber Clothing Co. 37 Milk (see front leaves)
Tappan McBurney & Co. 47 Summer (see front leaves)

India Rubbers. *(Shoes.)*

Tapley Amos P. 84 Milk st. (see Boots and Shoes)

Advertisement for rubber supplier Nathaniel Hayward, *an associate of Charles Goodyear, who began his N. Hayward Co. in Lisbon, Conn. in 1843. He specialized in making rubber shoes, buckets, etc. from recycled India-rubber under Daniel Hayward's 1854 patent.*

Railroad fire bucket made of recycled rubber. *Stenciled "CRR," the bucket may have been from the Concord Railroad, which operated in Massachusetts from the 1850s to 1889. These buckets were used in the passenger cars in case of fire and by the locomotive crew for drinking water. Remains of markings on the base: "D. Hayward's Patent 1854 - Manufactured By The N. Hayward Co. Stoneham, Mass."*

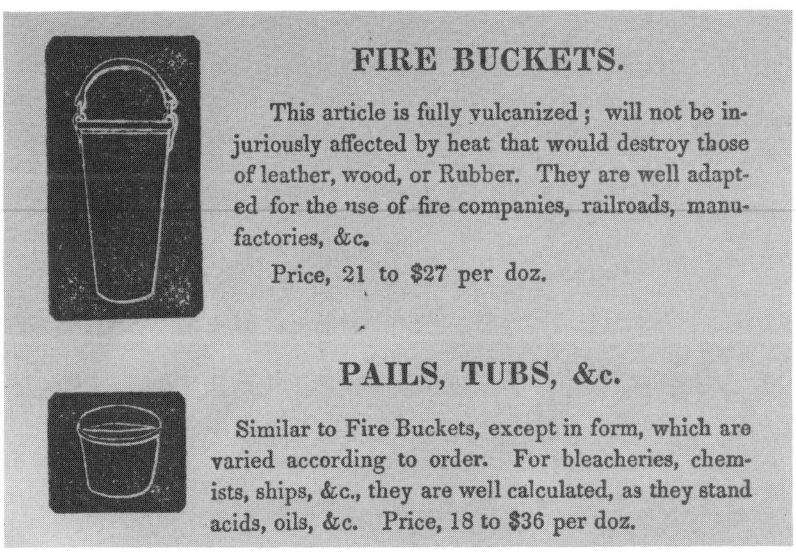

FIRE BUCKETS.

This article is fully vulcanized; will not be injuriously affected by heat that would destroy those of leather, wood, or Rubber. They are well adapted for the use of fire companies, railroads, manufactories, &c.

Price, 21 to $27 per doz.

PAILS, TUBS, &c.

Similar to Fire Buckets, except in form, which are varied according to order. For bleacheries, chemists, ships, &c., they are well calculated, as they stand acids, oils, &c. Price, 18 to $36 per doz.

Rubber buckets advertised in the 1854 catalog of the North American Gutta Percha Company. *Note the iron band at the top of each container.*

Camp scene *showing various pots, kettles, and, on the far left, a bucket that appears to be identical to the Hayward's patent rubber bucket recovered at Bull Run. (Library of Congress)*

NOTES - Buckets

---

[1] Gutta-Percha Advertising Brochure, Circa 1860s; NA, RG 92, Entry 225, CCF, "Gutta-Percha."
[2] Helaine Fendelman, "What is It? What is It Worth?," *Country Living,* (July 1998): p. 74.
[3] NA, RG 92, Entry 225, CCF, "Goodyear's Metallic Rubber Company."
[4] Gutta-Percha Advertising Brochure, Circa 1860s; NA, RG 92, Entry 225, CCF, "Gutta-Percha."
[5] NA, RG 92, Entry 225, CCF, "Gutta-Percha."
[6] Norman Wiard, *Wiard's System of Field Artillery* (New York: Holman, 1863) p. 21.

BUTTONS

"...they were equipped to the black gutta-percha button on the coat, for their was nothing of brightness allowed about them to attract the attention of an enemy..."

First Sgt. George Springer, 122nd Reg't, PA Volunteers, describing the uniforms of Berdan's Sharpshooters

(While specimens of the military buttons depicted in this section are pictured and described in popular books on military buttons,[1] most comments are based on the author's experience and samples in his collection.)

In addition to the hundreds of different hard rubber civilian buttons made during the Civil War era, several military Army and Navy styles were also manufactured and used in the War. Both shanked and sew-through types were made by the Novelty Rubber Company and the India Rubber Comb Company. Along with the manufacturer's markings, most buttons of the period also bear the date of the hard rubber patent, May 6, 1851. Some early hard rubber buttons, not commonly encountered, also have the

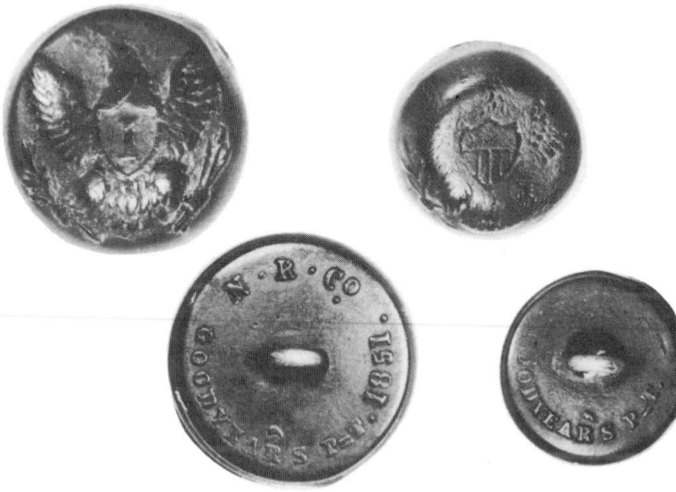

Some collectors theorize that this variety of hard rubber eagle button *represents the true Berdan's Sharpshooter's button. These scarce examples, distinguished by the high, uneven sides, are very similar to the prototype buttons on the Berdan's coat in the Smithsonian. While manufactured by the Novelty Rubber Company, it appears they were made in a crude mold. The blouse size eagle "I" was recovered at Aquia Landing.*

CDV photograph of C. R. Mead wearing the distinctive green frock coat with black hard rubber buttons of Berdan's Sharpshooters. *Mead belonged to the 10th Vermont Volunteers which joined the 1st U.S. Sharp Shooters. (Courtesy Mike McAfee)*

"1849" date of Nelson Goodyear's filing of the caveat for patent.

The best known of the military buttons are the hard rubber blouse and cuff size eagle buttons in both the plain "General Service" and Infantry "I" varieties. Some have speculated that these buttons may have been worn by some regular units, but they are most often associated with Berdan's Sharpshooters. In the *Concise History of the Camp and Field Life of the 122nd Regiment, Penn'a Volunteers*, diarist George F. Springer, First Sergeant, Company K, records an encounter with one of Berdan's Sharpshooters, who were attached to their brigade:

> "...in fact, they were equipped to the black gutta-percha button on the coat, for their was nothing of brightness allowed about them to attract the attention of an enemy, or the rays of the sun-the

Union Navy buttons and backmarks. *One reason for the popularity of these buttons may have been hard rubber's total resistance to any corrosion from the salt spray of the sea.*

gun barrels, too, being all bronzed or blackened."[2]

Colonel Hiram Berdan, founder of Berdan's Sharpshooters, appears to have originally considered metal buttons for the uniforms. However, a letter to General McClellan dated October 22, 1861 regarding a contract with the Seamless Clothing Company for the first uniforms specifies "...gray overcoats, gutta-percha buttons, and with green trimmings. The dark buttons and trimmings I deemed important for my Corps."[3]

There are two distinct shapes of hard rubber eagle buttons that has caused some discussion among historians. The most common type closely resembles the brass button in its shape with the convex face and beveled sides. The rarer example, which some say is the true type used by Berdan's, has a relatively flat face and slightly angled sides. Collector's sometimes refer to this style as the "pill-box" eagle button. The Smithsonian displayed a Berdan's coat with all buttons being this type. The only excavated hard rubber eagle button (the blouse "I" variety) known to the author is also this style.

Two styles of Navy overcoat buttons are found in hard rubber, the larger style having a tilted anchor and the smaller an upright anchor. Both are 4-hole sew-through types made by the Novelty Rubber Company.

A rare Confederate item is the Navy overcoat button imported through the blockade by Courtney

Union eagle buttons. *Worn by Berdan's Sharpshooters so they would not glisten in the sun. These may also have been worn by other officers and enlisted men. The backmarks are "N. R. Co. GOODYEAR'S P=T. 1851."*

and Tennent. This four-hole sew-through button, made of a composition very similar to hard rubber, contains the backmark of Confederate importer "Courtney & Tennent - Charleston, S C," and "Manton's Patent." On November 26, 1860, John Shakespeare Manton and button maker Thomas Islip were awarded British patent #2907 for "Certain improved compositions useful for many purposes in the arts & manufactures." This patent is listed under the subject matter index of patents under "Casting Plastic" and "Buttons," and is undoubtedly the patent referred to on the Confederate backmark.

In 1973, a button collector referenced what he termed a "Standard [Confederate] Rifleman button, Block 'R,' made of rubber, with no backmark and brass eyelet." He claimed the button had been found in Bannerman's store on Broadway around 1958. Another collector told of a rubber "M" Confederate Marine button he had seen. The author has not seen either button or heard of any similar examples that have surfaced.

There are hundreds of styles of hard rubber civilian buttons in sizes ranging from ¼ inch to 1 ½ inches in diameter. The varieties include shanked and

Political buttons. *Hard rubber Seymour and Blair button from the presidential campaign of 1872. Backmark is N. R. Co. Goodyear's Pat. 1851. Grant and Colfax used similar buttons in this campaign. Also shown is a composition Lincoln button.*

Confederate Navy composition button *imported from England through the blockade by Courtney & Tennent. Backmark: "Courtney & Tennent - Charleston, S C - Manton's Patent" Reverse also bears a small swan. Unlike brass Confederate Navy buttons, the face shows only "C N," despite a letter from Commander James H. North, representative of the Confederate Navy, to G. B. Tennent of Courtney & Tennent, stating, "...if our brass buttons have 'C. S. N.' on them, the overcoat button should also have 'C. S. N.' on them, or, in a word, whatever ornament is on one, the other should have also."[5] While this button was made under a British patent for a composition, it closely resembles hard rubber.*

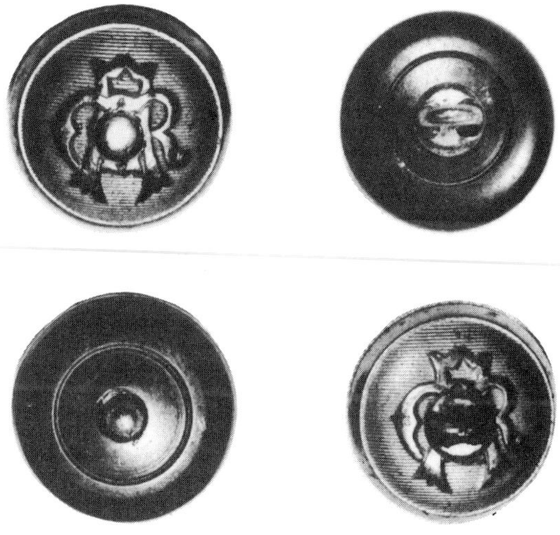

Reversible G. A. R. mourning buttons *with the standard brass front and removable shank. By screwing the shank into the brass face, the button was converted to a black hard rubber button for funeral services.*

Excavated hard rubber button. *This unusual button, dug at Aquia Landing, was made by the Novelty Rubber Company for the Beverly Rubber Company, a wartime manufacturer of soft India-rubber goods. It likely came from a cloak or raincoat.*

sew-through, round and square, plain and fancy ones with floral and geometric designs. Among the more detailed are the "sportsman's" buttons with a running deer, flying bird, or the head of a wolf or boar. Also of interest are the patriotic "Miss Liberty" buttons with 13 stars on her headband.

A Berdan's forage cap in the Smithsonian collection has plain domed hard rubber buttons on the strap.[4] Both military and civilian styles have been found in campsites indicating the latter were used as replacement buttons (as were wood, china, and metal buttons). Sometimes soldiers passed the time by carving rings and other items out of these buttons. Partially carved buttons have been excavated at prisoner-of-war camps as well.

Carved hard rubber button. *From the possessions of Captain J. Byron Brooks, Co. H, 4th Vermont Volunteers, who was discharged as a result of wounds received May 4, 1864. Backmark: N. R. Co. GOODYEAR'S P=T. 1851. The face is carved with a cross and initials "J.B.B.." (Courtesy Marius Peladeau)*

An interesting button and shank was patented in 1864 by Phillip W. Gengembre of Boston, who had at least three other button-related patents to his credit. His patent #41,292, titled simply "Button," employs India-rubber and fastens to the garment. The patent specification contains the following description:

> The button is secured to the cloth by means of a catch, which passes through the button, the inner end of the catch terminating in a device, by turning which, the button is fastened. The shank of the catch is surrounded by a tube of India-rubber which acts as a spring to tighten the locking device.

Composition buttons *taken from the uniform of a dead Confederate soldier at Gettysburg.*

Later buttons with a link to the War and Reconstruction period are the hard rubber buttons made for the Presidential campaigns of 1868 and 1872. These depicted Grant and Colfax or their opponents Seymour and Blair. There are also some post-war hard rubber Lincoln buttons and G.A.R. "mourning buttons," with a shank that unscrewed and converted the normal brass G.A.R. face to the solid black hard rubber face on the reverse.

Black rubber cuff button*, one of the styles associated with surviving Berdan's clothing. Brass wire shank secured through face of button and looped in back. "N. R. Co. GOODYEAR'S P=T." backmark.*[6]

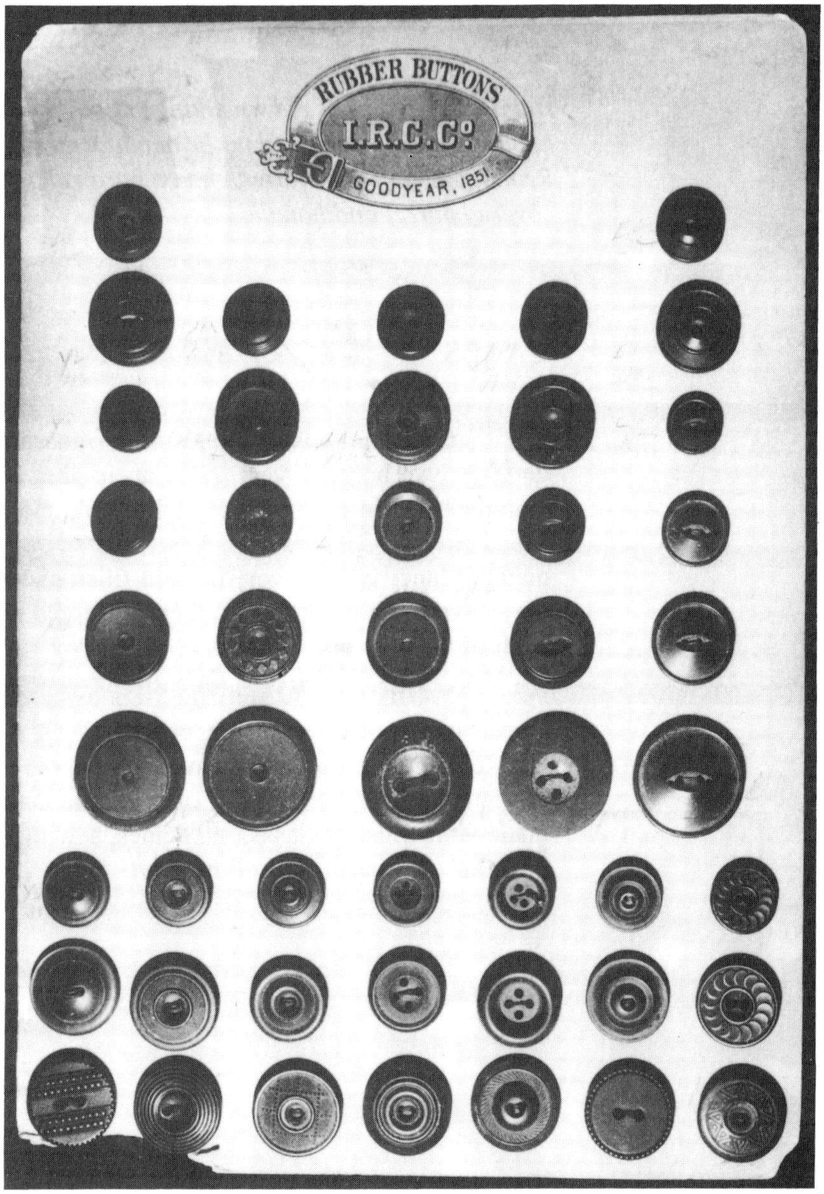

Sample card for hard rubber buttons *made by the India Rubber Comb Company. Each button design has a style and size number associated with it.*

(Also see **Section 3** and **Personal Items**.)

NOTES - Buttons

---

[1] Alphaeus H. Albert, *Record of American Uniform and Historical Buttons* (Boyertown, PA: Boyertown Publishing Company, 1977); David F. Johnson, *Uniform Buttons - American Armed Forces 1784-1948* (Watkins Glen, NY: Century House).
[2] First Sergeant George F. Springer, *Concise History of the Camp and Field Life of the 122d Regiment, Penn'a Volunteers* (The New Era Steam Book Print, 1885) p. 266.
[3] Roy M. Marcot, "The Uniforms & Field Equipment of the Berdan U.S. Sharpshooters," *North South Trader's CIVIL WAR*, Vol. 17, No. 6, (1990): pp. 30-31.
[4] *Ibid.*
[5] Alphaeus H. Albert, *Buttons of the Confederacy* (Boyertown, PA: Boyertown Publishing Company, 1963) p. 13.
[6] Information from Berdan collector Don Troiani based on surviving examples.

questioned and is probably best described by the following extracts from *Hardtack and Coffee*, by Union veteran John D. Billings:

> I still have in my possession the remnants of a water-filterer in which I invested after enlistment. There was a metallic mouth-piece at the end of a small gutta-percha tube, which latter was about fifteen inches long. At the other end of the tube was a suction-chamber, an inch long by a half-inch in diameter, with the end perforated and containing a piece of bocking as filter. Midway of the tubing was an air chamber. The tubing long since dried and crumbled away from the metal. It is possible that I used this instrument half a dozen times, though I do not recall a single instance, and on breaking camp just before the Gettysburg Campaign, I sent it, with some other effects, northward.
>
> I remember another filterer, somewhat simpler. It consisted of the same kind of mouthpiece with rubber tubing attached to a small conical piece of pumice stone, through which the water was filtered. Neither of these was ever of any practical value.[26]

(Also see **Section 6** and **Section 7**.)

Remains of an India-rubber water filter *from the steamer Maple Leaf, sunk in the St. John's River on April 1, 1864. While many thought these filters useless, some soldiers apparently still carried them late in the War. (Courtesy St. John's Archeological Society)*

S.H. Russell,

Canteen,

Patented Aug. 20, 1861.

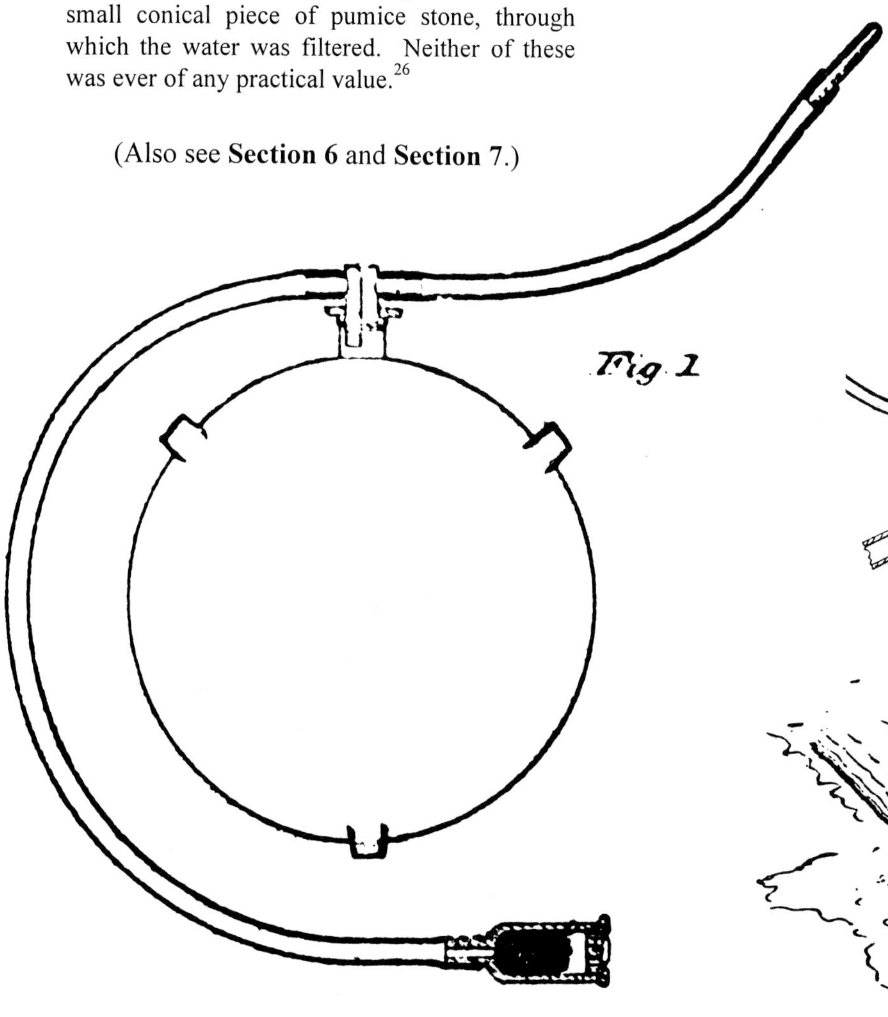

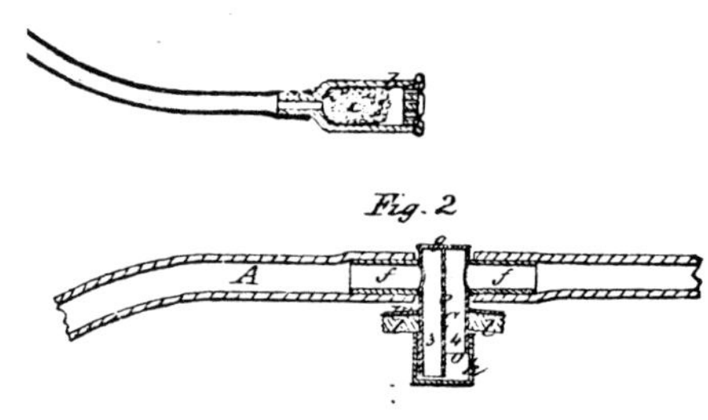

Patent drawing for Stephen Russell's Canteen-Filler*, patented August 20, 1861. The India-rubber tube was equipped with a hard rubber filter containing a sponge and could be used independently as a drinking tube or to fill the canteen.*

Hard rubber filter tube *designed to fit the regulation 1858 pattern canteen. It consists of separate sections of decreasing diameter glued together with what appears to be a contemporary rubber cement. Tiny holes in the bottom combined with an internal filtration device (sponge, charcoal, pumice, etc.) to cleanse field water. This was one of several versions of filter tubes widely used in both U.S. and British canteens during the Civil War period. (Courtesy Janet and Bedford Hayes)*

Flexible gray rubber spout *applied to a tin drum canteen to assist filling. The canteen's sling loops have been removed and a fabric loop used to support it for household purposes or as a powder can. (Courtesy Dave Mark)*

Regulation tin canteen with an India-rubber or gutta-percha adjustable strap. *The cloth-covered gum strap has an early repair. The wool cover on the canteen has the letter "O" and other illegible letters.*

Tin drum canteen with homespun strap and India-rubber stopper. *The stopper appears to have been used with the canteen but may have been larger at one time—possibly cut from a different canteen or bottle. The name "Woodman C. Coombs" and the date "1861" or "1867" are scrawled into the worn silver paint. (Courtesy of Paul Johnson)*

Close-up *of India-rubber stopper.*

NOTES - Canteens and Water Filters

[1] Charles Goodyear, *Gum-Elastic and Its Varieties* (New Haven, CT: Published for the author, 1853 and 1855), Vol. 2, p. 209.
[2] *Ibid.*, p. 134.
[3] National Archives, Record Group 92, Entry 225, Consolidated Correspondence File, "North American Gutta Percha Company."
[4] *Ibid.*, "Goodyear's Metallic Rubber Co."
[5] *Ibid.*
[6] NA, RG 92, Entry unknown, "Orders for Army Supplies, No. 1, June 17, 1846 - Feb. 15, 1847."
[7] *Ibid.*, Entry unknown, Vol. 33 (of 52), "Invoices of Stores...to be Delivered to Units."
[8] War Department General Orders No. 64, 29 November 1848, as copied in the Institute of Heraldry, Department of the Army, *U.S. Army Equipments, 1834-1870.*
[9] Author's opinion based on noted surviving examples.
[10] George G. Furber, *The Twelve Months Volunteer or Journal of a Private in the Tennessee Regiment of Cavalry in the Campaign, in Mexico. 1846-7* (Cincinnati: J. A. and U. P. James, 1850) p. 127.
[11] NA, RG 92, CCF, "Gutta Percha."
[12] *Ibid.*
[13] *Ibid.*, "Crosman."
[14] *Ibid.*, "North American Gutta Percha Company."
[15] NA, RG 92, Entry 332, Maj. G. H. Crosman, "Letters Received, 1835 - 62." Abstract dated August 19, 1856.
[16] *Ibid.,* Entry 1003,"Register of Letters Received," OQMG, Clothing and Equipage.
[17] NA, RG 92, CCF, "Gutta Percha."
[18] E. J. Coates and F. C. Gaede, Unpublished Compilation of U.S.
Army Quartermaster Department Contracts, 1861-1865, From NA, RG 217, Entry 236.
[19] Based on documented examples, advertisements, and soldier's letters quoted in this and other sections.
[20] *The Scientific American*, 22 June 1861.
[21] *Diary of George D. Carrington of Company B, 11th Illinois Infantry*; translated and edited by Ken Bauman; Original in Chicago Historical Society.
[22] *Diary of Sergeant Thomas Booth*, Box 4, Booth papers, Ohio State Historical Society, p. 12.
[23] *The Scientific American*, 11 May 1861.
[24] *Ibid.*
[25] *Ibid.*, 8 June 1861.
[26] Philip Van Doren Stern, *Soldier Life in the Union and Confederate Armies* (New York: Bonanza Books, 1961) p. 194.
[27] Author's discussions with two experienced collectors.

DENTAL APPLICATIONS

"...Mr. J. A. Bishop, of New York, ingeniously adapted a plate of vulcanite to both jaws that would supply the deficient front teeth and afford support to the lips."

Dental restoration detailed in *The Medical and Surgical History of the War of the Rebellion*

In 1847, Edwin Truman introduced gutta-percha as a denture base.[1] Around the same time, it was used as a temporary filling, soon receiving praises like these 1859 excerpts:

> Gutta-percha, with which some mineral substance, such as powdered silex or glass, has been incorporated, makes a remarkably good temporary filling, capable of lasting some months.
>
> Gutta-percha is useful as an experimental filling where there is doubt whether nerve can be preserved; In the case of the tooth aching, it can be removed by the patient if necessary.
>
> Of all temporary fillings in use, gutta-percha is the most reliable.
>
> If metal filling causes discomfort at hot and cold beverages, gutta-percha can be used for a few months, then replaced with gold.[2]

While the use of unvulcanized gutta-percha for interdental splints met with only limited success, one dentist was to later write, "Few substances have in so short a time acquired more importance in our profession than this."[3]

In 1848, gutta-percha was combined with quicklime, powdered quartz, feldspar, metal filings, and coloring and used as a permanent filling known as "Hill's Stopping." With subsequent improvements, it was used for that purpose until at least 1876.[4] Gutta-percha is still used in modern-day root canal therapy.

The saliva pump, with a hollow India-rubber air chamber to achieve suction, was devised by Robert Arthur in 1854. In 1865, William H. Dibble patented a device with an elastic suction cup and tube that held the jaw open and kept the tongue away from the teeth U.S. Patent #50,461). Dr. S. C. Barnum's introduction of the rubber dam in 1864 earned him gold medals, a gold watch, and a cash award.[5]

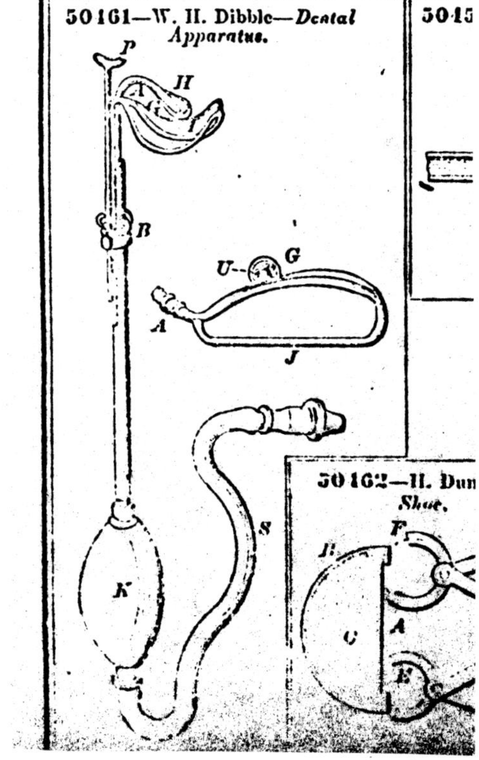

Patent drawing of William Dibble's saliva pump, *patented October 17, 1865. The saliva pump utilized a rubber bulb to create suction.*

The Samuel S. White Company, with depots in Philadelphia, New York, Boston and Chicago, listed the following dental supplies in their 1862 catalog:[6]

> Vulcanized Rubber Syringe, with gold points for injecting iodides and acids into abscesses...$5.00
>
> Vulcanizers and articles for vulcanizing work.
>
> Saliva Pump...$.75
>
> Preparations of India-rubber:
>
> American Rubber Company's Gutta-percha...per pound, $2.00
>
> American Rubber Company's Vulcanizing Base (sold in the U.S. to licensees only)...per pound, $3.00
>
> India-rubber Toothpicks in form of thread...per dozen, $1.50
>
> India-rubber Tubing, French, our own importation, seven different sizes...per foot, $.25

An interesting tool called "Taft's Extension Thimble" appears in White's 1867 catalog. The thimble fits over the finger and allows the dentist to place the thin, curved extension into the mouth where the fingers can't easily reach. It was used to hold napkins for blotting saliva or gold fillings in place until they were "...made fast in the proper position." The price for the tool in silver was $3. In "black vulcanite," the cost was a mere 40 cents.[7]

"Red vulcanite," a compound made of vulcanized India-rubber, sulphur, vermillion, and red sulphuret of mercury to achieve the color, was used largely in dentistry for artificial dentures. In 1864, it was blamed for kidney problems in a patient, although it was thought that some disorder was already present when he got the teeth.[8]

Dentures And Artificial Teeth

In an article on gutta-percha at the Great Exhibition in London in 1851, *The Illustrated London News* noted, "Mr. Truman contributes artificial teeth made from pure gutta-percha, which is said to be not only more durable, but, of course, more comfortable to the wearer, on account of [more] softness than the

Advertisement CDV for rubber dental plates. *Hard rubber (vulcanite) revolutionized the manufacture of plates and dentures.*

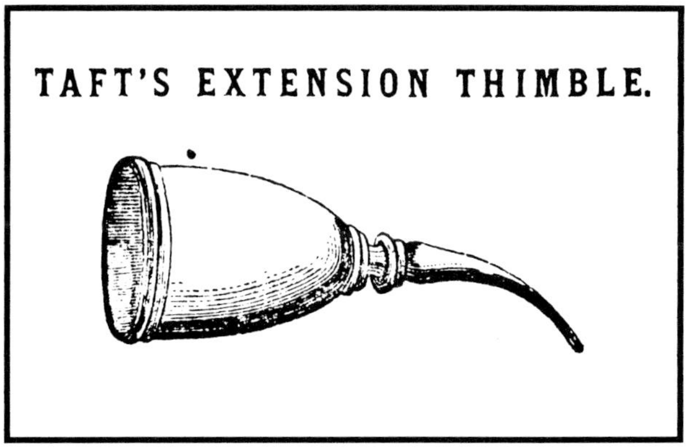

Taft's extension thimble, *as pictured in the 1867 catalog of the S. S. White Company. (Courtesy S. S. White Company)*

material ordinarily used."[9] The use of hard rubber for dental plates was included in a caveat of patent filed by John Cummings in 1852. Charles Goodyear, Jr. obtained a British patent for this purpose in 1855. In his patent, the first to embody both the hard rubber and the mold, the teeth were secured to the plate before the compound was vulcanized. Dr. C. S. Putnam worked with the young Goodyear on attempting to improve the color of the material and introducing the material and process to the dentists of New York in 1855 and 1856.[10] Hard rubber soon became the preferred material for the manufacture of replacement teeth. N. B. Griffin & Bros., "The largest and cheapest dental establishment in the United States," offered "Full or partial sets of teeth made of rubber at $10.00" in the classified section of *The New York Times* of April 12, 1861.

During the Civil War and Reconstruction years, a great battle took place within the dental profession regarding the use and license fees associated with the manufacture of the enormously popular hard rubber dental plates. Patents issued to John Cummings (U.S. Patent #43,009) and several others for improved artificial dental plates utilized the hard rubber process that was protected and licensed under Nelson Goodyear's patent of May 6, 1851. Dentists using small vulcanizers to make artificial teeth and plates had to pay licensing fees for the hard rubber and the specific molding process they used. The Goodyear Dental Vulcanite Company of Boston acquired the rights of the Cumming's patent and the Goodyear hard rubber patent as it pertained to dentistry, and strictly enforced the licensing and associated fees.[11]

Various court battles around the country

failed to change this situation and the fees apparently ceased only when the patents expired in the 1870s and 1880s. A chief advocate of the dentists in the more prominent cases was Dr. Samuel S. White, founder the modern-day S. S. White Company.[12] Dr. White himself held several patents for improved dental plates, including two patents that specified recesses in the artificial teeth so the hard rubber became firmly attached (U.S. Patents #38,072 and 38,073).

Restorations

Operations to restore facial features damaged by wounds and other injuries were performed during the Civil War and involved the use of hard rubber. In August of 1862, twenty year old Private Carleton Burgan of Purnell's Maryland Legion was admitted to the hospital at Frederick, Maryland. After being treated for pneumonia, he developed a jagged ulcer on his tongue, which eventually extended into the gum, cheek and roof of the mouth, destroying parts of the nose, mouth, lip, and cheek bone. A surgeon reported that "...the great loss of substance on the right side of the face caused frightful deformity."[13]

Reconstructive surgery on Burgan was reported in the *Medical and Surgical History of the War*, which noted the following:

> Before commencing the operation, dental fixtures, partly temporary, were ingeniously constructed by Mr. Thomas B. Gunning, of New York, and fitted to the cavity of the mouth to afford solid support to the soft parts that were transferred for the reconstruction of the mouth, and the closure of the cheek and nostril. Fixtures, in three parts, were made of vulcanite, two principal and one supplementary. The upper piece filled out the right half of the nose; the lower piece formed an artificial palate; the third part connected with the palate piece by bent spiral wire; the patient wore them two weeks prior to the first operation.[14]

The first in a series of five operations took place on March 26, 1863. In June of 1864, it was reported that Burgan enjoyed good health, and was working as an assistant nurse at the New York Hospital. He had been fitted with an artificial eye. The only dental fixture he wore was the principal piece, which covered the roof of the mouth and supplied missing teeth.

> It was worn constantly with entire comfort, and was removed and replaced at pleasure. Mastication of all descriptions of food was performed with facility. The speech, which, without the dental fixture, was hardly intelligible, scarcely betrayed any defect when it

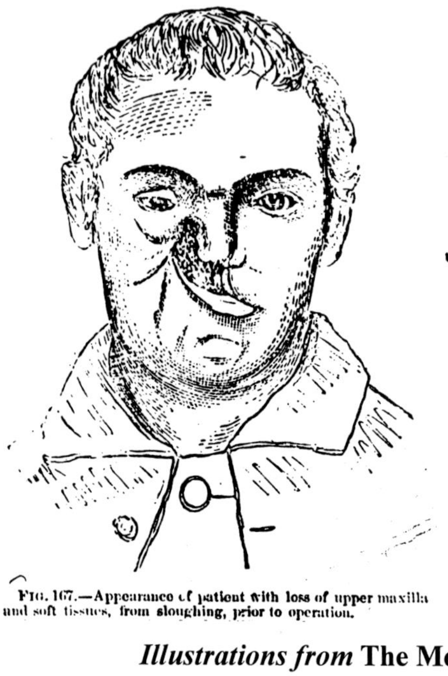

FIG. 167.—Appearance of patient with loss of upper maxilla and soft tissues, from sloughing, prior to operation.

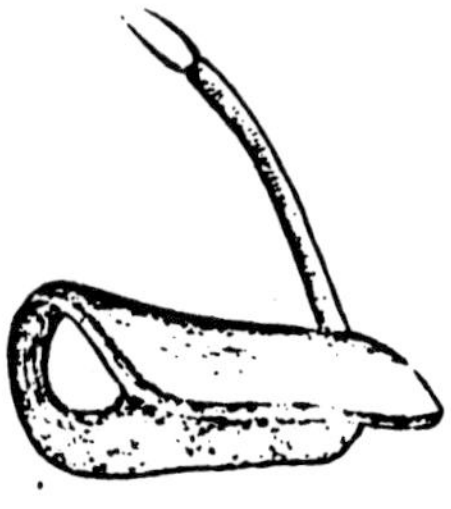

FIG. 168.—Cap for holding up roof-piece.

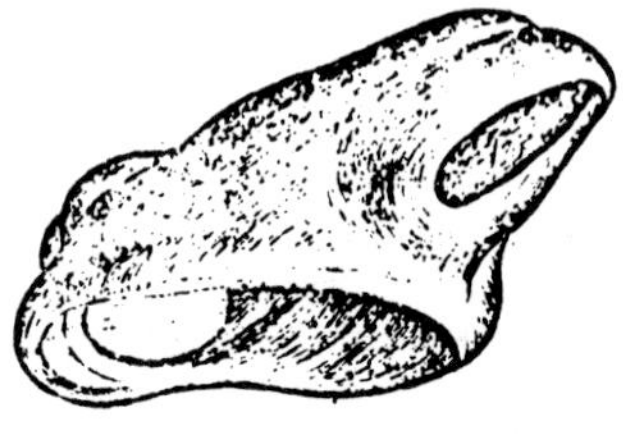

FIG. 170.—Nose-piece devised by Mr. T. B. Gunning.

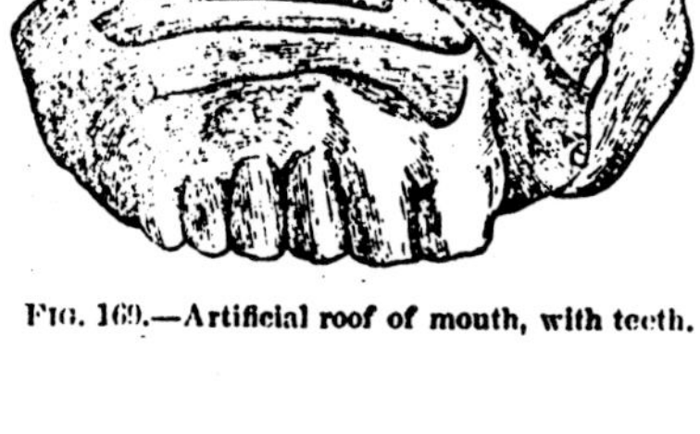

FIG. 169.—Artificial roof of mouth, with teeth.

FIG. 172.—Appearance of Burgan after the fifth and fi operation.

***Illustrations from* The Medical and Surgical History of the War of the Rebellion** *showing the hard rubber teeth, nose piece, and roof support designed by Thomas Gunning for reconstructive surgery on Private Carleton Burgan of Purnell's Maryland Legion.*

was worn. ...Bergan was a pensioner in December, 1871.[15]

Another case involved Private Elbert Hewitt of Company C, 6th Vermont Volunteers, who was wounded at Third Winchester on September 17, 1864.

> A fragment of shell struck his mouth, carrying away the upper and lower front teeth, lacerating under the lip and laying open the right cheek from the mouth to near the angle of the jaw.
>
> On December 12, 1865, he was induced to enter the New York Hospital, with a view of having an operation performed for the improvement of the mouth; preliminary to which, Mr. J. A. Bishop, of New York, ingeniously adapted a plate of vulcanite to both jaws that would supply the deficient front teeth and afford support to the lips.[16]

It was later reported that, "...Everything went on favorably after the operation, and the result was highly satisfactory to the patient as well as to the surgeon."[17]

Secretary of State William Seward suffered unfortunate fractures of the jaw when he fell from his carriage in early April, 1865. He was attacked and severely cut while asleep on the day following Lincoln's assassination. Mr. Gunning was summoned from New York and, following some disagreements concerning his procedures with Seward and his surgeon, began a series of operations on May 2. Three vulcanite splints were inserted and held in position by screws placed into the natural teeth. After four months, healing was satisfactory and the splints were removed.[18]

Similar operations in the Confederate States are attributed to Dr. James Bean, an Atlanta dentist.[19] Dr. Bean utilized a vulcanized India-rubber

Hard rubber mercury vial *used by dentists. This style could be a 20th century design.*

interdental splint, held in position with wires, compresses, and bandages, to treat fractured jaws of wounded soldiers. He worked gratuitously, with materials provided by E. N. Covey, the former Medical Inspector of the Confederate Army. Covey reported that patients with multiple fractures were able to eat soft foods within two months. He claimed the process was a "...very valuable improvement in surgery," and credited Dr. Bean with over forty operations in a six-month period, "...some of them, at first sight, apparently beyond remedy."[20]

(Also see **Section 3**, **Section 4**, and **Medical Applications**.)

NOTES - Dental Applications

[1] Arthur Ward Lufkin, *The History of Dentistry* (Philadelphia: Lea & Febiger, 1938) p. 210.
[2] John Tomes, *Dental Surgery* (London: J & A Churchill, 1859) pp. 329-330.
[3] Samuel S. White, *A History of Dental and Oral Science in America* (Philadelphia: White, 1876) pp. 71-72.
[4] *Ibid.*, p. 72.
[5] *Ibid.*, pp. 112-114.
[6] Samuel S. White, *Catalog of 1862* (Philadelphia: Lippencott & Co., 1862) A complete set of S. S. White catalogs can be found at the University of Pennsylvania in Philadelphia.
[7] Samuel S. White, *Catalog of 1867.*
[8] W. Braithwaite, The Retrospect of Practical Medicine and Surgery (New York: W. A. Townsend & Co., 1864) p. 249.

Two patriotic hard rubber pipes, *each featuring an eagle on one side and a shield on the other. Top example found in Tennessee. Bottom pipe recovered from the Fredericksburg, Virginia area.*

April, 1862 through the end of the war.

Archeological projects, under the direction of David R. Bush of the University of Pittsburgh's Center for Cultural resources, have yielded many prison artifacts. By studying the variations in artifacts found in latrines or sink holes over the period of the prison's operation, the project has been able to document changes in War department policy and treatment of the prisoners. For example, early latrines produced many beef and pork bones, as well as peach pits and grape seeds. No fruit pits or seeds were found in later latrines, and the bones recovered were bones of squirrels, rodents and migratory birds.[17]

The large number of carved hard rubber items and carefully cut fragments of hard rubber prompted Dr. Bush to do further research on hard rubber. He found that, as many prisoners documented their activities in letters and diaries, they often mentioned carving articles out of gutta-percha. One such reference, from a July 31, 1862 diary entry of Confederate Major James T. Poe, relates that, "There are men here making almost any and everything that can be possibly made of Girtipurture, (note presumable a type of wood common to the area)...."[18]

Many items were carved from hard rubber buttons, as partially-carved buttons have been recovered. Unfinished hard rubber buttons, with no drilled holes, markings, or ornamentation have been found, leading to speculation that the prison sutlers were selling them solely to be carved into other objects. Other small hard rubber fragments found at Johnson's Island are actually pieces cut from a chart rule or writing guide. Hard rubber rings and other jewelry items were carved by artisans, some of whom earned up to $3 a day. Coin silver inlays with initials inscribed were set into the hard rubber with silver studs.[19]

(Also see **Buttons** and **Personal Items**.)

These two non-military items frequently turn up in Civil War campsites. *The mess knife has the initials "G.A.A." carved into the hard rubber handle. The toothbrush bears the maker's name "Rubber Brush Co., N.Y." and "Goodyear Patent 1851" markings.*

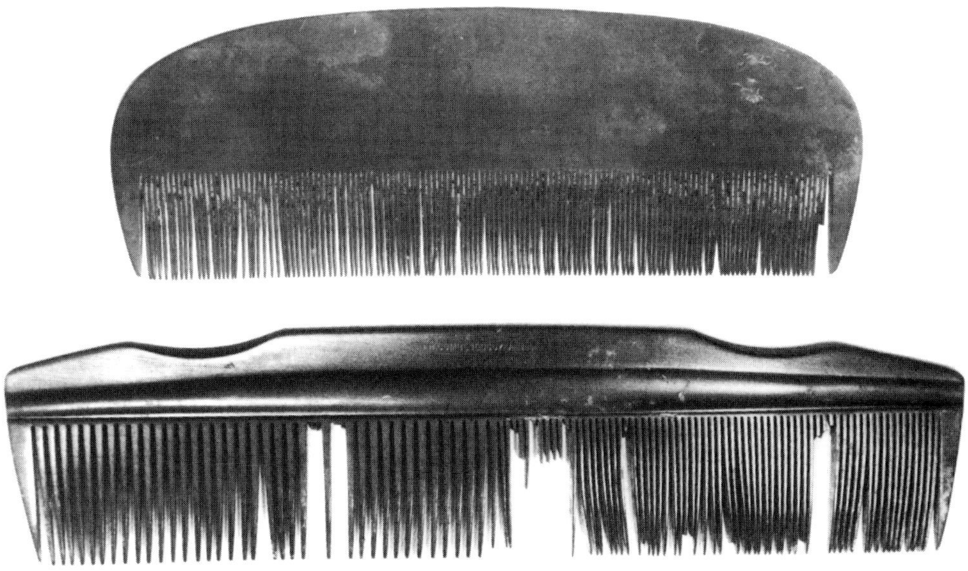

Hard rubber combs *excavated in Civil War era trash pits near Vicksburg, Miss. The smaller comb is a lice comb. The larger hair comb is marked "I. R. Comb Co. - Goodyear 1851."*

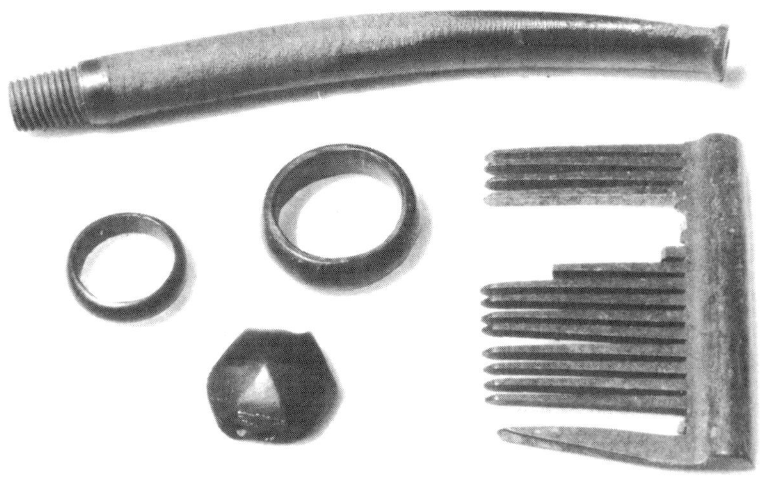

Selection of hard rubber artifacts *including pipe stem, comb, rings, and button found in a Union camp at Bridgeport, Alabama. (Photo courtesy of Charles Harris)*

Two excavated hard rubber pipes *marked "Goodyear's Patent May 6, 1851. (Courtesy of Stanley Phillips, from his notable book* Excavated Artifacts and Campsites of the Civil War 1861-1865, *published in 1974.)*

internal diameter of the case was to be the same as the general bore of the rifle so the ball would fit tightly into it without having to be tied. The patent also stated that a lubricating material on a disk of leather can be fit into the case so that, upon firing, the disk will lubricate the inside of the barrel for the next ball.

In 1861, B. C. English patented a cartridge case described as "...an improvement upon the cartridge case patented by Gilbert Smith" (U.S. Patent #33,429). English's cartridge case added "...a metallic bushing to the vent or point of ignition, for the purpose of preventing the cartridge case from being rendered unserviceable by the enlargement of the vent, consequent upon the action of the fire, at every discharge." This improvement allowed the case to be reused.

Musket Covers

Musket covers had been suggested as early as 1833, when Colonel Bomford recommended their use to save millions of dollars lost through corroded weapons.[11] Gun Covers were used successfully in the Mexican War and similar covers for artillery pieces were tested at the same time.[12]

Charles Goodyear suggested musket covers as a protection to both firearm and man. Goodyear wrote: "When...inflated by the self-acting valve tube, they are not only sufficiently buoyant to float the gun or rifle, but will answer in good degree as a life-preserver to the sportsman or soldier, in crossing a river." He added that the musket cover, when combined with an India-rubber canteen, equipped with a valve, "...is quite sufficient to buoy up a man in the water."[13]

Powder Flasks

The American Hard Rubber Company manufactured powder flasks of hard rubber during the 1850s, although it constituted only a small portion of the firm's overall business.[14] These flasks were made in several sizes and styles in both polished and unpolished varieties.[15] Most are faintly marked at the throat of the hard rubber body "Patented - Goodyear's May 6, 1851" and may also have a flask manufacturer's markings on the brass top. More scarce flasks of the decade have markings of the Goodyear patent and the Beacon Dam Company, both on the brass top.

A British patent (#875) was issued to A. J. Hoffstaedt and S. Blackwell in 1855 for the use of vulcanized India-rubber springs for powder flasks and shot belts or pouches. Another British patent (#339)

Hard rubber rifle powder flasks*, initially made in the 1850s, were oftentimes cracked by rough use in the field. (Left) Marked on the brass top and in fine letters on the neck: "Beacon Dam Co.- Goodyear's Patent-1851." (Center) Marked on the neck in fine letters: "Beacon Dam Co. - Goodyear's Patent." The brass top is marked "Dixon & Sons." (Right) Likely made by American Hard Rubber Company, this flask is marked in very fine letters on the neck: "Goodyear's Patent May 6, 1851." Beacon Dam Company was the earliest hard rubber manufacturer under Nelson Goodyear's hard rubber patent. The flasks are about 7 ½ " high.*

was issued in 1857 to W. Green for applying a coating of rubber and/or gutta-percha solution to a sheet metal form to make an imitation leather powder flask. The surface was then grained with heated dies to simulate grained leather.[16] Similar flasks for whiskey, made of grained hard rubber over glass, were manufactured in the United States.

Tompions

A tompion (also tampion) is defined as "A wooden plug or stopper placed in the muzzle of a piece of ordnance when not in use, to keep out dampness and dust."

In the early 1850s, a tompion of "gum-elastic sponge" was suggested by a former Naval officer, Lieutenant Samuel Seely, for "keeping cannon dry and free from rust."[17]

On June 14, 1858, Captain P. V. Hagner wrote to Chief of Ordnance Colonel Henry C. Craig about the possible use of India-rubber tompions. Captain Hagner had written to a manufacturer in New York asking if they could "...furnish tubing of the diameter to fit our Musket barrels—so that by cutting it into proper lengths, these could be mounted as tompions, and fit accurately without stressing the Rubber." He also "...suggested that probably they could mould tompions solid-of prescribed form which would be of exactly uniform size and of equal quality."[18] The manufacturer agreed to "...exert themselves to furnish precisely what might be required" and the officer asked for orders to proceed to New York with the pattern moulds he had prepared for both options saying he felt the "...solid Rubber Tompions will be the best & cheapest for general service."[19]

Wilmott's tompion*, patented November 24, 1863, had an expanding ring of India-rubber that sealed the barrel of the firearm.*

On November 24, 1863, George Wilmott was issued U.S. Patent #40,720 for a "Tompion for Firearms," with an expanding ring or packing of India-rubber. Fortunately for collectors, these tompions were well marked with the patent date and surviving specimens, even those recovered from battlefields and campsites, are easily identified.

Later tompions included a wooden variety with an India-rubber "Muzzle-cap" patented by John Cleu on November 15, 1864 (U.S. Patent #45,020) and an expanding tompion utilizing elastic bands patented by Thomas Schermerhorn and Joseph Anderson on December 27, 1864 (U.S. Patent #45,641).

(Also see **Section 6**.)

NOTES - Firearms and Related Accessories

1 *Dodd's Curiosities of Industry* (Published in England circa 1851) pp. 4, 5.
2 Charles Goodyear, *Gum-Elastic and Its Varieties* (New Haven, CT: Goodyear, 1853 and 1855) Vol. II, p. 137.
3 Letters to the author from experienced collectors.
4 Goodyear, *Gum-Elastic and Its Varieties,* p. 111.
5 Various references in Section 6.
6 Based on author's discussions with researchers and low serial numbers on surviving weapons.
7 Raymond G. Rieser, "The Elastic Weapons of R. E. Hodges," *Journal of the Society of Archer-Antiquaries*, Vol. 31, (1988): pg. 19.
8 Norm Flayderman, *Flayderman's Guide to Antique American Firearms* (Northbrook, IL: DBI Books, Inc., 1990. Krause Publications, 700 East State Street, Iola, Wisconsin. 1-800-258-0929) p. 152.

[9] National Archives, Records Group 92, Entry 225, Consolidated Correspondence File, "Goodyear's Metallic Rubber Company." Letter dated January 25, 1844.
[10] Francis A. Lord, *Civil War Collector's Encyclopedia* (Harrisburg: Stackpole Books, 1965) p. 341.
[11] *The Army and Navy Chronicle,* 22 Sep. 1836, Vol. 3, No. 12, Page 177; NA, RG 92, Entry 225, CCF, "Goodyear's Metallic Rubber Company."
[12] NA, RG 156, Office of the Chief of Ordnance, Entry 21, Letters Received, R-132 1847.
[13] Goodyear, *Gum-Elastic and Its Varieties*, Vol. II, p. 130.
[14] Ray Riling, *The Powder Flask Book* (New York: Bonanza Books, 1953) pp. 101-102.
[15] *Ibid.*, p. 196.
[16] *Ibid.*, p. 156.
[17] Goodyear, *Gum-Elastic and Its Varieties*, Vol. II, p. 147.
[18] NA, RG 156, Office of the Chief of Ordnance, Entry 21, Letters Received.
[19] *Ibid.*

GAITERS AND LEGGINGS

"...a long leggin...is designed only to be worn in the cold and snow, for which purpose it cannot be too highly recommended...."

Charles Goodyear, Writing in
Gum-Elastic and Its Varieties, 1855

The words "gaiters," "leggins," and "leggings" have been used to denote similar items of clothing. Coverings for the ankle and instep, as well as shoes with elastic insertions at the sides, are sometimes called gaiters or congress shoes.[1] Gaiters are also defined as outer coverings for the legs, extending from the ankles to the knees.[2] Higher leg coverings are generally called leggins or leggings.[3] Charles Goodyear wrote that leggins "...are used for walking, or riding on horseback, in muddy traveling."[4]

India-rubber and gutta-percha gaiters and leggings were worn by individual soldiers, while gaiters of leather and canvas were worn by some units.[5] Medical Department personnel viewed gaiters or leggins as an important protection from dust and dirt, but most soldiers found them to be a nuisance and soon disposed of them.[6]

Vulcanized India-rubber gaiters with a backing of knit web of textile fabric were patented (U.S. Patent #34,140) by Henry Goodyear in 1862. Short overshoes or gaiters with elastic rubber sides were found among the cargo of the recovered Steamer *Bertrand.*

India-rubber legging found in a North Carolina barn *near a Confederate cavalry camp. The lengthwise cotton border has 9 composition buttons.*

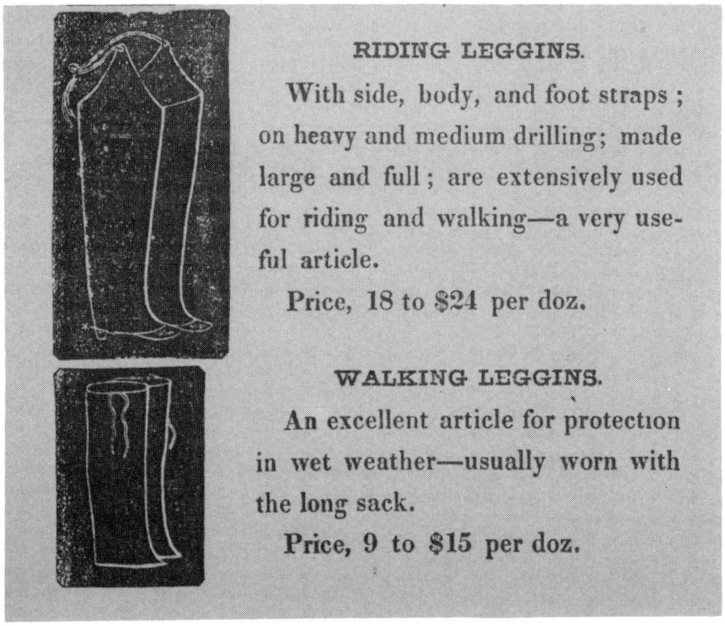

RIDING LEGGINS.

With side, body, and foot straps; on heavy and medium drilling; made large and full; are extensively used for riding and walking—a very useful article.

Price, 18 to $24 per doz.

WALKING LEGGINS.

An excellent article for protection in wet weather—usually worn with the long sack.

Price, 9 to $15 per doz.

Excerpt from the 1854 catalog of the North American Gutta Percha Company *advertising two styles of leggins. This New York City firm manufactured goods under the patents of Rider and Murphy.*

(Also see **Section 6**, **Boots and Shoes**, and **Excavated Relics and Archeological Artifacts**.)

India-rubber "riding" leggings owned by General Ulysses S. Grant. *Traces of the Goodyear patent markings and the Union India Rubber Company markings remain on the inside. (Courtesy Mike Cunningham)*

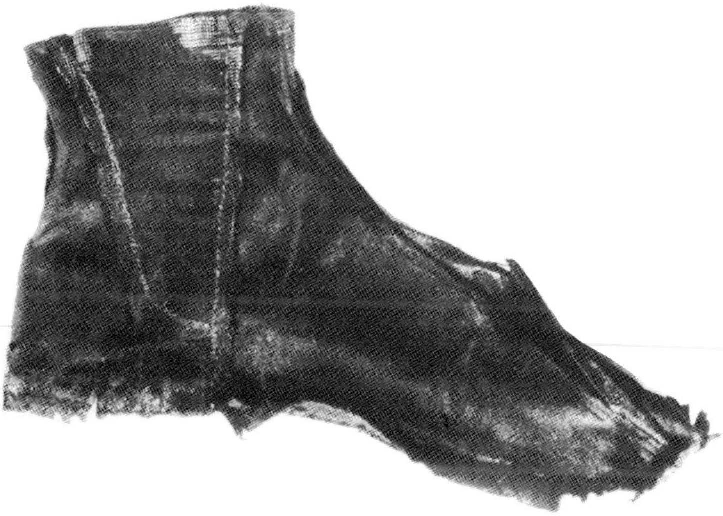

Textile gaiters with elastic gores *recovered from the sunken steamship Bertrand. Traces of the Goodyear patent markings, including dates of June 15, 1844 and June 15, 1858, remain on the elastic. (Courtesy Bertrand Conservation Laboratory, U. S. Department of the Interior.)*

NOTES - Gaiters and Leggings

---

[1] *Merriam Webster's Collegiate Dictionary* (Springfield, Mass: Merriam-Webster, Inc., 1993) p. 476; Charles Goodyear, *Gum-Elastic and Its Varieties* (New Haven: Goodyear, 1853 & 1855) Vol. II, p. 319.

[2] *Merriam Webster's Collegiate Dictionary,* p. 476.

[3] *Merriam Webster's Collegiate Dictionary,* p. 665.

[4] Goodyear, *Gum-Elastic and Its Varieties,* Vol. II, p. 309.

[5] Anthony Tafel, "The Leather Gaiter," *North South Trader*, Vol. 5, No. 2, (Jan.-Feb. 1978): p. 18.

[6] *Ibid.*

HAVERSACKS

"...a rubber haversack, a most useful article for every soldier and traveler."

Quoted from *The Scientific American,* June 22, 1861

Enlisted man's haversack *made entirely of India-rubber over duck or canvas. A brass button holds the flap closed.*

India-rubber haversacks were used by the military in the Mexican War period and carried by some individuals and volunteer troops in the Civil War.[1] Gutta-percha haversacks were included in the military trials of 1855 and some may have been used in the Civil War.[2] Haversacks, "...of gutta-percha, tin, or other material, and of pattern to be prepared by the Quartermaster Department, after making the requisite experiments," were specified in the Army regulations of 1861.[3] This specification was an error that was meant to refer to the canteen. The revised regulations specified only a "...haversack of painted canvas."[4]

Dressed for field duty stands an unidentified northern state militiaman. *Note his 1858 pattern canteen, full pack, and large non-regulation waterproof haversack. (Courtesy Art Marchand)*

In an article on The Rubber Clothing Company, The Scientific American of June 22, 1861 noted:

> ...a rubber haversack, a most useful article for every soldier and traveler. They are undoubtedly vastly superior to the kind in ordinary use, which are made of cotton duck; the cotton soon gets saturated with grease from the rations which the soldiers are often required to carry in them and they not only look very filthy, but they soil the uniform of the wearer. In the use of india-rubber haversacks, such evils are entirely overcome.[5]

In addition to the standard waterproof haversack, Charles Goodyear suggested a perforated variety that allowed for the circulation of air.[6]

(Also see Sections 6 and 7.)

NOTES - Haversacks

---

[1] National Archives, Records Group 92, Entry unknown, Vol. 33 of 52, "Invoices of Stores to be Delivered to Units, June 8-Dec. 14, 1846" and "Orders for Army Supplies, No. 1, June 17, 1846-February 15, 1847." Also Civil War period advertisements and documented surviving examples.
[2] NA, RG 92, Entry 225, Consolidated Correspondence File, "Gutta-percha." Possible Civil War use is author's opinion based on period advertisements and surviving examples.
[3] War Department, *Regulations for the Army of the United States-1861* (New York: Harper & Bros., May 1, 1861).
[4] War Department, *Revised Regulations for the Army of the United States-1861* (Philadelphia: J.G.L. Brown, Philadelphia, August 10, 1861).
[5] *The Scientific American,* 22 June 1861.
[6] Charles Goodyear, *Gum-Elastic and Its Varieties* (New Haven: Goodyear, 1853 & 1855) Vol. II, p. 136.

HORSE EQUIPMENT

"...Some of the 26th are on guard,...are sewing an India rubber house for my horse."

Major Fredrick C. Winkler,
January, 1863

Various accessories for horses were patented and used including improved horseshoes, blankets, and curry combs. John Jones' "Elastic Horseshoe" (U.S. Patent #9,173) consisted of a sheet of India-rubber between two plates to aid the horse. The rider was assisted by A. Livingston's patented stirrup with an India-rubber footpiece to prevent slipping (U.S. Patent #33,354). A predecessor to the bicycle seat was the elastic saddle, consisting of a seat suspended on India-rubber springs, patented by Lewis Bishop in 1859 (U.S. Patent #22,480).

India-rubber horse covers or blankets were suggested as early as 1845 as a means of sheltering the animals from wind and rain.[1] On October 28, 1862, Lt. Colonel and Chief Quartermaster Rufus Ingalls wrote to Quartermaster General Meigs:

Bridle rosettes made of hard rubber *with brass attachments. The back of the rosettes is marked: "Goodyear 1851."*

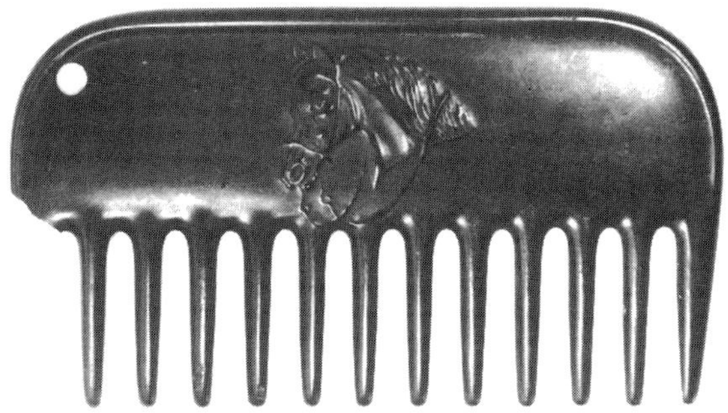

Hard rubber horse mane comb. *The comb is marked on the reverse: "I. R. Comb Co. - Goodyear 1851."*

> I directed Colonel Sawtelle to consult you on the expediency of providing rubber or other kind of waterproof horse blankets for our cavalry, artillery, and even wagon horses. I am of opinion that the measure would prove an economical one, particularly for cavalry.[2]

E. L. Perry of New York patented a horse cover on October 12, 1864 that was an improvement over ordinary India-rubber or gutta-percha blankets (U.S. Patent #44,654). Perry's cover was "self-ventilating," containing apertures that permitted air to circulate and perspiration to escape without letting rain or snow enter. The use of the blanket was suggested as a preventive measure to protect horses against the elements:

> Many persons thoughtlessly expose them to inclement weather, when a little forethought and small expenditure would have saved the services of a veterinary surgeon and the animal would have been in condition for use instead of being laid up in ordinary like a naval vessel under repair.[3]

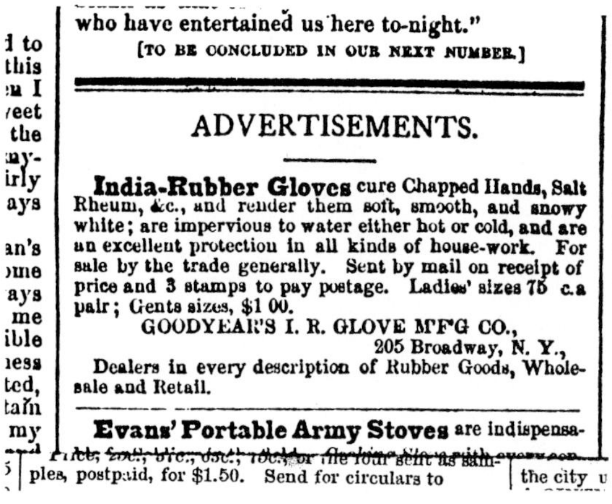

who have entertained us here to-night."

[TO BE CONCLUDED IN OUR NEXT NUMBER.]

ADVERTISEMENTS.

India-Rubber Gloves cure Chapped Hands, Salt Rheum, &c., and render them soft, smooth, and snowy white; are impervious to water either hot or cold, and are an excellent protection in all kinds of house-work. For sale by the trade generally. Sent by mail on receipt of price and 3 stamps to pay postage. Ladies' sizes 75 c. a pair; Gents sizes, $1 00.

GOODYEAR'S I. R. GLOVE M'F'G CO.,
205 Broadway, N. Y.,
Dealers in every description of Rubber Goods, Wholesale and Retail.

Evans' Portable Army Stoves are indispensa-

ples, postpaid, for $1.50. Send for circulars to

Advertisement for India-Rubber Gloves taken from Harper's Weekly, *December 20, 1862. Rubber gloves were not used for surgery until the 1890s.*

HOUSEHOLD ARTICLES

"Clothes brushes, covered bottles, table mats, preserve jars, wash tubs, baskets, furniture…."

> Household articles listed by Charles Goodyear in *Gum-Elastic and Its Varieties.*

While many household articles could be classified as personal items, certain products and the associated advertising were clearly targeting domestic necessities of daily life. Some of these items are highlighted here.

Gloves

While rubber gloves were not used for surgical purposes until the 1890s, they were in use by workers and for household purposes. Gloves and other household and industrial goods were made during the Civil War by Goodyear's India-Rubber Glove Manufacturing Company.

(Also see **Medical Applications**.)

Canning Accessories

The Scientific American of August 12, 1865 printed the following application:

> INDIA-RUBBER CAN COVERS
>
> D. B. of New Hope, Pa., now uses India-rubber (gum cloth) tied over jars and wants two dozen covers this minute to seal peaches with.[1]

Shirt Collars

The Scientific American ran the following article on April 30, 1864:

> INDIA-RUBBER SHIRT COLLARS
>
> Linen, cotton, paper, and steel collars are now made, and to these vulcanized india-rubber collars have been added by W. J. Smith of Sale, England, who has taken out a patent for them as a new article of manufacture. He states that suitable patterns may be painted or printed on the collars, either before or after they are cut from the sheet; and they may be made white, or colored, or embossed. Cuffs and wristbands may be made also of the same material.[2]

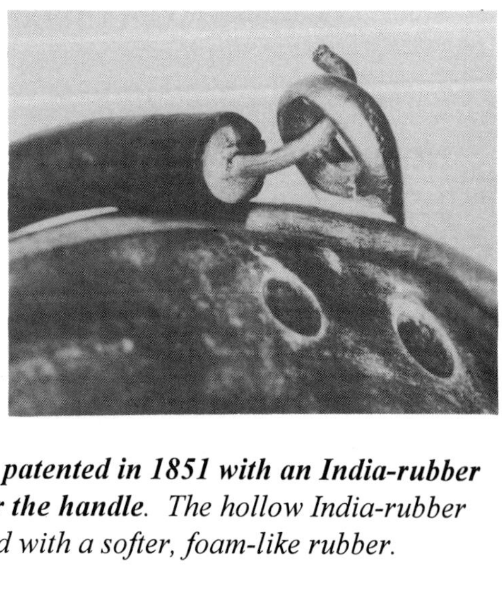

Brass bucket patented in 1851 with an India-rubber covering over the handle. *The hollow India-rubber tubing is filled with a softer, foam-like rubber.*

The hard rubber handles *on these knives and forks were permanently vulcanized directly to the steel shafts. The plain set has no markings. The set with bone tips is marked "Meridan Cutlery Co. - Goodyear's Patent May 6, 1851 Ext'd." This marking indicates the knives were made after the 1851 hard rubber patent was extended on May 6, 1865. Other knives and forks were made by the Hard Rubber Cutlery Company and marked "Goodyear's Patent May 6, 1851."*

Clothes Wringer

An important application for households was the India-rubber clothes wringer, patented by Whitney Washington in 1861 (U.S. Patent #33,861). *The Scientific American* commented: "One of the most useful inventions that has made its appearance to confer blessings upon toiling womankind is the clothes-wringing machine, in which rubber rollers are employed. It saves much hard labor, and much wearisome twisting."[3]

Hard Rubber Items

Hard rubber items of all types, including combs, shuttles, thimbles, eyeglass frames, napkin rings, etc., were prevalent in households. Knife and fork handles of hard rubber, vulcanized directly to the steel utensil, readily endured hot water, cleaning, and constant use.[4]

(Also see **Sections 2 - 5**.)

Hard rubber thimble and shuttle *for tatting (fine lace work). It is a sign of the times that the thimble was found in a Civil War campsite near Clear Spring, MD.*

NOTES - Household Articles

[1] *The Scientific American*, 12 Aug. 1865.
[2] *The Scientific American*, 30 Apr. 1864.
[3] *The Scientific American*, 19 Nov. 1864.
[4] Author's observations, based on various period references and surviving artifacts.

***Illustration of Confederate submarine** from* Harper's Weekly *of November 2, 1861. Note the India-rubber suction plate and air tube.*

INDUSTRIAL AND COMMERCIAL USES

"By the use of a thin sheet of vulcanized India rubber…a joint is produced that is absolutely and permanently perfect."[1]

Rev. Bradford K. Pierce,
Charles Goodyear biographer, 1866

Creative applications of India-rubber and gutta-percha resulted in many improvements and innovations in commerce and industry. The use of gutta-percha in driving bands or belts in machinery of all types is documented as early as 1847.[2] In 1860, *The Edinburgh Journal* was quoted as saying:

> New applications of India rubber to mechanical purposes are being discovered almost every month; in springs to lift the saw in sawing machines, and with considerable economy in power; in springs for cables, or for moorings, proof against any strain to which they may be subject; and in a new code of signals introduced into the navy at Plymouth, comprising a series of flexible cones.[3]

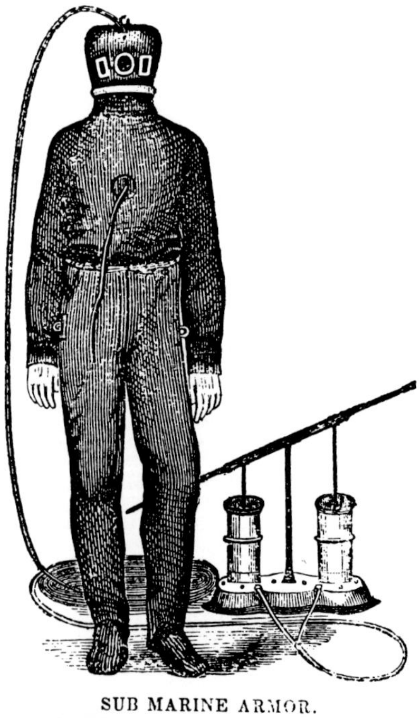

***Rubber diving suit** with a copper helmet marketed as submarine armor by the Rubber Clothing Company in 1861. It featured a rubber ensemble including rubber hoses and parts of the air suction pump.*

While the use of India-rubber for washers, gaskets, and packing may be considered trivial, the flexibility and durability of these parts made them ideal for use in machinery and even steam engines. Addressing the use of rubber in high temperature environments, *The Scientific American* commented:

> On the issue of heat with India-rubber gaskets used with iron parts, etc.: Vulcanized rubber may be successfully used at temperatures not exceeding 230 degrees F.; At that point the sulphur fuses and destroys the union with the rubber and itself.[4]

In 1860, layers of vulcanized India-rubber were employed as gaskets of sorts in the joints of girders in new bridges.[5]

As the railroads and telegraph became more of a necessity for the United States, India-rubber and gutta-percha contributed to improved performance and increased efficiencies. The military also used accessories made of the substances in their newest wartime experiments. A few examples of these products and applications are highlighted below.

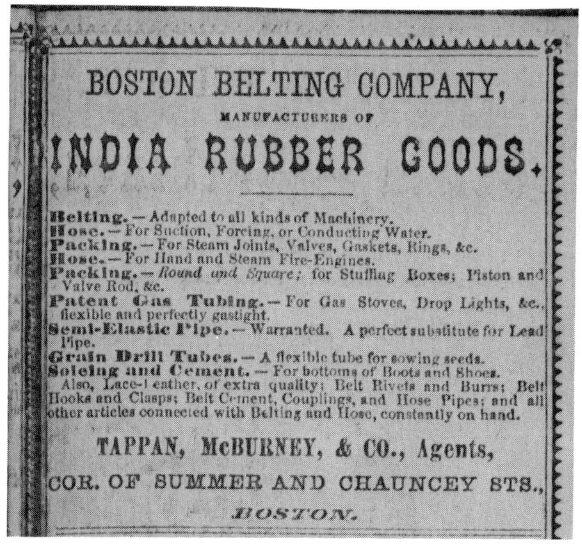

BOSTON BELTING COMPANY,

MANUFACTURERS OF

INDIA RUBBER GOODS.

Belting.—Adapted to all kinds of Machinery.
Hose.—For Suction, Forcing, or Conducting Water.
Packing.—For Steam Joints, Valves, Gaskets, Rings, &c.
Hose.—For Hand and Steam Fire-Engines.
Packing.—*Round and Square;* for Stuffing Boxes; Piston and Valve Rod, &c.
Patent Gas Tubing.—For Gas Stoves, Drop Lights, &c., flexible and perfectly gastight.
Semi-Elastic Pipe.—Warranted. A perfect substitute for Lead Pipe.
Grain Drill Tubes.—A flexible tube for sowing seeds.
Soleing and Cement.—For bottoms of Boots and Shoes.

Also, Lace-Leather, of extra quality; Belt Rivets and Burrs; Belt Hooks and Clasps; Belt Cement, Couplings, and Hose Pipes; and all other articles connected with Belting and Hose, constantly on hand.

TAPPAN, McBURNEY, & CO., Agents,

COR. OF SUMMER AND CHAUNCEY STS.,

BOSTON.

Boston Belting Company advertisement *from the* Boston Almanac *of 1861. Listed specialties included India-rubber hoses.*

HOSE.

India-rubber hoses included fire hoses, *as illustrated above, and those employed with gas generators to inflate Professor T. S. Lowe's balloons for U.S. Army aerial surveillance. (National Archives)*

Military Balloon and Submarine Appendages

Goodyear's Rubber Belting and Packing Company supplied hose, rubber pipes, packing, etc. for the gas generators used by Professor Lowe's Balloon Corps.[6] A Confederate submarine, described and illustrated in *Harper's Weekly* of November 2,

1861, details components including an India-rubber suction plate and air tube.[7]

Paints and Coatings

Many paints and coatings containing India-rubber or gutta-percha, to stop rotting and decay of wood in ships, were patented in the 1840s, 1850s, and 1860s.[8] Roof cements and coatings containing the gums were also patented and advertised. *The New York Times* of April 10, 1862 advertised "Leaky Roofs Cured - Ellery's Patent Rubber Paint-one cent a foot and warranted."

Rail Car Springs

As the railroads of the nation and world expanded, one of the more prosperous divisions of the rubber industry became the manufacture of railroad car springs. Charles Goodyear details the use of many kinds of springs in *Gum-Elastic and Its Varieties,* but highlights car springs "...of alternate discs of India-rubber and metal," made by Mr. Fowler M. Ray of New York. In 1849, he believed there was an average of one thousand pounds of gum per day processed into car springs. Subsequent improvements

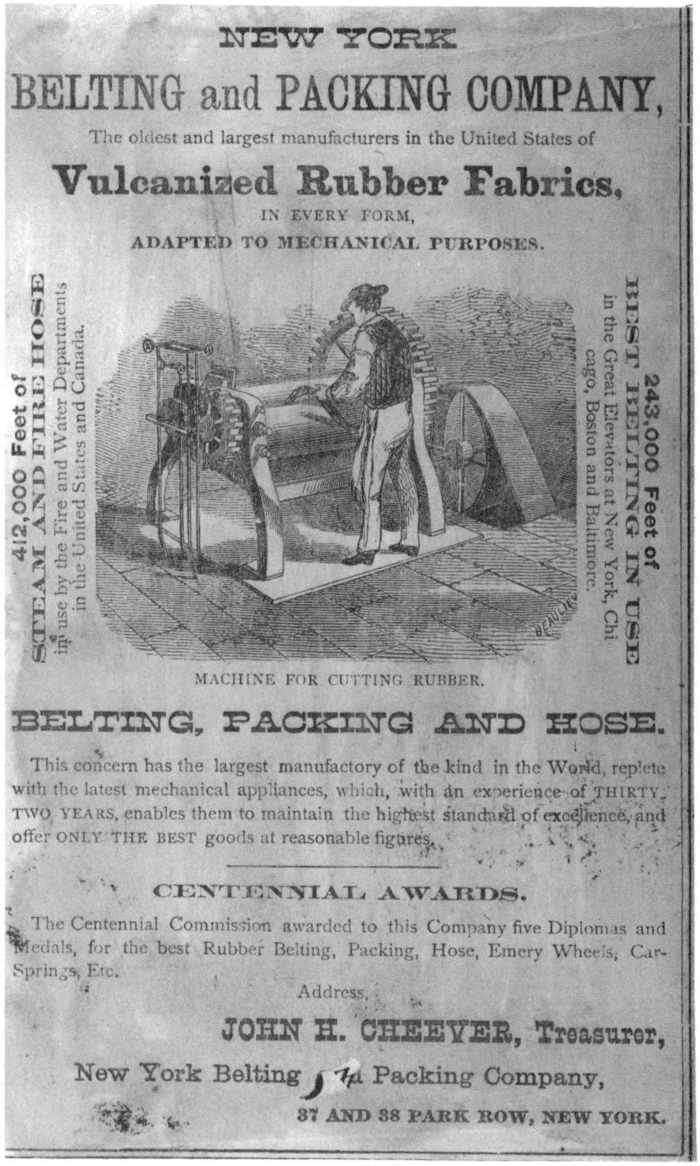

Advertisement for the New York Belting and Packing Company, *dating to 1877.*

under development when Goodyear's book was published in 1855 included air springs and blocks of rubber encircled with iron rings that allowed compression of the rubber.[9]

George Spencer was a British rubber pioneer with many inventions to his credit. He invented a mechanical method of testing the spring quality of rubber that he preferred over the contemporary method of "tooth and hand;" that is, biting the rubber or bending it in the hand. His railroad car spring, which consisted of a vulcanized India-rubber moulding in the shape of two truncated cones with their larger diameters together, was patented in over a dozen countries, including the United States. The Spencer conical rubber springs, encased in malleable cast iron rings to limit their lateral expansion, were

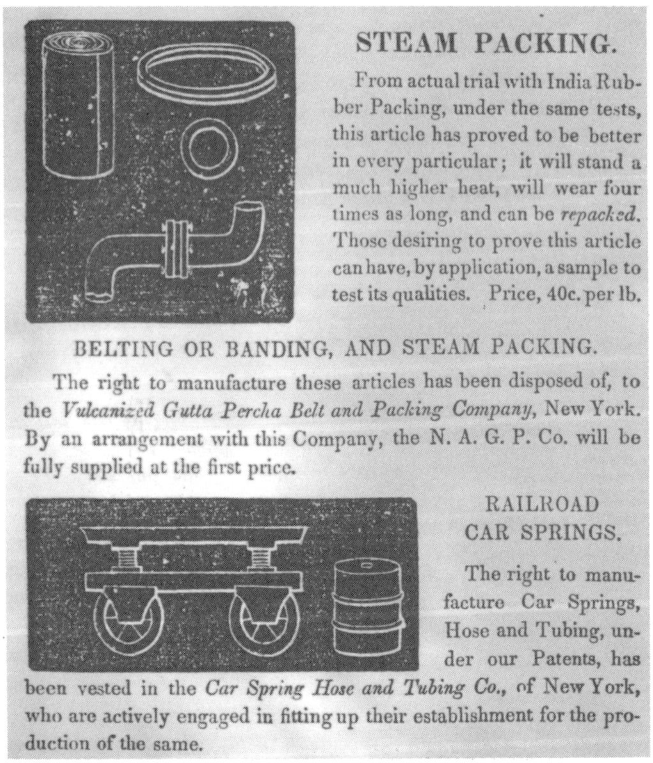

Industrial products using gutta-percha *advertised in the 1854 catalog of the North American Gutta Percha Company. (National Archives)*

Upper portion of sheet music from 1858 *showing a cross-section of the first transatlantic cable with three coatings of gutta-percha insulation.* ***Piece of the first transatlantic cable,*** *placed by the* Great Eastern *in 1858. The surplus cable was purchased in reels by Tiffany's and cut up for sale as souvenirs. A small brass tag on the cable reads: "Atlantic Telegraph Cable Guaranteed By Tiffany's & Co. Broadway-New York."*

far superior to contemporary steel springs. The rubber springs were not liable to fracture during sudden concussions, required no lubrication, were lighter, quieter and required far less maintenance than steel springs.[10] Although popular in Britain and other parts of the world, they never sold well in the United States, likely due to high tariffs and freight costs and because other competitive rubber springs were being produced and sold in this country.[11]

Telegraph Wire, Cable, and Insulators

One of the earliest applications for gutta-percha was as insulation for telegraph wires in the late 1840s. When the first transatlantic telegraph cable was placed by Cyrus Fields and the ship *Great Eastern* in 1858, it was three coats of the finest gutta-percha that provided the insulation between the telegraph wire and the external protecting cable.[12]

Early in the transatlantic cable project, *The Scientific American* had commented on the value of gutta-percha:

> The rapid progress of the submarine telegraph, setting aside other important commercial uses, calls for fresh supplies. If no other purpose had been subserved by this Indian gum than that of encasing the telegraph wires, mankind would have reason to be eminently grateful to the discoverers.[13]

Hard rubber telegraph insulators, originally patented in England, replaced traditional glass insulators in some terrestrial applications.

The military also made use of India-rubber and gutta-percha for telegraphic purposes. A. H. Caldwell, writing to Major T. T. Eckert from the Headquarters of the Army of the Potomac on May 29, 1864, detailed the destruction of the battery wagon by Rebel fire and added, "Dover wants you to send quantity of rubber slips for covering places where connections have been made in field Wire."[14] Lt. Colonel William Stokes of the 4th South Carolina Cavalry reported with obvious excitement the capture

KNAPSACKS

"...the proposals for knapsacks ranged all the way from 2.07 to 2.50 dollars, For what I call an apoligy [sic] for a knapsack. My knapsack will out-ware [sic] two and give comfort to the Soldier on a march."

John Rider to Quartermaster General Meigs
February, 1864

India-rubber knapsacks were used by the United States Army as early as 1846, as evidenced by contracts for almost 20,000 being issued that year.[1] Many of these were rejected by the government due to inferior quality of the knapsack or the India-rubber itself. As the industry matured through the years, the overall quality of all India-rubber goods improved and were better received. India-rubber knapsacks were employed by some individuals and volunteer units during the Civil War.[2] Gutta-percha knapsacks were used on a trial basis by the military in the 1850s and some were employed during the Civil War. These alternatives to the tarred knapsack kept their contents dry but were uncomfortable in summer's heat.[3] A painted rubber knapsack, manufactured in England and bearing the "Isaacs and Campbell" name, was among the cargo of a captured blockade runner, which at least confirms their intended use in the Confederacy.[4]

The *Regulations for the Army of the United States*, published during May of 1861, do not specify a material for knapsacks.[5] The "*Revised Regulations*," printed in August, 1861, show the approved knapsack being made of "painted canvas," likely referring to the familiar tarred knapsacks.[6] But the India-rubber and gutta-percha companies were making some sales. On May 3, 1861, John Rider of the Gutta Percha Manufacturing Company proposed gutta-percha knapsacks to the Quartermaster General and indicated they were already being used by New York State troops.[7] A bill for $200, still due on June 26, 1862, confirms the delivery of gutta-percha knapsacks to Colonel Dwyer's regiment on May 21, 1861.[8]

On June 21, 1861, Rider sent a letter to Secretary of War Simon Cameron, informing him he had sent a sample of a "...white gutta-percha knapsack, which I would most respectfully submit for your consideration." Rider went on to say "The

CDV of a private in Co. A, 151st New York Vol. Inf.
With Sharps rifle in hand and his white gutta-percha knapsack in the foreground. Note "Bowen's IND. RIFLES" painted on flap. (Courtesy Mike McAfee)

knapsacks can be made almost any colour you may desire and are perfectly waterproof. Can be washed with ease. They are very durable and a strong non conductor of heat and electricity, which I should consider important to the soldier's health & comfort in a hot climate."[9]

Several letters between Rider, writing from Willard's Hotel in Washington, D.C., and Quartermaster General Meigs reveal Union purchases of gutta-percha knapsacks early in the War. On July 17, 1861, Rider proposed the following to Meigs:

> The samples of Gutta-Percha Knapsacks submitted for your consideration yesterday I will furnish at the following prices:

Blanket & Shawl Knapsacks		$ 3.75 each
White French pattern	"	2.75
California -----------	"	3.00
Army regulation -----	"	3.75

White fabric knapsack and straps lined with gutta-percha, *worn during the Civil War by E. S. Yergason, Co. B, 22nd Conn. Volunteers. A large circular marking on the inside of the front flap reads: "The Gutta Percha Manufacturing Company - 139 Broadway - New York - Rider and Murphy's Patents June 1, 1852 and May 30, '54." The buckles are stamped "Patented Feb. 18, 1862." (Courtesy of Don Troiani)*

> I am satisfied that either of the above will give better satisfaction & comfort to the soldier as well as economy to the Army than the present ones in use. They are now in use by Col. Baker's regiment & the Garibaldi Guards & others, and they speak in the highest praise of them. I would respectfully ask you, log in me an order to furnish the U.S. Army with sufficient for a fair trial.[10]

On July 18, General Meigs placed the following order for gutta-percha knapsacks:

> The Department will take from you one thousand and ten gutta percha knapsacks of the blanket & shawl pattern and the same number of the white French Pattern - the latter at \$2.75 ea. and the former at \$3.75 ea. Have them made with a strap to secure the mess tin, and deliver them to the Military Storekeeper in this City.[11]

On August 5, 1861, General Meigs advised John Rider that the accounts of the Gutta-Percha Manufacturing Company for knapsacks & haversacks furnished to Colonel Baker's California Regiment of Volunteers had been referred to Colonel D. D. Tompkins for payment. On the same day, Meigs wrote to Colonel Tompkins directing him to pay the bill:

> You will pay the New York Gutta Percha Manufacturing Company for fifteen hundred Knapsacks at $ 2.87½ each, and for the same number of havresacks [sic - contemporary from French *havresac*] at 87½ cents each, and invoice them to Lieut. Francis G. Young, Regtl. Qr. Master, California Volunteers, now in this city.[12]

The following letter to Quartermaster General Meigs gave a glowing report on the use of gutta-percha knapsacks in the field:

> Head Qrs. 5th Maine Regt.
> Camp Franklin Va. Feb. 1st '62
>
> Brig. Gen'l Meigs
> Qr. Mr. Genl.
> U.S. Army
> Sir:
>
> Agreeable to your request I furnish you with the following report upon the "Gutta Percha French Grey Knapsacks" issued to this Regt.
>
> These Knapsacks possess a great advantage over any I have ever seen, in the ease with which they can be packed, strapped, and carried, and their general finish and appearance.
>
> We have used them for nearly three months, and have found them durable and waterproof. In the way of ornament they are a great addition to the Army Uniform. In short, in every respect, they surpass any article of the kind that has yet come under my observation.
>
> Respectfully,
> Your Obt. Servt,
> N. J. Jackson
> Col. Comdg. Regt.[13]

On August 1, 1862, A contract was awarded to John Rider for 20,000 vulcanized gutta-percha knapsacks at $2.75 each. Later in the month, Rider informed Quartermaster General Meigs that he had delivered 10,000 gutta-percha knapsacks to Colonel Vinton, who would not approve the contract for payment without Meigs' authorization. Colonel Vinton was in charge of the New York Office of Army Clothing and Equipage, one of the main Quartermaster procurement offices in the country. Rider asked Meigs to approve the bills so he can meet payments due on materials. He indicated they were making 1,000 a day and asked Meigs if he will need more. He requested a prompt reply as "...cotton is continually going up and we do not want to discharge our hands."[14]

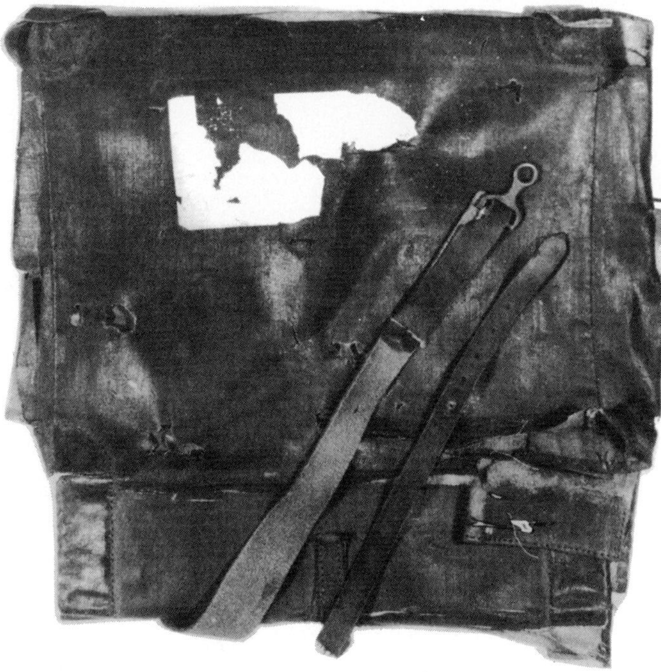

This knapsack, made of India-rubber painted on canvas, *was bound for the Confederacy on a blockade runner. Fragmented remains of an old typed label glued to the front tells that the knapsack "...was on a cargo of British goods...captured...later bought by Dr. Morse....used it many miles of camping trips." Marked "Isaacs and Campbell." The wooden frame and several straps are missing.*

On December 12, 1862, Colonel George Buckingham, Acting Brigadier General and Commander of "Seymour Brigade," wrote to Meigs requesting his special approval that they might be issued, "Riders Patent Vulcanized Gutta Percha Knapsacks." Buckingham stated:

> Most, if not all, the Regt, which have recently left this city for the "Seat of War" have the new Knapsack, which I understand, unless special authority from your Dept is given, the old tar canvas satchels are issued—they are unfit for field service in my judgement.[15]

Deputy Quartermaster General Vinton's December 22 reply on behalf of Meigs noted that "There are none of them within described knapsacks 'on hand' at this Depot [New York]. The difference in cost of them and the regulation Knapsack is sixty-one cents."[16]

Price was still an issue in September of 1863,

as indicated in Vinton's response to Meigs' question of why contracts were not awarded to John Rider for Gutta-percha knapsacks. Vinton stated:

> "...I was limited in price in your letter of the 15th of July to $2.75....His bid...was $2.75....The lowest bids under that advertisement were 700 at 90 cents, 100,000 at $1.93, 25,000 at $2.09 and 25,000 at $2.12 each to which the awards were made before the receipt of your letter."[17]

The issue of price versus quality apparently carried on throughout the War. The following letter, written by John Rider on a letterhead as the President of the Goodenough Horse Shoe Manufacturing Company, presented interesting points:

> New York, Feb'y 26, 1864
>
> Gen'l M. C. Meigs
> Qr. Mr. Genl. U.S.A.
>
> Sir:
>
> I am informed at the Qr. Mr. Dept. in this city, the proposals for knapsacks ranged all the way from 2.07 to 2.50 dollars, For what I call an apoligy [sic] for a knapsack.
>
> My knapsack will out-ware [sic] two and give comfort to the Soldier on a march.
>
> A Connecticut-return regiment passed through here on Saturday, equipped with my Knapsack, they had been in service two years. They were apparently in as good condition (with the exception of soiled) as when they were issued. And the soldiers said it was the best & only knapsack they saw, that the soldier could march with ease in. Col. Vinton informs me he is not authorized to pay over $2.90 for my knapsack. They have cost me over $2.75 Cash & from all information I can obtain they are the best & cheapest in the field.
>
> Will you be good enough to say by return mail the most you are willing to allow me for the 7600.40.
>
> They are intended for Soldiers use, & I must sell them for that purpose.
>
> Respectfully, Yr Sevt,
> John Rider[18]

The response (and a reference to the "7600.40") may be found in a contract with John Rider dated March 5, 1864 for 7,640 gutta-percha knapsacks at $2.09 each.[19]

***India-rubber knapsack** with wooden interior box frame that was likely made under a local state contract. Stenciled "MASS. 26" on front and "CO. I 26" on side.*

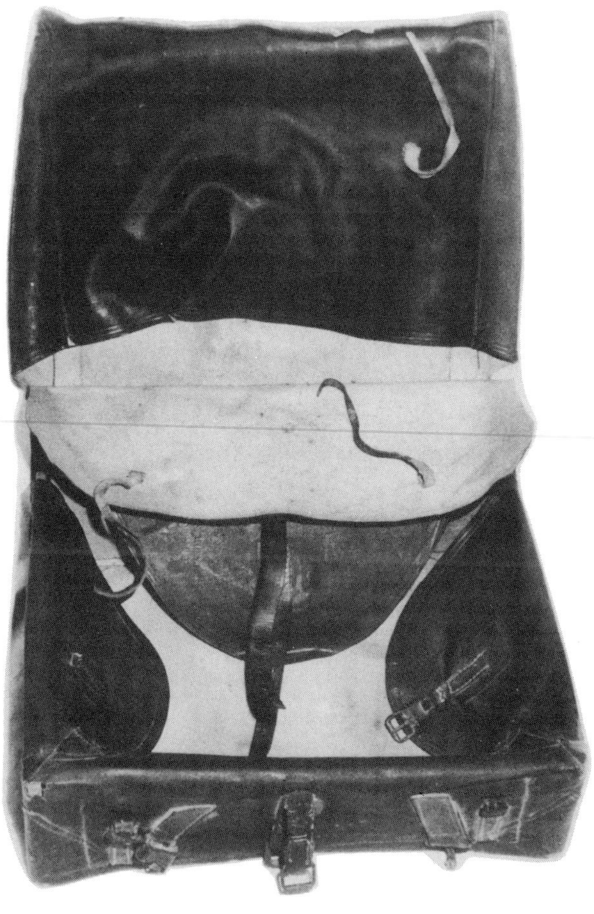

Combination Knapsacks

A combination knapsack was patented on April 12, 1859 by William B. Johns, an officer in the United States Army (U.S. Patent #23,582). Johns' knapsack is sometimes referred to as "Rider's Tent-knapsack," as the knapsack was manufactured under an agreement with John Rider and the North American Gutta Percha Company. On April 12, 1859, a Board reported that 1,000 of these gutta-percha knapsacks were to be ordered from Rider for testing by the Army.[20]

The tent-knapsack consisted of a sheet of gutta-percha 5¼ feet by 3¾ feet, with brass studs and button-holes along three edges and straps and buckles on the fourth side. The unit weighed three pounds and could be used as a knapsack, a rubber blanket, or, when combined with three others, a 5-man tent. A promotional brochure stated:

> The tent knapsack may be applied to many other useful purposes, as a water-proof cape or talma, as a bed cover, as a pack cover, as a dry shelter for ammunition, and other uses, where protection against rain and moisture is desired.[21]

The Rubber Clothing Company manufactured a combination knapsack, cloak, and camp blanket.[22] Augustus Clark's Patent #37,936 of March 17, 1863 for an "Improved Knapsack" was assigned to this company.

Corporal Granville Abbott of Company I, 31st Indiana Infantry invented the "Patent Hammock Knapsack" which he claimed, "Combined the Infantry Knapsack, Cavalry Valise, Officers' Portmanteau, Camp, Hospital and Field Bed and Shelter Tent, Litter, Ambulance, etc." A booklet describing the hammock knapsack also touted its use for "...hunting, fishing, excursion, engineering, exploring, and emigration parties over the plains to the Pacific coast." The unit consisted of a woolen blanket, an India-rubber cloth, canvas sacking, and cords. There was a pocket for storing personal articles that could be used as a pillow. A portable frame to be used for the hammock or stretcher, was also available. The unit without the frame weighed sixteen pounds and could be slung in the ordinary manner or carried by a handle like a valise.[23]

A sample of the hammock knapsack was first submitted to Quartermaster General Meigs for inspection on September 17, 1863 with a request that it be shown to the Secretary of War. Among the benefits highlighted was the fact that the soldier could sleep above the ground, adding to his comfort and health.[24]

On September 29, 1863, Corporal Abbott's commander, Captain James Hallowell wrote to Lt. Colonel C. Goudard, requesting Abbott be granted a sixty day furlough so he could procure a patent for the hammock knapsack. Hallowell sang the praises of

The Tent Knapsack is essentially a Gutta Percha sheet, 3 feet 8 inches wide by 5 feet 3 inches long, double edges on

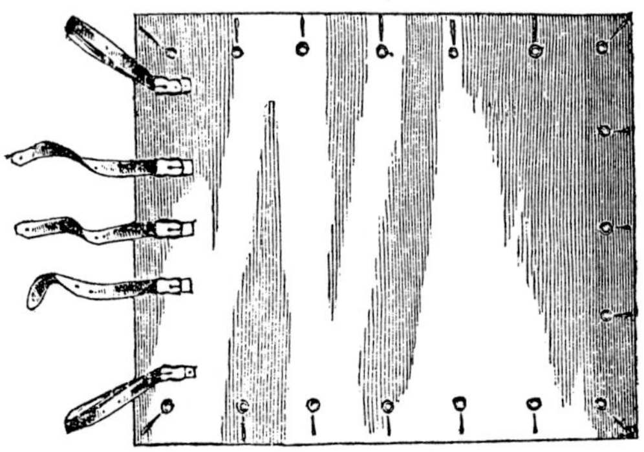

one side, with brass studs and button holes along two edges, and straps and buckles on the fourth edge; the whole weighing three pounds .

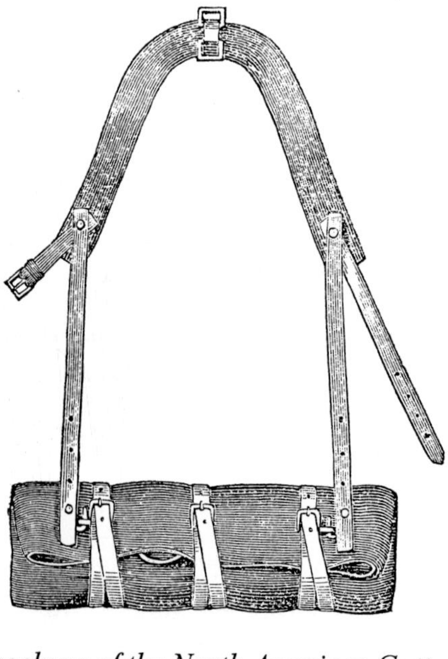

The Gutta Percha Tent Knapsack, *as illustrated in an advertising brochure of the North American Gutta Percha Company, circa 1859. Marketed to the Army in 1859 and during the Civil War, it served as a shelter half and was easily rolled up for individual transportation. (National Archives)*

Fig. 2.

Fig. 2 represents the infantry hammock knapsack rolled up, ready to be slung on the shoulders, showing the straps and the handle for carrying it as a valise in the hand, at the side.

In this form, containing the hammock, rubber and woolen blankets, with every article allowed the soldier to be carried in a knapsack, it weighs 16 pounds.

Fig. 3.

Fig. 3 represents the soldier with the knapsack slung in the ordinary manner ; also with it unrolled, and twisted in the form of a scarf, and thrown over the shoulder, as is often done with the ordinary blanket ; a very easy and convenient method of transportation in warm weather, and on forced marches.

***Diagrams and descriptions of Corporal Abbott's Patent Hammock Knapsack**, as depicted in a wartime advertising brochure. The combination knapsack was manufactured at The Wiard Ordnance Works in Trenton, NJ, and was sold at several stores. (National Archives)*

Fig. 4.

Fig. 4 represents the cavalry hammock knapsack rolled up, ready to be attached to the crupper or carried in the hand. In this shape it is admirably adapted for transportation by officers and on shipboard.

Abbott, stating that he had served faithfully for two years without absence, and fought in the battles of Donaldson, Shiloh, Stone River, and Crawfish Springs. His hammock knapsack was seen and approved by Generals Rosecrans, Cullenden, Thomas, Palmer, and others.[25]

On November 27, 1863 Major F. E. Trotter of the 1st Regiment Invalid Corps reported to Lt. Colonel and Assistant Adjutant General J. A. Hardie on the weights of knapsacks, with the following results:

"Old style knapsack" containing:
1 Woolen Blanket
1 Rubber Blanket
1 Shirt
1 Drawers pair
1 Socks
1 Blouse
Weighs 13 ½ lbs.

"Abbott's Patent Knapsack & Hammock" contains:
Same as above
Weighs 15 ¾ lbs.[26]

Corporal Abbott wrote directly to Secretary of War Stanton on December 2, 1863 offering 200 hammock knapsacks, complete with a wool blanket and a rubber blanket, at a cost not to exceed $22 each. The units were to be made by December 20, "…of as good material as can be found for that purpose" by Mr. V. Wiard of Trenton, New Jersey. The Wiard Ordnance Works would later manufacture and sell the hammock knapsacks directly and through distributors.[27]

On December 11, 1863 Acting Quartermaster General Charles Thomas advised Abbott that "The Secretary of War authorized the purchase of one hundred of your "Soldier's Improved Knapsacks and

---

The shelter may be pitched also in another manner, by fastening one edge to the ground and attaching the opposite one to the rope inside.

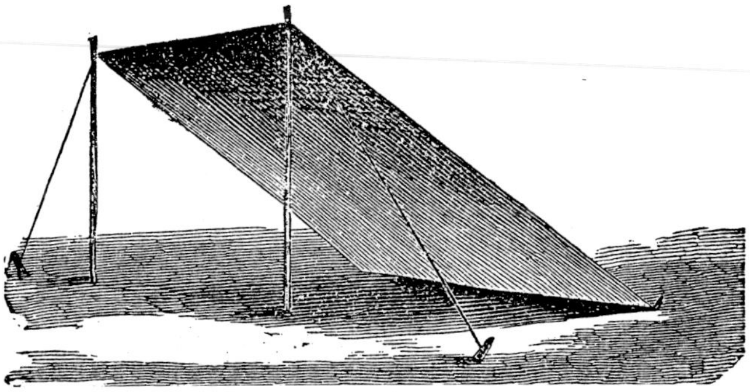

***An advantage of the gutta-percha Tent Knapsack,** patented by William Johns in 1859, was its value as a waterproof cover for the soldier. Four knapsacks combined to make a shelter tent. (National Archives)*

***Corporal Abbott's Hammock Knapsack**, advertised for use in camp and hospital to provide extra comfort. The Union Army tested Abbott's invention, but rejected it as "...too heavy and complicated for enlisted men." (National Archives)*

Haversacks," the cost not to exceed $20 each." He directed Abbott to confer with Colonel A. J. Perry in the Clothing Branch of the Quartermaster's Office as to delivery directions.[28]

Lt. Colonel G. H. Crosman, the Assistant Quartermaster General in Philadelphia, wrote to General Meigs on December 17, 1863 advising him that Corporal Abbott had stopped by with a hammock knapsack on his way to New York to arrange for manufacture of the 100 units. Besides the fact that he saw no way that Abbott could meet his promised delivery date of December 20, he also stated that the "...blanket shown me...is entirely too light in weight and thickness for a soldier in the field." He recommended that Abbott use a regulation five pound blanket with the hammock knapsacks.[29]

The lack of acceptance of Abbott's Patent Hammock Knapsack can be summed up by a letter from 1st Lieutenant A. Edman, written on April 8, 1865, probably to General Meigs. Edman says he had distributed twenty of thirty-nine hammock knapsacks to the Reserve brigade and the 1st U.S. Engineers and the reports were that they were "...too heavy and complicated for enlisted men," and were recommended for mounted officers. The hammock knapsacks weighed fifteen to sixteen pounds as compared to eleven pounds for the common knapsack, blanket, and rubber poncho. Edman's closing comment was, "I have still nineteen on hand, which it will be difficult, if not impossible, for me to dispose of."[30]

(Also see **Section 6** & 7, and **Blankets and Ponchos**.)

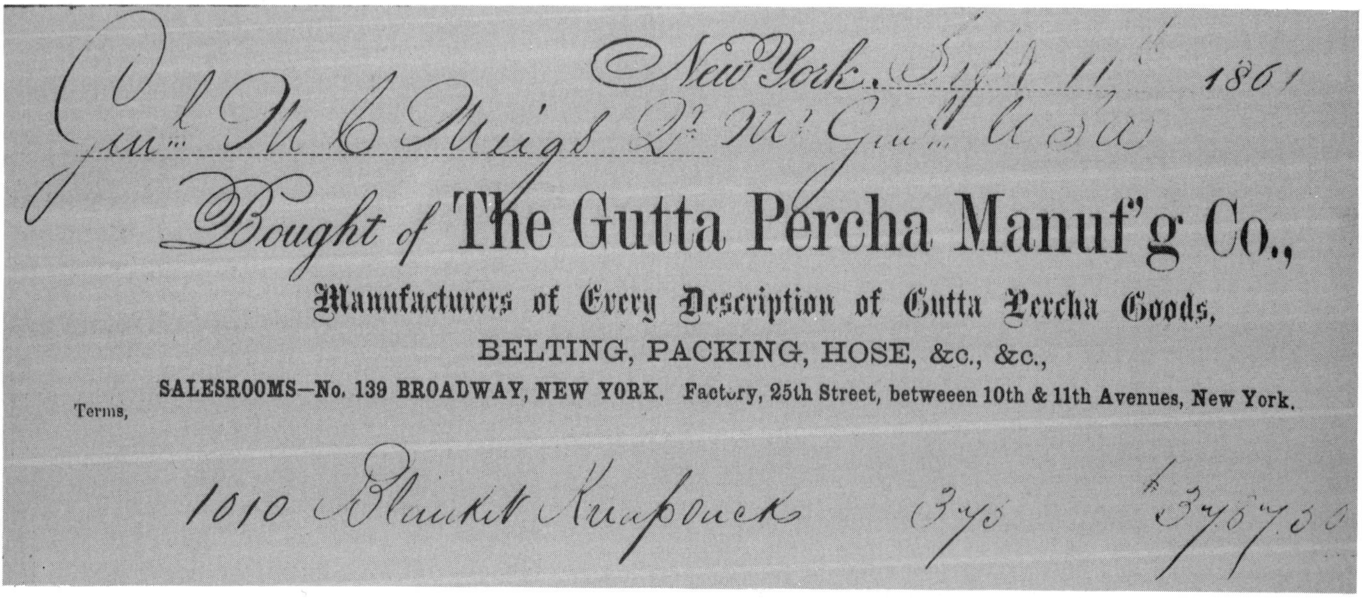

New York, Sept 11th 1861

Genl M C Meigs Qr Mr Genl U S A

Bought of The Gutta Percha Manuf'g Co.,

Manufacturers of Every Description of Gutta Percha Goods,

BELTING, PACKING, HOSE, &c., &c.,

Terms, SALESROOMS—No. 139 BROADWAY, NEW YORK. Factory, 25th Street, betweeen 10th & 11th Avenues, New York.

1010 Blanket Knapsack 3.75 $3787.50

***Invoice for 1,010 Blanket Knapsacks** at $3.75 each, sent to Quartermaster General Meigs by the Gutta Percha Manufacturing Company early in the Civil War. (National Archives)*

NOTES - Knapsacks

[1] National Archives, Record Group 92, Entry 25, "Orders for Army Supplies, No. 1, June 17, 1846 - February 15, 1847."
[2] Based on period advertisements and surviving examples.
[3] Francis A. Lord, *Civil War Collector's Encyclopedia* (Harrisburg: The Stackpole Co., 1965) p. 144.
[4] Author's collection. See accompanying illustration.
[5] War Department, *Regulations for the Army of the United States-1861* (New York: Harper & Bros., May 1, 1861).
[6] War Department, *Revised Regulations for the Army of the United States-1861* (Philadelphia: J. G. L. Brown, August 10, 1861).
[7] NA, RG 92, Entry 225, Consolidated Correspondence File, "Gutta-percha."
[8] *Ibid.*
[9] *Ibid.*
[10] *Ibid.,* "Knapsacks."
[11] NA, RG 92, Entry 999, Office of Quartermaster General, "Letters Sent (Clothing & Equipage)," Vol. 18, p. 171.
[12] *Ibid.,* Vol. 18, p. 219.
[13] NA, RG 92, Entry 225, CCF, "Knapsacks."
[14] *Ibid.*, "Gutta-percha."
[15] *Ibid.*
[16] *Ibid.*
[17] *Ibid.*
[18] *Ibid.*, "Knapsacks."
[19] "Reports of the Secretary of War," *Congressional Serial Set 1230*, p. 181.
[20] NA, RG 92, Office of Quartermaster General, "Register of Letters Received (Clothing)," Vol. 11.
[21] NA, RG 92, Entry 225, CCF, "Knapsacks."
[22] *The Scientific American,* 22 June 1861.
[23] NA, RG 92, Entry 225, CCF, "Knapsacks."
[24] *Ibid.* Letter to Meigs. Signature illegible.
[25] NA, RG 92, Entry 225, CCF, "Knapsacks."
[26] *Ibid.*
[27] *Ibid.*
[28] *Ibid.*
[29] *Ibid.*
[30] *Ibid.*

Fig. 10.

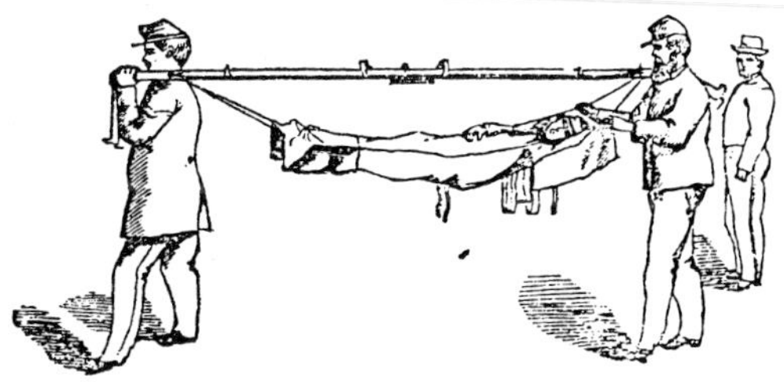

Fig. 10 represents the hammock suspended to the carrying pole, transporting a wounded soldier from the field of battle.

Abbott's Hammock Knapsack. *(National Archives)*

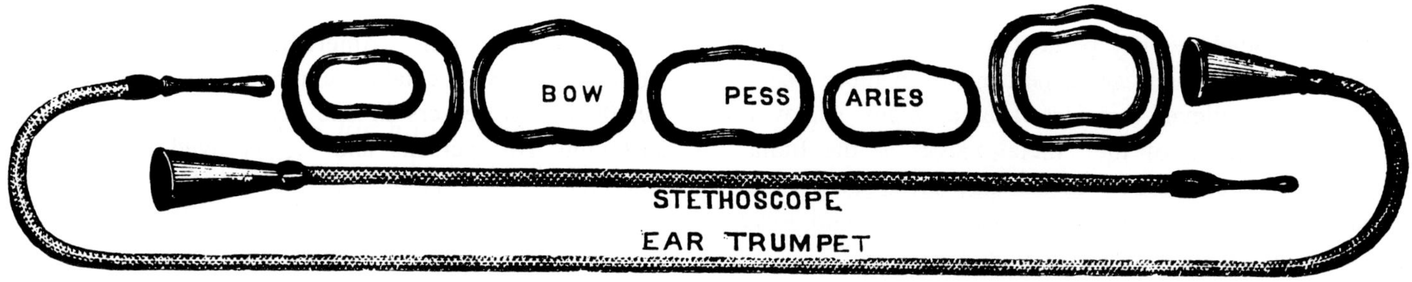

Illustrations from the W. A. Weed & Co. catalog of 1872.
(Courtesy Dittrick Museum of Medical History)

MEDICAL APPLICATIONS

"An evening with Venus, a month with Mercury."

> Sailor's warning on sexual affairs and the local treatment of venereal diseases by caustic injections

India-rubber, gutta-percha, and hard rubber have been used in a wide variety of medical items. South American natives fabricated bulb syringes of rubber centuries before the discovery of vulcanization.[1] Charles Goodyear, in *Gum-Elastic and It's Varieties,* suggests various articles such as bandages, syringes, cushions, supporters, hot water bottles, vericose stockings, hospital sheets, and water beds.[2]

The use of liquid caoutchouc as an elastic plaster was described in the *New York Journal of Medicine and Collateral Sciences of 1855.*[3] In 1859, the Royal Orthopaedic Hospital used moulded gutta-percha splints, applied hot and shaped to the body, in treating diseased hip joints.[4] A simple tourniquet, consisting of an "...endless band or tube of vulcanized rubber with or without pads," was patented by F. Bond in 1863 (U.S. Patent #38,442) and Lea & Lambert's tourniquets utilized elastic and non-elastic straps.[5] T. E. Gordon's patented crutch contained an India-rubber spring to reduce concussion (U.S. Patent #45,709).

Novel and creative uses that took advantage of the inherent properties of India-rubber were not uncommon. In 1855, Dr. Gunning S. Bedford addressed the case of a North Carolina woman, who suffered from, "...irritation about the bladder, giving rise to a frequent desire to micturate." He diagnosed "prolapsation of the womb" due to an, "...unusually large pelvis...and increased capacity." Perplexed as to how to remedy this disorder, he wrote, "All I did was to introduce into the vagina a soft India-rubber ball, for the purpose of giving gentle support to the uterus, and thus relieve the bladder from pressure." He went on to explain his partiality to the India-rubber ball, as it, "...forms an admirable support to the uterus," and how he instructed the patient to remove, clean and replace it at least every 24 hours.[6]

Various bags, balloons, and animal bladders that had been employed for induction of labor and treatment of hemorrhage and uterine disorders were now made of India-rubber. The *ballon excitateur,* introduced in 1862 by Etienne Tarnier and generally called "Tarnier's balloon," was a soft rubber bag inserted through the cervix on the end of a catheter and inflated to the size of a hen's egg to dilate the canal and stop hemorrhage.[7] Bushe's Compressor was an India-rubber bladder introduced into the

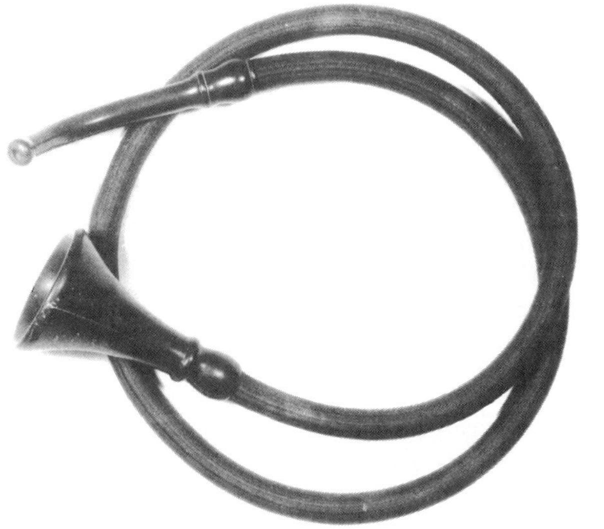

A predecessor to the hearing aid*, this hearing tube's hard rubber ear-piece and composition speaking mouthpiece are connected by an India-rubber tube.*

rectum and dilated by ice water for treating hemorrhage.[8]

Military uses in the Civil War were extensive. Early in the War, the Gutta Percha Manufacturing Company of New York sent Secretary of War Simon Cameron a sample of an Army cot which they claimed was, "...used by the French & English Armies for Soldiers when in quarters or on the Sick List." The company claimed they were very strong and "...could be put together or out the way when the room is wanted for some other purpose." The price quoted to the Government was $4 each.[9]

The Medical and Surgical History of the War of the Rebellion contains many references to the use of India-rubber and gutta-percha for medical applications, including the use of air cushions[10] and the treatment of a patient with bedsores using a water bed.[11] Other uses detailed in these volumes include:

- The use of vulcanized rubber rings with automatic catches proposed in an operation to unite separated intestines.[12]
- Employment of India-rubber in splints for treating fractures, including reference to an article in the *American Journal of Medical Science* on the use of India-rubber in treatment of wounds of the femur.[13]

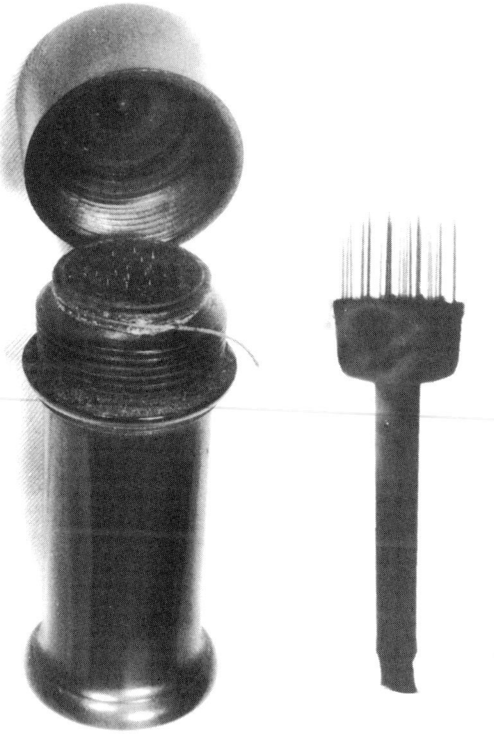

__Traveling hard rubber vaccinator__ with screw-on cap. An X-ray of the instrument reveals a heavy metal shaft along with the piercing needles. (X-ray courtesy of Gordon Dammann)

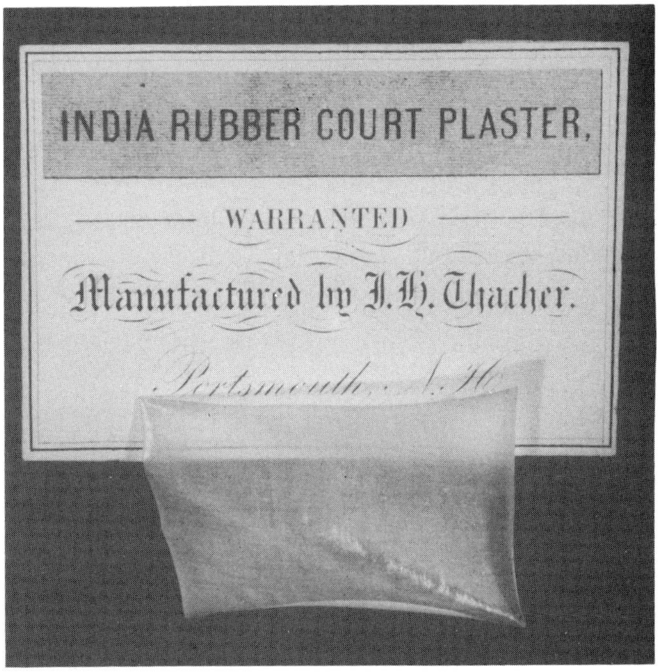

***India-rubber court plaster**, applied as a hot poultice on sore or inflamed skin.. Thacher is listed in the* 1861 New Hampshire Apothecary Directory. *The envelope contains a sheet of now brittle unvulcanized rubber.*

- An ambulance, constructed by Union Army Assistant Surgeon B. Howard, "...utilized blocks of rubber in strategic places to provide for patient comfort." [14]
- Stretchers in a 30-stretcher hospital railroad car suspended on rubber rings that acted as springs.[15]
- Private T. M. Peterson, a patient in a Union hospital suffering from a fistulous urethra allowing urine to escape from two openings at the base of the penis, was compelled to wear a rubber bag on account of constant dribbling.[16]
- A patient with a plastic operation on his mouth was given nourishment by mouth through a funnel with a long gutta-percha tube.[17]

Hospitals, medical wagons and medical kits contained various items of India-rubber and gutta-percha. The contents of Perot's Medicine Wagon included the following:

1 8 oz. Hard rubber syringe
1 Self-injecting rubber syringe
8 Gutta-percha bed covers
2 Yards gutta-percha cloth[18]

Autenreith's Medicine Wagon contained rubber blankets, gutta-percha bed covers, and 6 gum-elastic catheters. Regimental surgeon's medical outfits were equipped with a gutta-percha pouch and gum-elastic catheters # 1, 3, 5, 7, 9 & 11.[19]

The list of articles and quantities of supplies purchased during the war by the Medical Department of the Union Army shows the following entries:

106,011 Yards of gutta-percha cloth
39,551 Gutta-percha bed covers
1,144 India-rubber water beds
6,486 Rubber cushions with open centre
11,724 Small rubber cushions for air or water[20]

Hard rubber enema syringe with tip. *Holds 14 oz. Length is 15 inches. Marked 3 times on body near the plunger end, "xxxGoodyear's Patent May 6, 1851xxx."*

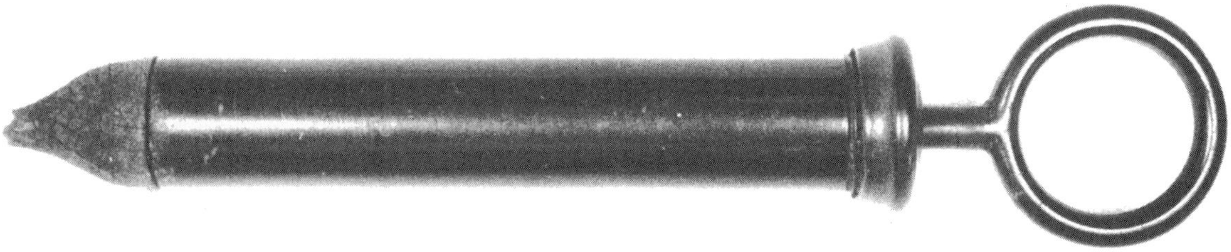

Hard rubber syringe with red soft rubber tip. *Possibly a post-War variation of ear or male gonorrhea syringe. Length is 4¾ inches. No markings.*

MEDICAL DEPARTMENT SUPPLY TABLE

INSTRUMENTS TO EACH MILITARY POST	AMOUNT
Syringes, Enema (1-4 oz, 1-12 oz.)	2
Syringes, Penis (2-glass, 6-metallic)	8
Syringes, Vagina (2-glass, 1-metallic)	3
Bedding:	
Gutta-percha cloth, yds.	4

Extracts from a medical supply table shown in the Regulations for the Army of the United States, *published in May of 1861, showing gutta-percha cloth and certain glass and metallic syringes.*[21]

MEDICAL DEPARTMENT	QUANTITIES FOR ONE YEAR FOR COMMANDS OF					
	100-200	200-300	300-400	500	1000	
Syringe, Enema	3	3	3	3	6	1-Davidsons; 1-4 oz, 1-8 oz
Syringe, Penis, glass	2	4	6	8	16	
Syringe, Penis, metallic	6	12	18	24	36	
Syringe, Vagina	3	3	3	3	6	Hard India rubber-1
Gutta-percha cloth, yds	4	6	8	10	16	
Oiled silk or gutta-percha tissue or India rubber tissue, yds.	4	6	6	8	12	

For amputating, exsecting and general operating kits, a gutta-percha pouch was specified.

STANDARD SUPPLY TABLE FOR FIELD SERVICE

ARTICLE	Regiment 3 months	Battery 3 months	Command 3 months	
Syringe, Enema	4	2	1	1-Davidson's; 1-Hard rubber, 6 oz
Syringe, Penis, glass	8	4	2	
Syringe, Penis, India rubber	8	4	2	
Gutta-percha cloth, yds.	8	4	2	
Gutta-percha bed covers, no.	8	4	2	So constructed as to form, when united, a continuous spread or covering
Oiled silk or gutta-percha tissue or India rubber tissue, yds.	8	4	2	
Pans, Bed	2	1	1	Of hard India rubber or other material

In the Revised Regulations for the Army of the United States, *published in August of 1861, the overall list of authorized equipment is greatly expanded. These tables of extracts show many more specific applications of India-rubber, hard rubber, and gutta-percha.*[22]

***Excavated hard rubber male gonorrhea syringe** found at Berkely Plantation (Richmond), Virginia. Length is 5 ½ inches. Marked "Goodyear's Patent May 6, 1851."*

Artificial Limbs

The Civil War took its toll on limbs, as 29,661 amputations are detailed, with 7,074 resultant deaths. Discounting the 9,121 survivors of finger and toe amputations, a potential 13,466 veterans sought artificial arms, legs, hands, and feet.[23] The number of patents for artificial limbs increased as the War progressed. In 1861 and 1862, only one patent for artificial limbs was issued each year; In 1863, 11 patents were issued, 18 in 1864, and 24 in 1865. Many of these patents utilized elastic straps and contained blocks, springs, and other parts made of India-rubber.[24]

The lower part of Thomas White's Artificial leg, patented June 24, 1862 (U.S. Patent #35,723), was covered with a case made of India-rubber. In 1863, Amasa Marks patented artificial limbs of sponge rubber and hard rubber combined with wood (U.S. Patent #40,763).

In an order to Medical Directors dated May 13, 1865, United States Army surgeon C. H. Crane authorized certain manufacturers to provide artificial limbs to mutilated soldiers. The maximums paid were $75 for legs, $50 for feet, and $50 for arms. Among the manufacturers authorized to provide legs was A. A. (Amasa) Marks of New York City, whose price was $65.[25] Marks continued his work with India-rubber components, as he published *A Treatise on Artificial Limbs with Rubber Hands and Feet* in 1902.

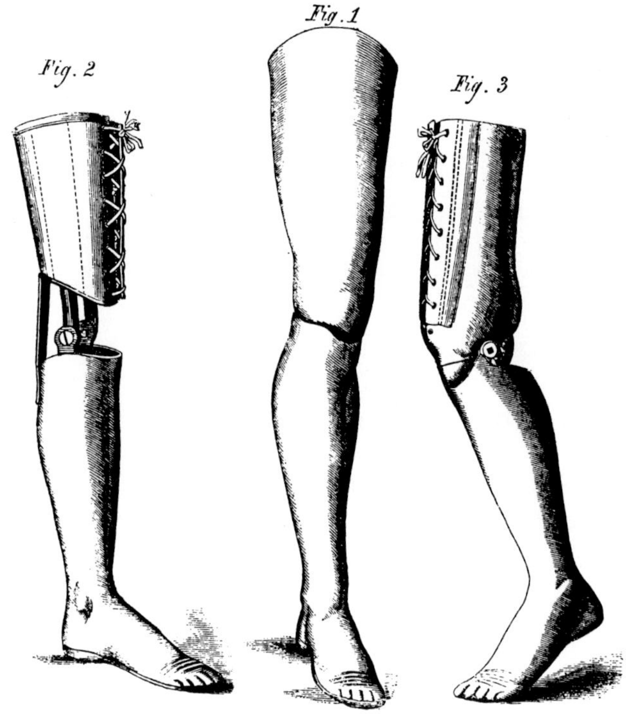

Above and below: ***Marks' artificial limbs****, as depicted in* The Scientific American *of April 15, 1865. Wartime casualties created a large market for various improved designs.*

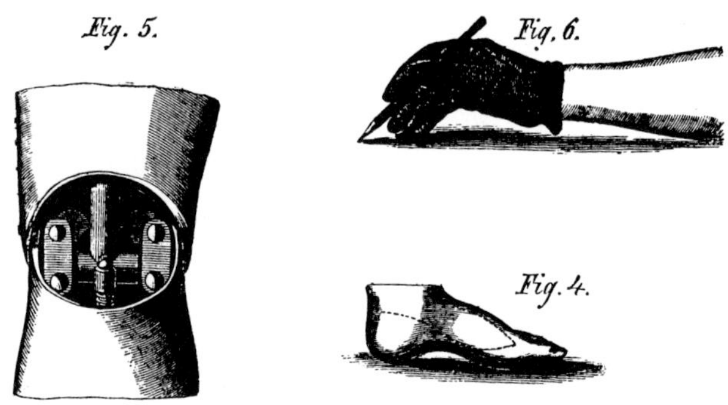

MARKS'S ARTIFICIAL LIMBS.

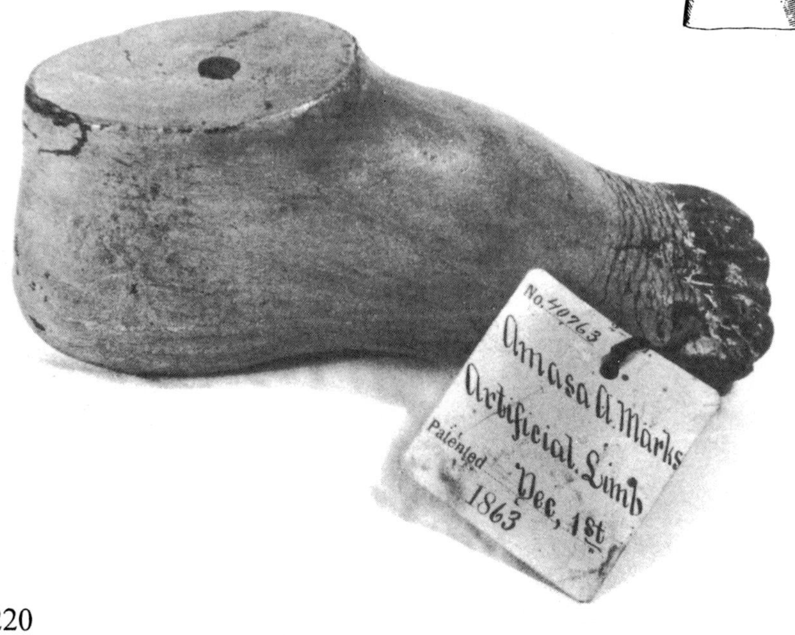

Life-size patent model of an India-rubber artificial foot. *United States Patent #40,763 was issued December 1, 1863 to Amasa A. Marks, for the use of sponge rubber and hard rubber combined with wood for various artificial limbs, including hands, arms, and feet. The original tips of the toes have been replaced.*

Bandages

As noted in the history of gutta-percha, Dr. Montgomerie introduced and advocated its use for surgical bandages around 1842. Ten years later, elastic bandages of India-rubber appeared in the catalog of Arnold Company, a surgical instrument maker. When excessive warmth from these bandages became a problem, perforated bandages were recommended.[26]

A "Bandage/apparatus," which consisted of India-rubber tubing through which cold water could be circulated for cooling parts of the body, was patented by Orange M. Fuller in 1864. (U.S. Patent #43,679) The specification detailed a bandage for the head for reducing fever. An elastic catamenial bandage was patented by D. F. Robinson in the same year (U.S. Patent #45,523).

Bougies

Bougies of metal, wood, and elastic-gum were used in contraction of the anus and stricture of the rectum to restore its caliber.[27]

Catheters

Silver or bone tubes had been used since ancient times to drain fluid or pus and in 1851 a gentleman named Chassaignac introduced a perforated rubber tube for this purpose. Catheters were made of metal until Auguste Nelaton invented a soft rubber catheter in 1860.[28] Ordinary gum catheters, used as drainage tubes, facilitated the discharge of fluids in cases of empyema, a collection of pus in the lung.[29]

Catheters were employed in the treatment of wounds of the urethra. The Gelis catheter is often referred to in such operations. Catheters of gutta-percha or balata (a substance having qualities between gutta-percha and India-rubber) are recommended in *The Medical and Surgical History of the War of the Rebellion*, which also noted:

> In the last few years caoutchouc has again come in vogue in the shape of vulcanized rubber. Catheters of this substance are soft and pliable. They are very useful for those who have to catheterize themselves and in cases where the catheter is to be worn for several days; they are retained with less inconvenience than harder and less flexible instruments and long resist the action of the urine. They must be thick so as to not collapse under pressure.[30]

This hard rubber artifact, *excavated at Clear Spring, Maryland, appears to be a double-action medical suction or cupping device. If a thin rubber valve was placed in the instrument, suction could be created both by pushing the body onto the skin and pulling the plunger out.*

Inhalers

The use of inhalers for ether and chloroform was introduced in 1846-47. Facepieces, straps and mouthpieces for inhalers were made of India-rubber.[31]

Rubber Gloves

Rubber gloves had been manufactured for industrial and household purposes since the 1850s. Charles Goodyear suggested "dissecting gloves," combined with a "dissecting apron," to protect the wearer from the "...the fatal accidents that so frequently occur in the profession" and the "offensive effluvia of the dissecting room."[32] A brief history of their later introduction into surgery will exclude any thoughts of the use of these gloves in surgery prior to the late nineteenth century.

In 1889 or 1890, W. S. Halstead of Baltimore asked the Goodyear Rubber Company to make two sets of thin rubber gloves with gauntlets for a nurse who complained of dermatitis caused by the process of sterilizing the hands in a solution of mercuric chloride. They were tested and worked so well that soon the entire operating team wore them.[33]

F. L. Reichert, in the *International Abstract of Surgery of May, 1930,* noted:

> ...that the use of rubber gloves in surgery was not the result of an inspiration to eliminate the hands as a source of infection during an operation. Their use was a matter of slow evolution, first as a protection for the hands of the assistants from irritating solutions, then as an added protection on the part of the operator in exploring joints, later as an aid to the operative dexterity of those accustomed to gloves as assistants, and finally as a regular adornment to be worn invariably in all cases, clean and septic, by the operator and all members of the operating team.[34]

Stethoscopes

The idea of the stethoscope was first suggested in 1761.[35] Early examples were one-piece monaural instruments resembling small hearing trumpets and could have been manufactured in hard rubber or artificial ivory, as suggested by Goodyear.[36]

The Cammann stethoscope, which incorporated some hard rubber parts, was invented in America in the early 1850s and was the first commercial binaural stethoscope.[37] Another type was introduced at the Great Exhibition of 1851 in London. The Illustrated London News, reporting on gutta-percha displays at the exhibition, wrote that "Dr. O'Leared exhibits a double stethoscope for ascertaining the state of the lungs; single stethoscopes of gutta percha have been in use for the past three years."[38]

Syringes

A wide variety of medical syringes formerly made of other substances are known to have been made of rubber and hard rubber during the mid-1800s. Soft rubber was used in syringe bags, bulbs and tubing while hard rubber was used in pump syringes of various types as well as non-corrosive fittings on rubber syringes. With the exception of buttons, hard rubber syringes of many types represent the largest grouping of surviving Civil War era hard rubber items.[39]

Ear Syringes

Ear syringes with a small rounded tip were made with a protective head that prevented insertion too far into the aural cavity. A long-tubed injecting syringe was also used to inject solutions, such as diluted liquor potassae, through a catheter and into the nostrils for treatment of some middle ear disorders.[40]

Enema Syringes

Medical books of the seventeenth, eighteenth, and early to mid-nineteenth centuries are replete with illustrations of the plunger-type enema syringes found in campsites and hospital sites of the Civil War. Earlier plunger syringes were made of pewter,[41] while hard rubber varieties were introduced in the 1850s. In the *Hospital Steward's Guide* of 1863, Assistant

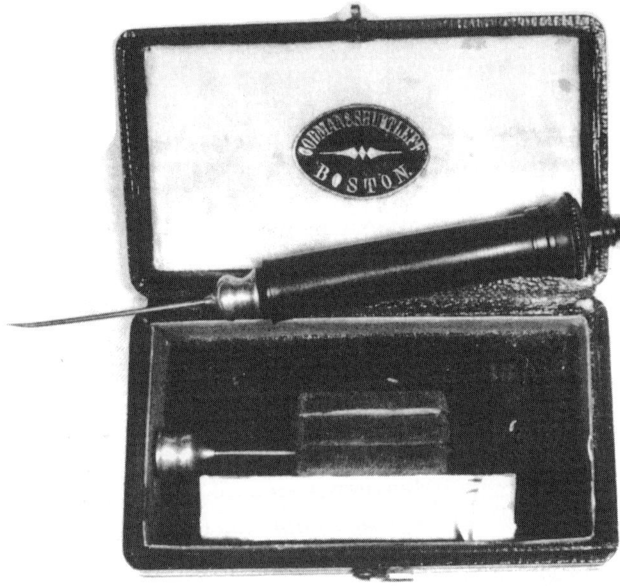

***Hard rubber hypodermic syringe** and two gold-washed needles in leatherette case. Case has an inner label reading: "Godman & Shurtleff-Boston." Syringe is marked "Goodyear's Patent May 6, 1851." Syringe size without needle: 2 5/8" long x 3/8" diameter. Hypodermic syringes were first used in Scotland in 1853.*

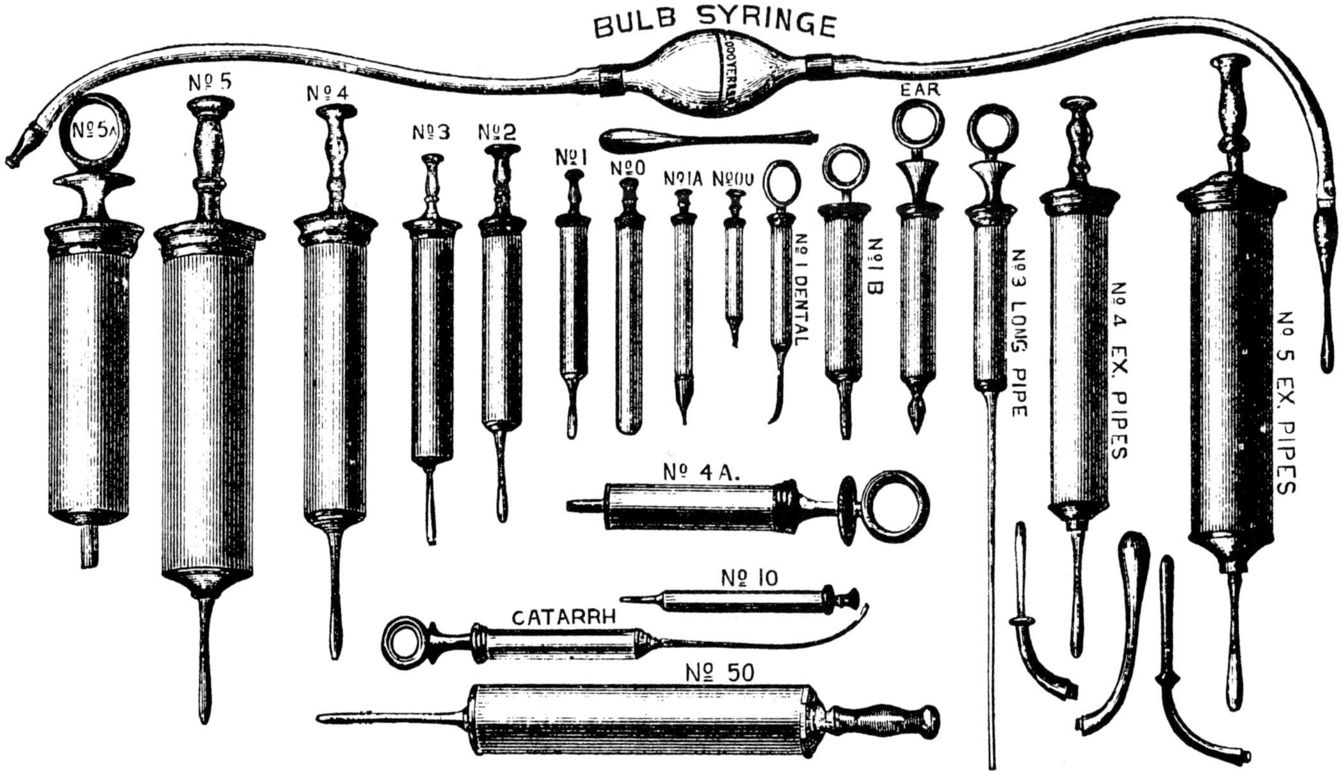

Pages from the W. A. Weed & Co. catalog of 1872*, showing various elastic and hard rubber syringes and numbered to match a separate price listing. Most wartime styles of hard rubber syringes were still being made late in the century. The bulb syringe depicted is a Davidson's patent syringe patented in 1857. These syringes, used for enemas and douches, functioned like a siphon. (Courtesy Dittrick Museum of Medical History)*

Surgeon Joseph J. Woodward described "...a large enema syringe made of hard India-rubber furnished with the medical supplies."[42] These instruments have sometimes been mistakenly identified as horse or veterinary syringes.

Bulb syringes consisting of a hollow rubber bulb with fittings and nozzles of ivory or hard rubber were commonly used for enemas, or clysters, as they were referred to at that time. While few of the popular "Davidson Syringes" have survived, the oval pasteboard boxes, with labels bearing the patent markings, appear frequently at collector shows. The elusive syringe, employing India-rubber tubing and a bulb that was squeezed to create the flow of fluid, was patented by Charles and Harman Davidson on March 31, 1857 (U.S. Patent #16,956).

While there were many other types of enema syringes employing pumps and valves, B. T. Babbitt's patented "Fountain Syringe" (U.S. Patent #16,822 dated March 17, 1857) set the design for the vulcanized rubber bag syringe still in use today. The patent drawing shows the simple method of utilizing hydrostatic pressure to deliver the liquid. While the patent belonged to Babbitt, soft rubber syringes of this type were in use prior to this patent. In *Gum-Elastic and Its Varieties,* Charles Goodyear refers to them as "self-acting syringes" and comments that they have "...for many years been made in Europe, of the Macintosh cloths."[43] Mr. Thomas Lewis of Philadelphia and Mr. Davidson of Charleton, Massachusetts manufactured this type of syringe for

"Fountain" and "Ball" syringes *advertised in the 1854 catalog of the North American Gutta Percha Co.*

vaginal use.[44] It is likely that the "One self-injecting syringe" listed among the contents of Perot's Medicine Wagon was of this type.[45] A modified standard Union tin canteen, with a soldered drainage spout to accept the rubber tubing, also appears to have been used for this purpose.[46]

Eye Syringes

One eye syringe or eye douche consisted of 4½ feet of India-rubber tubing from a container placed on a high shelf, allowing a jet of water regulated by pressure that was "...less irritating than cup-type douches."[47]

Gonorrhea Syringes

There were 73,382 cases of syphilis and 95,833 cases of gonorrhea among members of the Union forces during the Civil War.[48] Most of the smaller syringes found in collections today are some version of the gonorrhea syringe that was used to treat that disease. Male versions with different styles of "tips" and female versions resembling test tubes with small holes at the tip were sold at the time.

The *Hospital Steward's Manual* stated: "Injections by the urethra are frequently employed, especially in cases of gonorrhea. For this purpose small syringes of hard India-rubber or of glass, containing about ½ oz., are furnished with the medical supplies."[49] The Confederate Regimental Dispensing Kit recommended the inclusion of "3 penis syringes." Interestingly, these instruments are found in catalogs of the late 19th century and may have been in use as recently as 1930.[50]

Box for Davidson's syringe *complete with label and patent dates. These leatherette or cardboard boxes from Davidson syringes are sometimes offered for sale with incorrect hard rubber syringes inside.*

In his book, *A Treatise on Gonorrhea and*

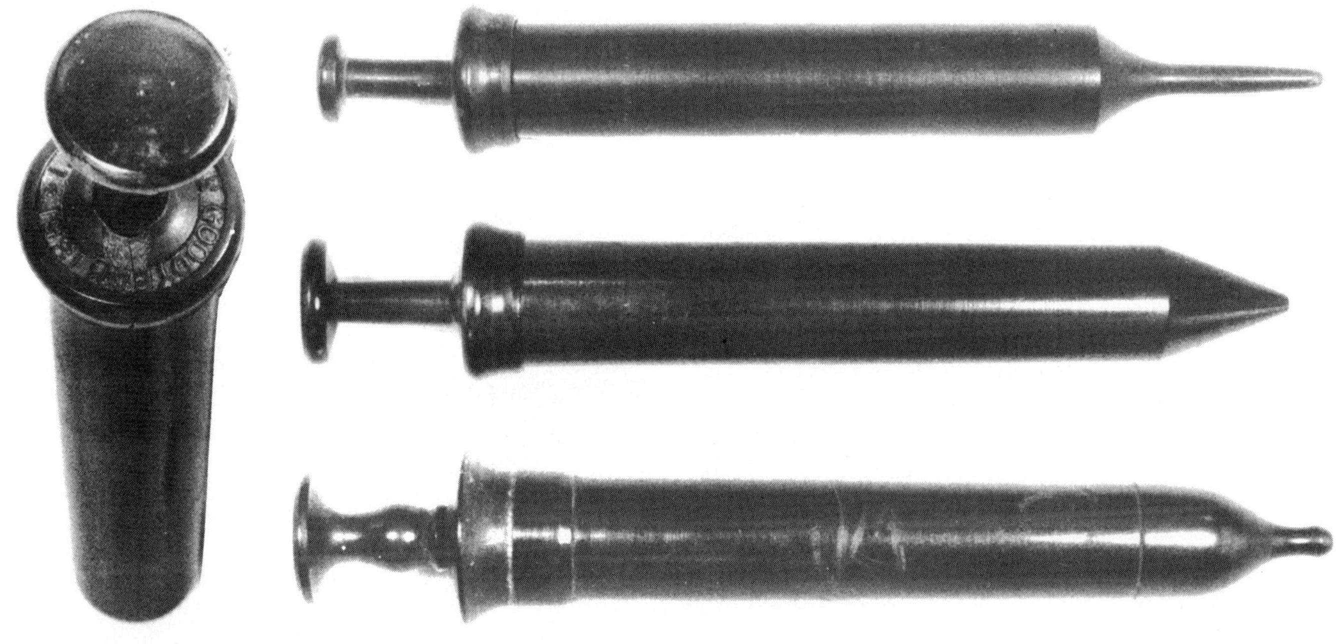

Group of male syringes *found in the Benecia Arsenal and San Francisco earthquake areas. (Top) Male gonorrhea syringe with pointed tip. Marked "xxx Goodyear's Patent May 6, 1851 xxx." (Center) Male gonorrhea syringe with conical tip. Marked "I. R. Comb Co. Goodyear 1851." Note close-up of patent marking at left. (Bottom) Male gonorrhea syringe with short bulb tip.*

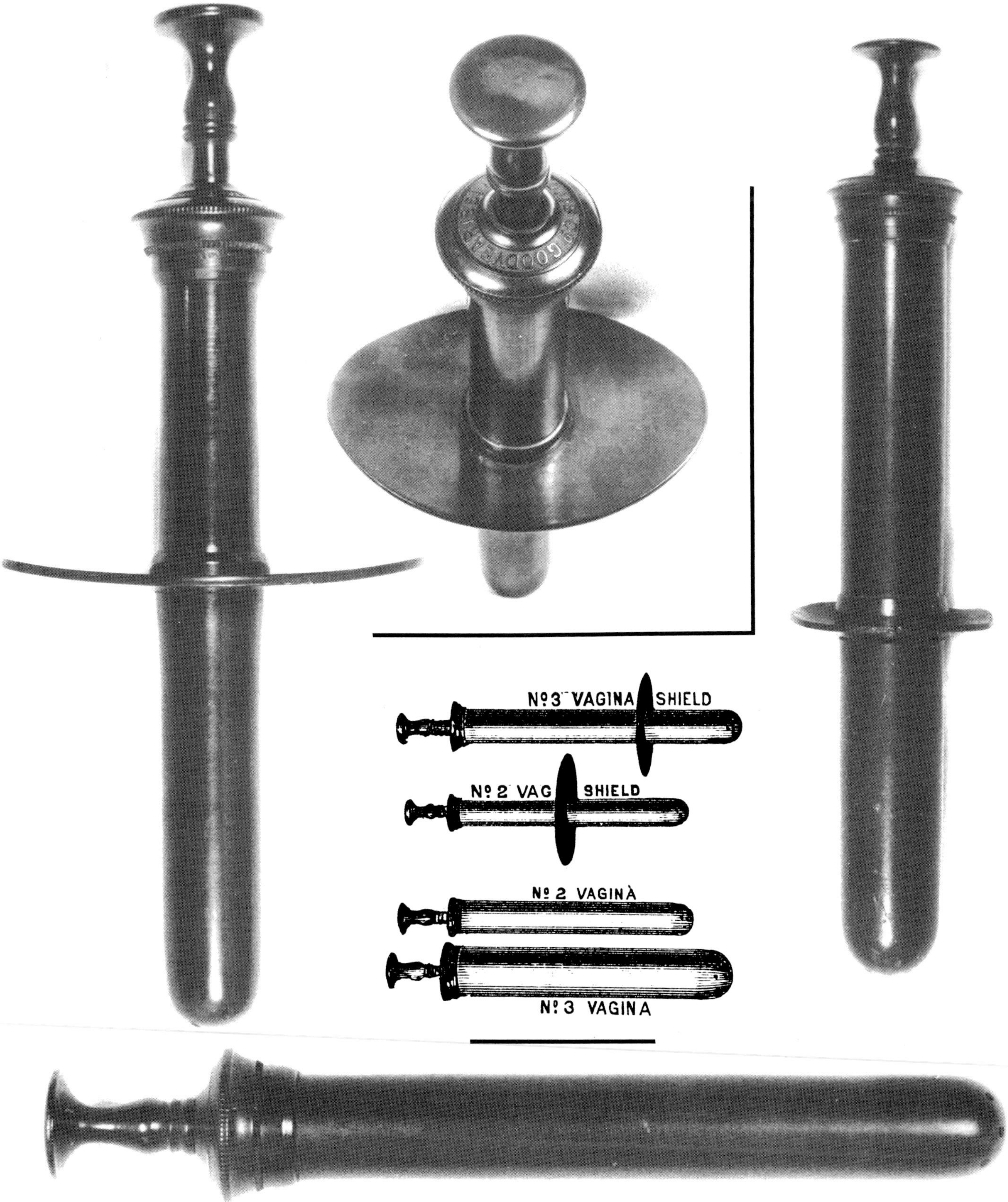

Three hard rubber female gonorrhea syringes and illustrations from the 1873 W. A. Weed & Co. catalog. *Each syringe is about 6½ inches long and holds approximately ¾ oz. (Left and close-up of patent marking on top) Syringe with hard rubber guard. Marked "I. R. Comb Co. Goodyear 1851." (Right) Syringe excavated from the Benecia Arsenal/San Francisco earthquake area. A replacement guard was recently added to original recess in syringe. Marked "xxxGoodyear's Patent May 6, 1851xxx." (Bottom) Syringe marked "Goodyear's Patent May 6, 1851." Also marked "Pat. 1860." This 1860 marking, also seen on later period hard rubber syringes, has not been successfully traced as an original patent marking.*

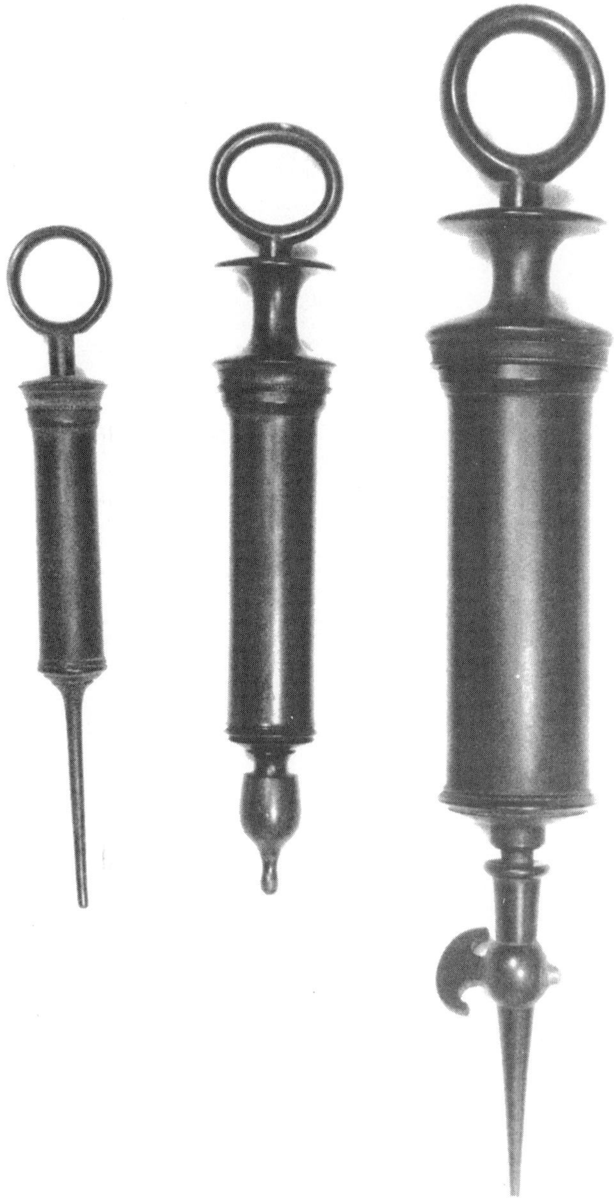

Hard rubber ear syringes. *The long thin tube on the syringe, at left, was used to slowly drip medications into the ear. The right syringe (length 9½ inches) has a removable tip with a shut-off. All three syringes are marked "Goodyear's Patent May 6, 1851."*

Syphilis, Dr. Silas Durkee admits that the use of local injections to cure disease in the male patient is a very controversial subject. He states that the syringe must be handled in the "most adroit and cautious manner by the surgeon."[51] Three or four grains of nitrate of silver with an ounce of rosewater is the solution he recommended. However, he also discussed an abortive method utilizing 10 grams of nitrate of silver to the ounce of rosewater. Other ingredients used in the injections were potash, red bark with nitric acid, pernitratis, subnitrate of bismuth, and mercuric or red wine solutions.[52]

Although this treatment had the sanction of eminent surgeons, it often caused ulcerations and inflammations of the bladder and urethra. In at least one case these reactions were so severe that the patient died within a week.[53] The treatment was extremely painful but many men chose this "quick cure" in order to maintain their normal domestic lives. While a popular sailor's warning about sexual affairs, "An evening with Venus, a month with Mercury,"[54] likely referred to this cure, it should be noted that various forms of mercury were also administered by mouth for treating syphilis and gonorrhea.[55]

Continued popularity of local injections was discussed by Dr. Felix von Niemeyer in *A Textbook of Practical Medicine,* published in 1876. The doctor mentioned possible complications developing because infected men were embarrassed to ask for a penis syringe at a drug store and purchased instead an ear syringe, used for treating inflammations and irritations of the ear. The doctor pointed out that the ear syringe was not designed in capacity or shape for use as a penis syringe.[56]

Treating females with these injections was apparently even more widely accepted in medical circles. Solutions used in female syringes included nitrate of soda, tincture of iodine, red bark with nitric

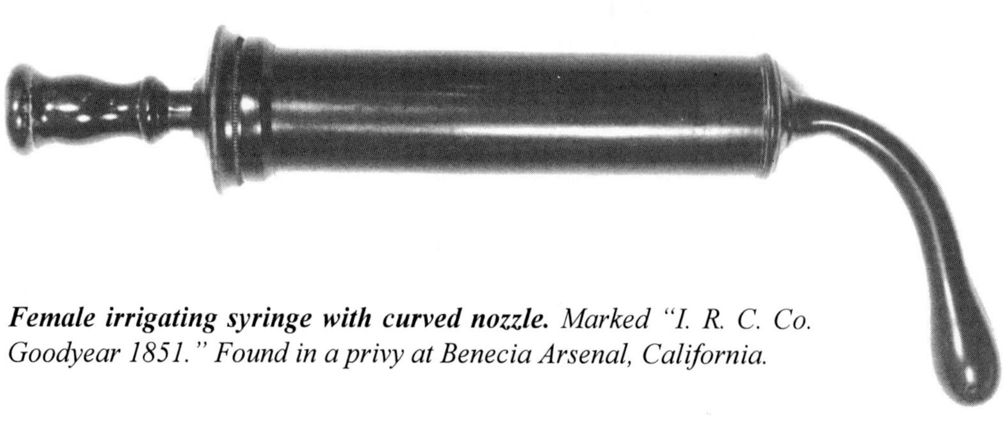

Female irrigating syringe with curved nozzle. *Marked "I. R. C. Co. Goodyear 1851." Found in a privy at Benecia Arsenal, California.*

acid, and chlorate of potash. In some cases, larger volumes were required and common pump syringes were employed. Dr. Durkee's opinion was that the solutions were strong as, "...the vagina will tolerate an injection of higher strength than would be proper for the male urethra."[57]

For centuries, condoms made of animal intestines were used for the control of diseases but were difficult to make, expensive to buy, and did not fit very well. The discovery of vulcanization led to the manufacture of an improved crepe rubber condom that was in widespread use by the 1870s.[58] An 1867 advertising booklet offered "Indian Rubber Safes...a perfect shield against disease or conception...$4 per dozen."[59]

Hypodermic Syringes

When one mentions syringes, the hypodermic syringe is generally the first to come to mind. These syringes made their appearance shortly before the Civil War. In 1853, Alexander Wood of Edinburgh, Scotland used a syringe-like instrument to inject a narcotic to treat neuralgia. Local anesthesia by injection was first used around 1858.[60] Early hypodermic syringes were made of metal, hard rubber, or glass with metal fittings. In recent years, at least one syringe with a glass barrel was listed as having been found at a Civil War site, although the all-glass syringe was not introduced until 1896.[61]

Syringe bottle for medicinal agents. *This combination medicine bottle, hard rubber syringe, and screw-on cap was patented by John Stull on March 25, 1856 (U.S. Patent #14,524). (Courtesy of Janet and Bedford Hayes)*

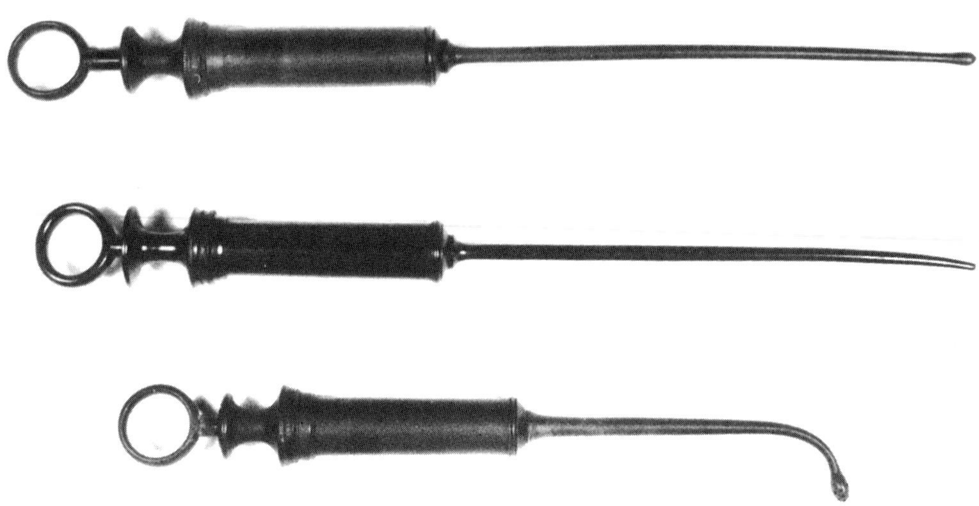

Intrauterine irrigating syringes. *(Top) Hard rubber intrauterine irrigating syringe with long, straight nozzle and one outlet hole. Marked "I. R. Comb Co. Goodyear 1851." (Center) Hard rubber intrauterine syringe with 7" long pointed tip with slight curve and one hole at end. Marked "I. R. C. Co. Goodyear 1851." (Bottom) Hard rubber intrauterine syringe with long, curved nozzle and several outlet holes. Marked "I. R. Comb Co. Goodyear 1851."*

Richmond Daily Dispatch of October 20, 1863 advertised "600 dozen India-rubber fine combs" among the cargo of the steamers *Ella* and *Annie*.

Two unusual combs were patented in 1865 by G. F. J. Colburn, the first a folding comb with graduations to use as a rule (U.S. Patent #45,909). The second was contained in a case that slipped on as a handle (U.S. Patent #46,082).

Drinking Cups

Tumblers of hard rubber were manufactured in the 1850s, but the most popular surviving examples of cups are collapsible hard rubber cups.[5] Pewter and silver collapsible cups were popular long before the appearance of hard rubber examples of the 1850s. Like early hard rubber collapsible cups, these were usually carried in leather or metal cases. Peter Niles of Boston patented a significant improvement in 1860, which added a removable cover and an attached base to prevent the cylindrical rings from slipping apart (U.S. Patent #28,597).

Hard rubber collapsible cup and tole tin case. *The tin lid is stamped "India Rubber Telescopic Tumbler made by The India Rubber Comb Co." The top section of the cup is marked: "***I. R. C. Co. Goodyear***1851***." In this early style of collapsible cup, nothing held the concentric sections together, so they could slip apart and easily be lost.*

Niles hard rubber collapsible cup. *The hard rubber cover is marked: "Niles Drinking Cup-Patented June 5th 1860." Top section of cup is marked: "***Goodyear's Patent May 6, 1851***." Peter Niles' improved cup added a base that kept the sections together and a self-contained lid, eliminating the need for a separate case. His detailed patent drawings are illustrated at right.*

P. H. NILES.
TELESCOPIC DRINKING CUP.

No. 28,597. PATENTED JUNE 5, 1860.

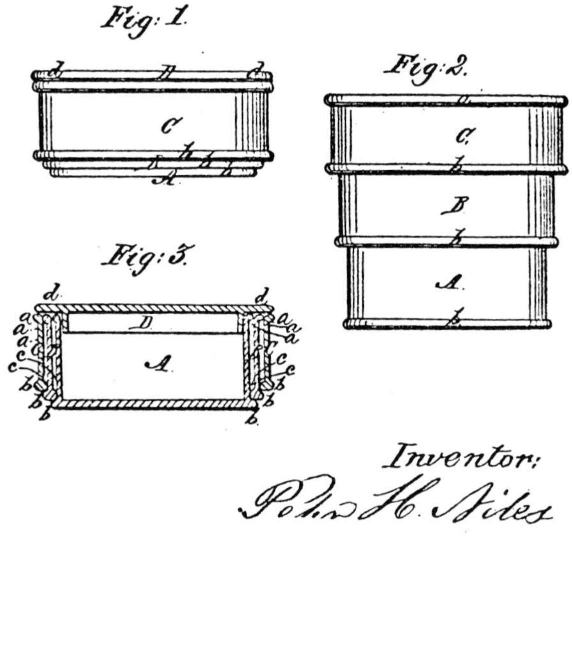

Hard rubber locket with raised 5th Corps design. *Loved one's photograph inside. (Courtesy of Art Marchand).*

Jewelry

In 1866, Goodyear biographer Reverend Bradford K. Pierce commented on the "rich jewelry" that was part of the inventor's display at the 1855 Exposition Universelle in Paris. This jewelry gained in popularity, as indicated by Pierce's reference to "...the immense trade that has since arisen in India rubber jewelry."[6] These articles included bracelets, watch chains, pendants, brooches, and lockets. Increasingly popular "mourning jewelry" was sometimes made of hard rubber.[7]

Patriotic and advertising tokens, some bearing a calendar for the year, are also found in hard rubber. The popular hard rubber buttons of the era were frequently carved into rings and other jewelry items. Soldiers and prisoners passed the time in this avocation. In a letter to Dr. J. H. Douglas, Associate Secretary of the Sanitary Commission, regarding the treatment of Confederate prisoners at Point Lookout Maryland, W. F. Swalm noted:

> Others employ themselves in making rings, chains, seals, &c., from bone and gutta-percha, and notwithstanding the complaint that they do not get enough to eat, you will find them on the main street, which they call The Change, gambling both for money and rations.[8]

Hard rubber rings. *Two feature silver inlays, one with initials, the other displaying symbols of the 3rd and 5th Army Corps. The other three rings appear finely made. Similar rings, some carved from hard rubber buttons, have been found in Civil War campsites and prisons.*

Hard rubber bracelet *with hinge and locking clasp.*

Military soap box ***showing a proud eagle holding a razor in its talons and a banner in its beak that reads "Morning Exercise." Inside the cover is marked: "Manufactured by the Novelty Rubber Co. New Brunswick New Jersey under Goodyear's Patent May 6, 1851." The author has been told of a matching straight razor with the same eagle motif, which once accompanied the soap box.***

Shaving Accessories

Many straight razors were encased in materials such as wood, bone, and ivory. While some cases or handles are made of hard rubber, most of this type are actually a black composition similar to the popular daguerreotype cases.[9]

Shaving soap boxes or cases made it convenient for travelers and military personnel to tend to their facial appearance. These containers were made to hold and preserve the round cakes of shaving soap that were lathered up by a shaving brush. Charles Goodyear recommended soap boxes, or shaving boxes, of hard rubber, "...for the reason that it is not injured by hot water, will not warp like wood or corrode like metal, will not break like china or glass, and admits of the box being made water-tight by screwing on the lid."[10]

Some shaving soap boxes contained a mirror on the underside of the lid and a brass ring for hanging it from a nail or branch. Shaving soap boxes with military motifs were available for soldiers and sailors.

Hard rubber shaving soap box with floral design. *The mirror in this box is missing. Marked: "Manufactured by the Novelty Rubber Co., New Brunswick, N. J. under Goodyear's Patent May 6, 1851."*

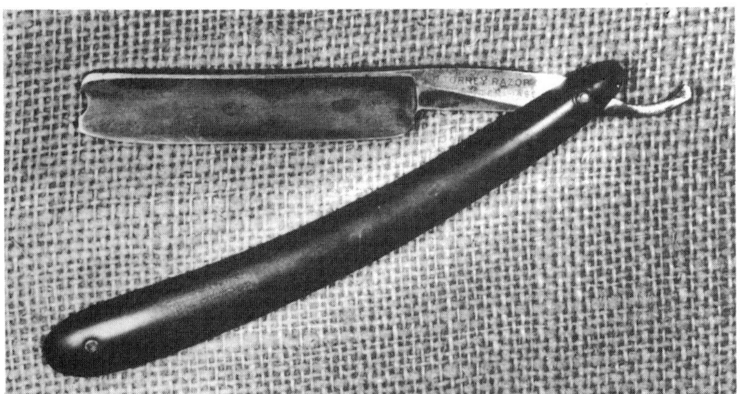

Straight razor with hard rubber handle*, a style that enjoyed wide use by soldiers in the Civil War. Note the marking of "Torrey Razor-London" on the blade.*

Navy shaving soap box with original soap. *Removable cover has ring for hanging and original mirror on inside. The top has a design of an anchor and "U S N - Don't Give Up The Ship." Marked on the base and inside cover: "Manufactured by the Novelty Rubber Co. New Brunswick, New Jersey under Goodyear's Patent May 6, 1851." The shaving soap inside has an imprinted starburst design similar to the lid of the soap box and the words "Naval Soap" in the middle. Also imprinted, partially smeared and illegible, is the soap manufacturer's name, "J. C. Hull & Son - New York."*

Decorative hard rubber shaving mirror, *an attractive lightweight style which could be carried by civilian and soldier alike. Note the ring for hanging from a nail or hook.*

Hard rubber cigar case. *Holds 4 cigars. Accompanied by a museum tag that reads: "Looted on southern battlefield during Civil War. Brought north by Lieutenant John T. Blake of Battery B R. I. Artillery." Case is marked: "Novelty Rubber Co. Goodyear's Patent May 6, 1851."*

Hard rubber Wellington-style pipe *with screw-on briar bowl.*

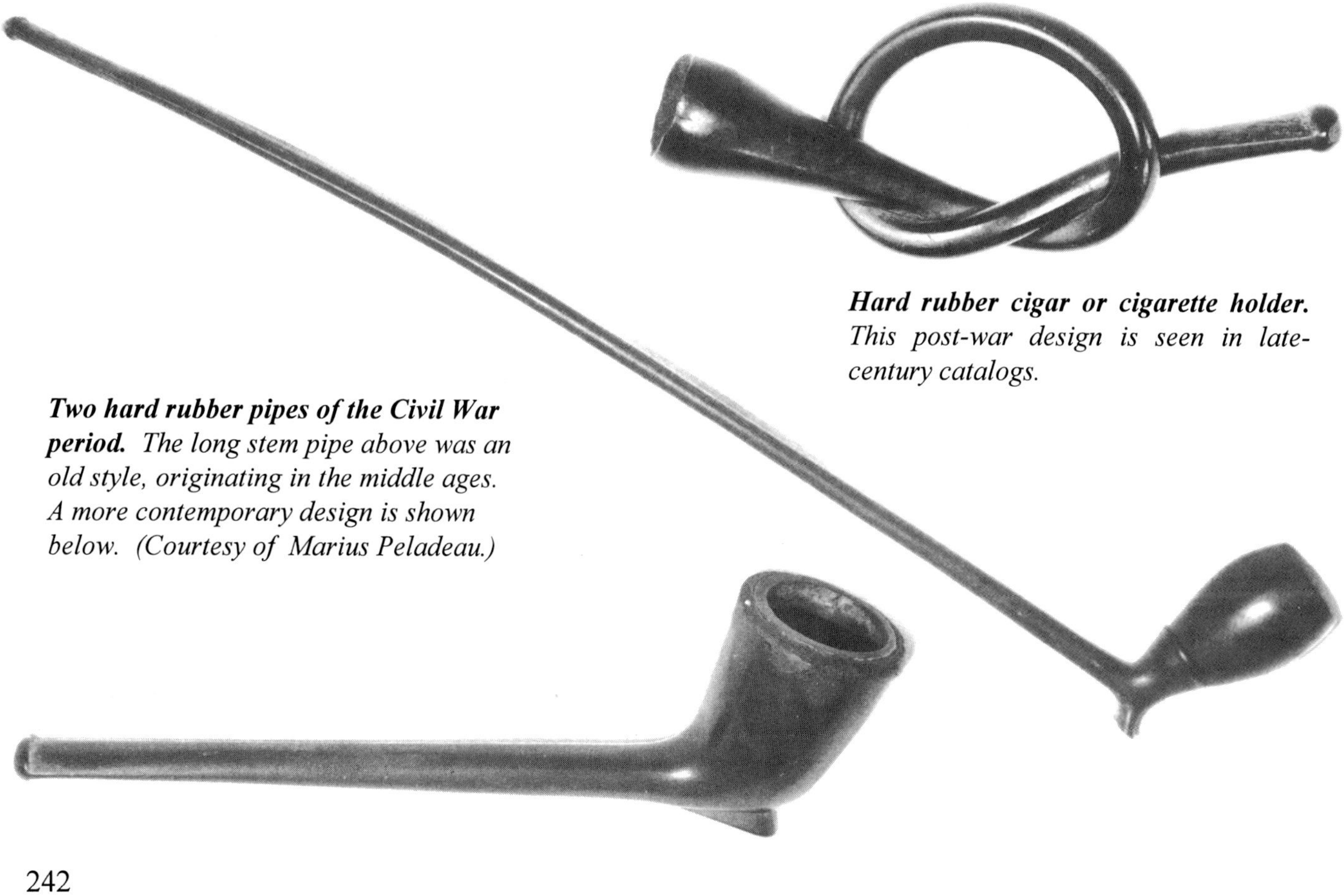

Hard rubber cigar or cigarette holder. *This post-war design is seen in late-century catalogs.*

Two hard rubber pipes of the Civil War period. *The long stem pipe above was an old style, originating in the middle ages. A more contemporary design is shown below. (Courtesy of Marius Peladeau.)*

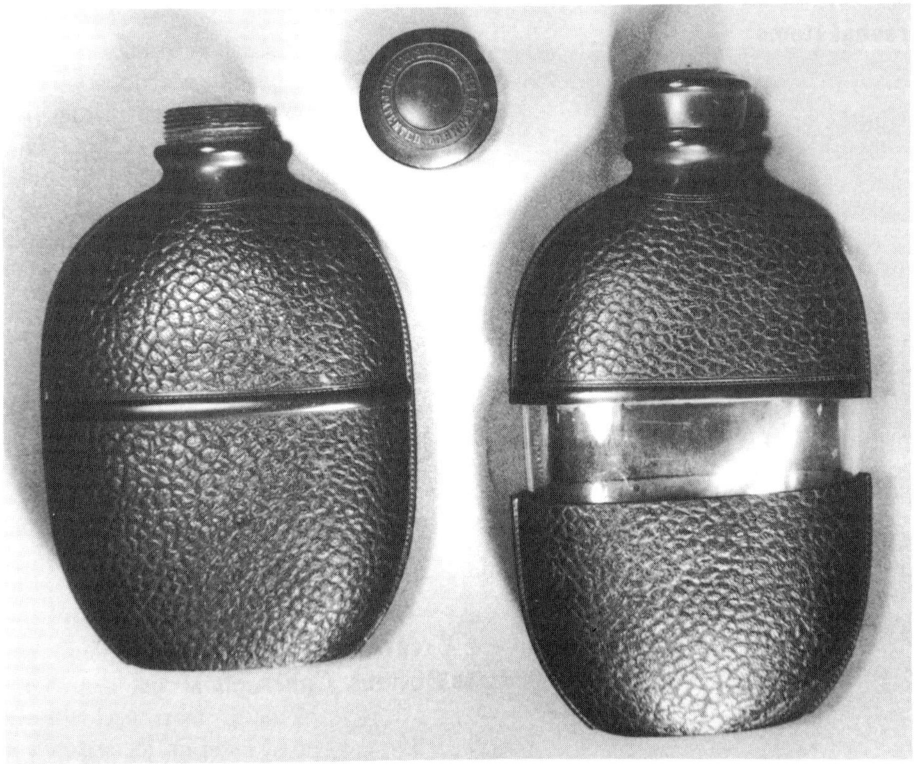

Hard rubber ½ pint whiskey flasks *with simulated grained leather body and screw-on cap. One style is one-piece and all hard rubber. The other is a glass advertising flask with a molded hard rubber cover and removable cup. The top of each cap is marked: "The I. R. Comb Co. Goodyear 1851 Patented." The advertisement in the base likely indicates that this flask is a post-Civil War item.*

Whiskey Flasks

A variety of hard rubber whiskey flasks with screw-off caps were manufactured by the India Rubber Comb Company, including ½ pint flasks with a grained finish to give the appearance and feel of leather. One style was a glass flask with a hard rubber cover and removable slip-off cup. Another flask looked the same, but was all molded hard rubber and had no removable cup.

Larger glass flasks were covered with a smooth-finish hard rubber and had the removable cup.

Miscellaneous

Other noted articles made from hard rubber and gutta-percha include folding fans, corkscrews, dice, dice cups, penholders, and inkwells.

balls, parlor balls, and "foot balls," which were round at that time, were also popular. A humorous event is alleged to have taken place during the 1855 Exposition Universelle in Paris, where the Goodyear display included a pile of round vulcanized foot balls stacked as in an artillery yard. Louis Napoleon, while touring the display, asked what they expected to do with India-rubber cannon balls. These foot balls were likely predecessors to the first round leather soccer ball with an inflatable India-rubber bladder, introduced in 1862.[9] Gutta-percha made it into the "Football Association Rules of 1863," as rule number 13 stated, "No player shall be allowed to wear projecting nails, iron plates, or gutta-percha on the soles or heels of his boots."[10]

Golf balls of gutta-percha, sometimes mixed with unvulcanized India-rubber and other substances, were introduced in 1848 as an alternative to the expensive "featherie" ball, made of stitched cowhide stuffed with boiled goose feathers.[11] At first, the smooth, molded, round ball was an aerodynamic disaster, skipping erratically across the ground. But, after being played a bit, the nicked balls became more controllable and took to the air very well. Around 1860, ball makers began using chisels to impart a pattern on the "gutty" and in 1875 the molds themselves were designed to produce a patterned ball.

India-rubber skate rollers*, as advertised* in Harper's Weekly *of July 12, 1862. Note similarity to modern "in-line" skates.*

The gutty remained a popular ball, even after the introduction of the rubber-cored ball in 1898.[12] Hand-patterned gutta-percha balls are extremely rare and much sought after by collectors of golf memorabilia.[13]

(Also see **Section 1**.)

NOTES - Toys and Sporting Goods

[1] Institution of the Rubber Industry, *History of the Rubber Industry* (Cambridge, England: W. Heffer, 1952) p. 250.
[2] Ralph F. Wolf, *India Rubber Man* (Caldwell, Idaho: The Caxton Printers, Ltd., 1939) pp. 41-42; *History of the Rubber Industry,* p. 250.
[3] *History of the Rubber Industry,* p. 250.
[4] Charles Goodyear, *Gum-Elastic and Its Varieties* (New Haven: Goodyear, 1853 & 1855) Vol. II, p. 192.
[5] *History of the Rubber Industry,* p. 251.
[6] Constance Eileen King, *The Collector's History of Dolls* (New York: Bonanza Books, 1981) p. 492.
[7] Goodyear, *Gum-Elastic and Its Varieties,* pp. 191-197.
[8] *History of the Rubber Industry,* p. 251.
[9] Tony Brown, *A Short History of Soccer* (http://www.innotts.co.uk/soccerstats/hist1.htm, 1996)
[10] *Ibid.*
[11] "Sports," *Science 81,* (July-August 1981): pp. 90-91. This source also notes that the featherie replaced the original carved boxwood golf balls in the 17th century. The spelling has also been seen as "feathery."
[12] Dale Concannon, *Golfing Bygones* (Buckinghamshire, UK: Shire Publications, Ltd., 1994) pp. 15-18.
[13] Rosemary McKittrick, "Antiques," *Pittsburgh Post-Gazette - Week End,* (26 July 1996).

TRIVIA

"I used to sport a diamond breastpin,
Patent leather boots
Take my bitters at the Carlton,
Coffee at Maschutts
Now the case is slightly altered,
Here we've got to work
On biscuits twice as hard as brickbats,
And gutta percha pork"

From "When This Cruel War is Over! No. 3"

This short section contains an assortment of entertaining applications, ideas, and references relating to India-rubber and gutta-percha. The author trusts that the reader will delight in them as much as he has.

An Unusual Entreaty

The following article appeared in the *Civil War Round Table Newsletter ISSN 0023-0146, Vol. XXVI, No. 5*, May 9, 1983:

> WERE THEY GOING TO STOMP THE YANKEES?
>
> The preceding report by Glenn Kerfoot on usage of rubber in the Civil War calls to mind a story about salty old Henry A. Wise, a Confederate brigadier who, in 1861, was on a recruiting expedition in Western Virginia.
>
> At every crossroads community, Wise would halt briefly and issue this stirring call to arms: "Come join me, and bring a rifle! If you have no rifle, bring a musket; if no musket, a shotgun; if no shotgun, a pistol; if no pistol, a gate hinge. No gate hinge? Then, by God, bring an India rubber shoe. But come!"

A Young Inventor's Reward

The following letter was printed in *The Scientific American* of October 29, 1859

> INDIA RUBBER SOLVENT
>
> Messrs. Editors:
>
> I was somewhat amused by reading an article on the above-named subject, in your paper of Oct. 8th. Mr. S. W. Ells cannot be posted, though he would have been if he had had my experience, which is this: About 12 years ago I undertook to dissolve some india-rubber in some turpentine, and succeeded very well. The rubber which I tried was a pair of old-fashioned overshoes, and I pretty effectually spoiled them by reducing them to a liquid form. My next operation was to daub the solution with a brush over a pair of fine calf boots, and the consequence was I spoiled them also; for it took them so long to dry, that the dust collected on them and could not be removed. To pay me for my trouble, I received a "most glorious thrashing" from my father, and thus ended my experiments in the india-rubber line.
>
> T. Middleton
> Chicago, Ill., Oct. 18, 1859[1]

HEARING TRUMPETS.

Hearing trumpets formed of vulcanized india rubber and made to fasten on a suitable cap, as represented in the annexed figure, might be used by military scouts for the purpose of detecting distant sounds unappreciable by the naked ear. It is of great importance to detect military movements made during night; hence soldiers who are quick in hearing should always be chosen to perform picket duty during hours of darkness, when sudden attacks or secret movements of the foe are expected. It is well known that some animals, such as dogs, can hear distant sounds when the human ear is at fault; it was therefore customary for Roman soldiers on guard in an enemy's country, to use a watch dog that was quick of ear so as to give warning when a foe was approaching with stealthy footsteps. It does not appear unreasonable that art may enable men to rival the most sensitive-hearing creature by such a device as is here represented.

Hearing Trumpets for Military Scouts

From The Scientific American *of February 1, 1862, this India-rubber invention would allow the wearer to "rival the most sensitive-hearing creature" in detecting enemy movements. No evidence of its wartime use.*

1,144 rubber water beds during the course of the War.[5]

While water beds were still being sold in the late 1800s, their popularity diminished for some time in this century.[6] In his 1939 biography of Charles Goodyear, rubber historian Ralph Wolf wrote:

It is surprising to find that almost every article mentioned in the inventor's chapter on medical and surgical devices is in common use today. An exception is the water bed, which consisted of a box filled with water and covered over with a rubberized cloth, supposedly very comfortable for the sick and lame.[7]

(Also see **Medical Applications**.)

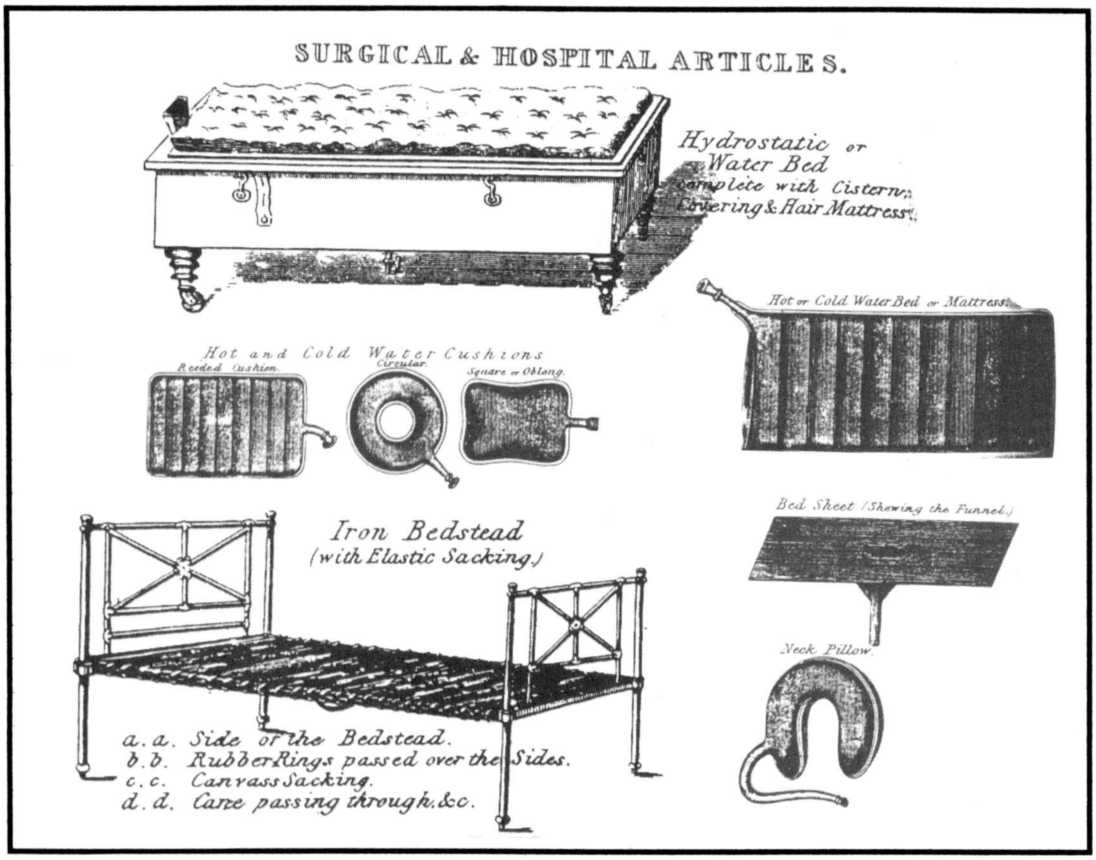

Water beds and accessories. *These engravings show an ornate "hydrostatic bed" and an iron bed frame with flexible rubber supports. Illustrated in Thomas Hancock's* Personal Narrative, *published in 1857, are various rubber sheets, pillows and mattresses intended for hospital use.*

NOTES - Water Beds

[1] Bradford K. Pierce, Trials of an Inventor: Life and Discoveries of Charles Goodyear (New York: Carlton & Porter, 1866) p. 206.
[2] *Dodd's Curiosities of Industry* (Published in England circa 1851) p. 8.
[3] Adolph C. Regli, *Rubber's Goodyear* (New York: Julian Messner, Inc., 1941) p. 194.
[4] Charles Goodyear, *Gum-Elastic and Its Varieties* (New Haven: Goodyear, 1853 & 1855) Vol. II, p. 168.
[5] Surgeon General, U.S. Army, *Medical and Surgical History of the War of the Rebellion* (Washington: Government Printing Office, 1883) Surgical Volume, Part III, p. 996.
[6] *Goodyear's Manual*, Catalog of goods issued by T. G. Thorne of the Brooklyn Depot for Goodyear's Rubber Goods, 1878, p. 23. Copy in author's possession.
[7] Ralph F. Wolf, *India Rubber Man* (Caldwell, Idaho: The Caxton Printers, Ltd., 1939) p. 28.

WHIPS

"...very nice whips do they make out of India rubber."

Quoted from the *Newark Daily Advertiser,* December 26, 1855

Team whips were made by the Newark India-Rubber Company and were thought to be somewhat more durable than leather ones, as they were not damaged by water. An article in the December 26, 1855 *Newark Daily Advertiser* made mention of this company and their product:

> In the whip shop they are making at the present time from 50,000 to 60,000 dozens a year, and very nice whips do they make out of India rubber, it only requiring the little wooden back bone on which to gum the rubber to make as durable and good a whip a wagoner ever cracked.[1]

Artificial gum-elastic ivory (i.e. hard rubber) was suggested by Goodyear as a suitable material for riding whips, coach whips, and switches, as well as for the mountings of team whips.

NOTES - Whips

[1] *Newark Daily Advertiser*, 26 Dec. 1855.

WRITING ACCESSORIES

"Elastic bands for papers, letters, etc., inkstands with elastic bottle, writing tablets, book covers, erasing rubber in squares and bottles."

Stationery articles listed by Thomas Hancock in his *Personal Narrative,* 1857

India-rubber was sold by stationers in the late 18th century to erase or "rub out" mistakes. This application led to the widespread acceptance of the name "rubber." On March 30, 1858, Hymen Lipman patented the combined lead pencil and eraser combination (U.S. Patent #19,783) that we still use today. A combined pencil, eraser, and stamp was patented by Eberhard Faber on June 16, 1863 (U.S. Patent #38,892) and a penholder and ink eraser by D. E. Holmes on May 17, 1864 (U.S. Patent #42,769).

India-rubber was also used extensively for rubber bands and elastic parts of improved inkwells. One improved inkwell was created by merely placing a rubber band around an inkstand to hold pens when not in use (U.S. Patent #33,160). Some inventions alleviated the strain and fatigue of continuous writing by providing support to the hands and fingers (U.S. Patents #34,983 and 41,495).

James Dorr of New York suggested hard rubber pen tips as improvements over quill and steel tips in the early 1850s.[1] Hard rubber writing accessories, such as penholders, inkstands, portable inkwells, rulers, and mechanical pencils, are frequently found among older personal belongings and soldier's effects. It is interesting that, while penholders of the era were long—like the quill pens

Hard rubber inkwell *showing distinctive markings of the India Rubber Comb Company, a leading producer of hard rubber items from the 1850s to 1898.*

***Hard rubber chart rules,** forerunners to modern-day rulers but made without numbers or graduations. Different company names appear, such as "I. R. C. Co.," "American Hard Rubber Company," and "Poppenhusen & Konig." Patent markings include "Goodyear's Patent May 6, 1851," and the rare early marking "Goodyear's Patent 1849-51." The close-up of the markings reveals the name "Konig," while printed references use the spelling "Koenig."*

that preceded them—early mechanical pencils were very short and thin.[2] Rulers or "chart rules" of the period contained no graduations of inches or other measurement tools.[3] While some patents for "fountain pens" existed as early as 1858 (U.S. Patent #20,741 dated June 29, 1858), hard rubber fountain pens examined by the author are, in his opinion, products of the 1880s.[4]

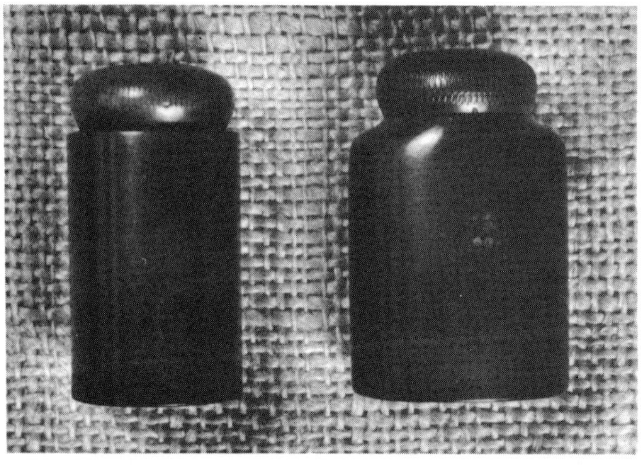

***Hard rubber traveling bottles** with screw-on caps. The oval bottle is marked "xxxGoodyear's Patent May 6, 1851xxx" and "U.S. 1861."*

Some India-rubber and gutta-percha stationery supplies were utilized by the military. The following entries are taken from "Contingent Expenses of the War Department for 1864" from the *Reports of the Secretary of War*:

EXPENDITURES:

Blanchard & Mohun (Stationers)

Gutta Percha Rulers .75 each

3 Gutta Percha Penholders-$ 1.20

2 Pounds India Rubber @ $ 1.00 per pound

1 Doz. Pieces India Rubber $ 1.00

1 Gutta Percha Inkstand .75

1 Long Gutta Percha Ruler $ 1.00

1 Gutta Percha Penholder .40

Philip & Solomons:

3 ½ Gross Rubber Bands @ 9.80[5]

(Also see **Sections 1 - 4** and **Section 6.**)

***Improved inkwell** patented by R. Gleason, Jr. on March 18, 1856 (U.S. Patent #14,451). The inkwell is hard rubber with an India-rubber washer and valve that allowed the user to pump ink to the upper reservoir. Marked on base: "Manufactured by L. Brandau & Co. under Goodyear's Patent May 6, 1851 - Patented January 16, 1856." The date of January 16 is likely the date that Gleason applied for the patent.*

***Barrel-shaped hard rubber sand shaker** marked "Novelty Rubber Co. Goodyear's Patent May 6, 1851." (Courtesy of Marius Peladeau)*

***Hourglass shaped sand shaker,** with "Novelty Rubber Co., Goodyear's Patent May 6, 1851" clearly marked on the base.*

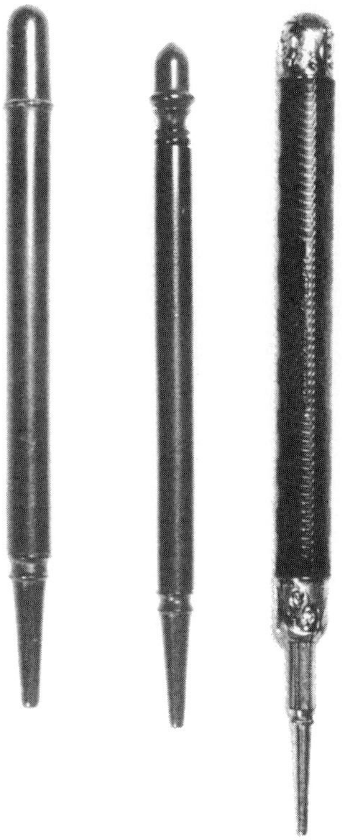

Hard rubber traveling inkwells *with screw-on caps. Like many other hard rubber items, these inkwells have no patent dates and may be post-Civil War.*

Hard rubber mechanical pencils. *The pencil on the left was excavated at the camp of the 11th Corps near Stafford Court House, Va. Marked: "A. G. Day's Patent Aug. 10, 1858." August Goodyear Day, a nephew and apprentice of Charles Goodyear, later patented his own hard rubber process (U.S. Patent #21,222). The pencil in the center is marked "Goodyear's Patent May 6, '51." The pencil on the right, with gold accents, is marked "Mabie's Patent Oct. 3, 1854" (U.S. Patent # 11,762).*

Unusual hard rubber inkwell *with pivoting closure top. Marked: "*I. R. C. Co. *Goodyear's 1851*."*

Attractive post-Civil War desk set *with two inkwells for different color inks. Marked "Novelty Rubber Company," this set is made entirely of hard rubber. The penholder, also pictured below, was found in a Civil War period portable desk. It is marked "Goodyear's Patent May 6, 1851."*

W. Burnet.
Inkstand.
No. 25,175. Patented Aug. 23, 1859.

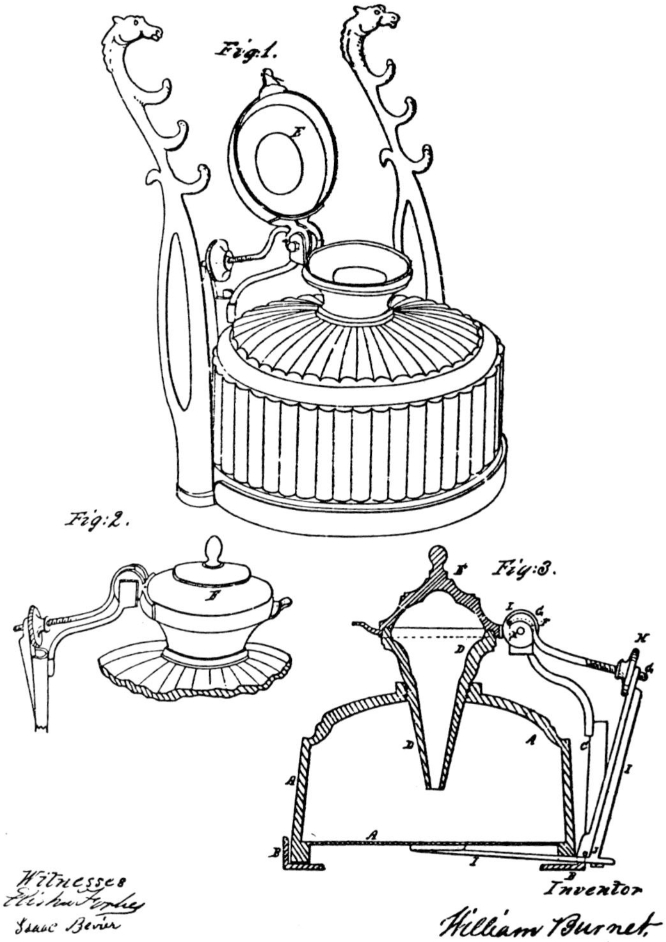

***India-rubber inkwell reservoir**, recovered in the area of Benecia Arsenal near San Francisco in the 1970s. The patented design included a lever that racheted under the flexible India-rubber reservoir base and forced ink to a glass well at the top. Marked: "Burnet Patent Pat'd Aug. 23, 1859 - Manuf'd. under Goodyear's Patent 1844." Though buried over one-hundred years, the India-rubber is still soft and pliable.*

***Patent drawing of Burnet's inkwell,** patented August 23, 1859 (U.S. Patent #25,175), showing a proposed elaborate base and frame with the mechanism for forcing the ink up to the top of the reservoir.*

NOTES - Writing Accessories

---

[1] Charles Goodyear, *Gum-Elastic and Its Varieties* (New Haven: Charles Goodyear, 1853 & 1855) Vol. II, p. 41.
[2] Author's observations of known examples.
[3] *Ibid.*
[4] Based on lack of "Goodyear 1851" markings and later manufacturer's markings, patent dates, etc.
[5] "Expenditures from the Contingent Fund of the War Department" Letter March 2, 1865, War Department, Washington, City to Speaker of the House of representatives, 38th Congress 2d Session, House Exe. Doc. No. 85.

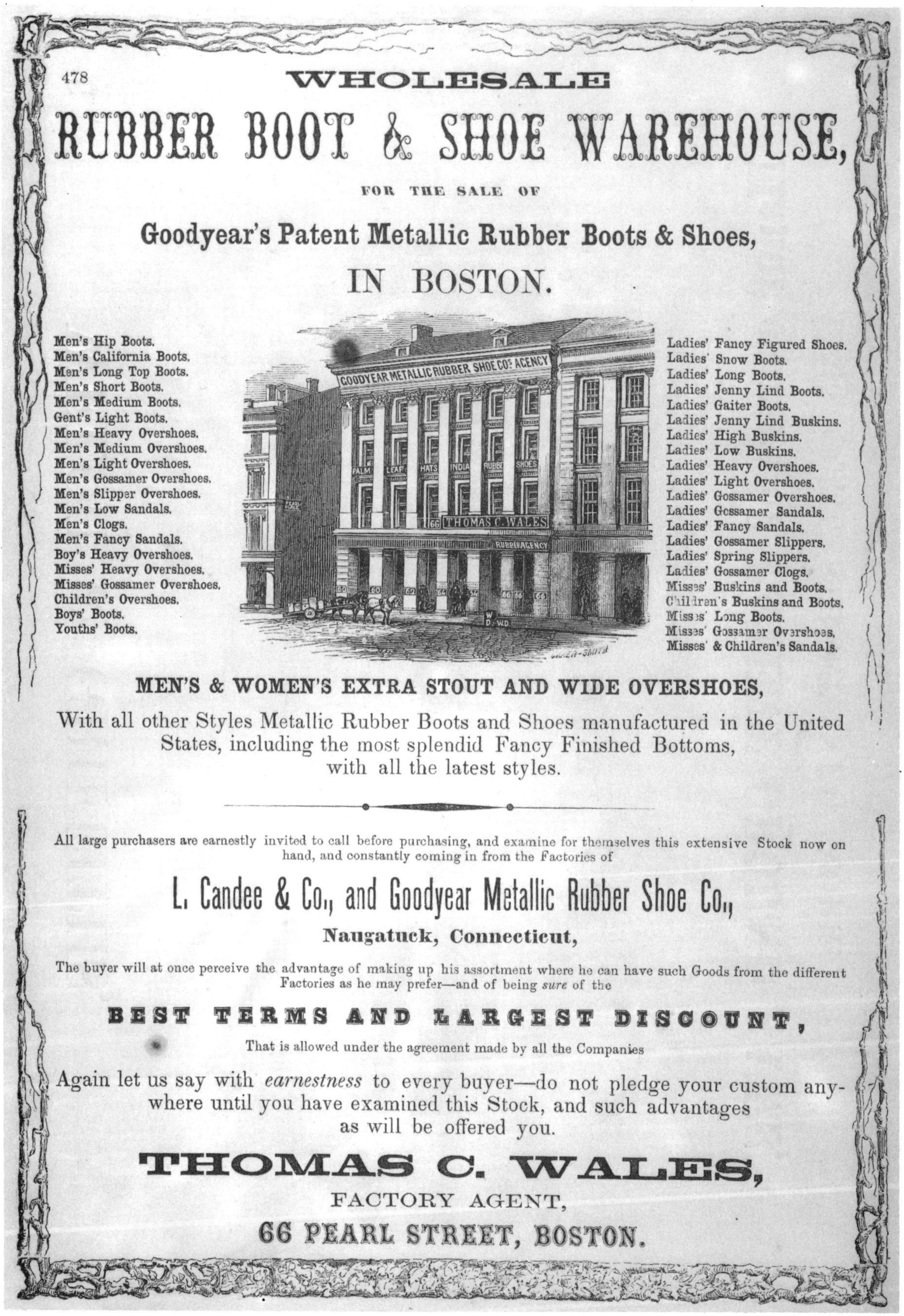

Rubber boots and shoes. *Manufactured by two of the earliest Goodyear licensees, L. Candee & Company and Goodyear Metallic Rubber Shoe Company. The ad is taken from the* Historical Picture Gallery, *published by John R. Chapin in Boston in 1856.*

PART THREE

ARTIFACT IDENTIFICATION AND REFERENCES

Dating, Preservation, Patents, and Manufacturers

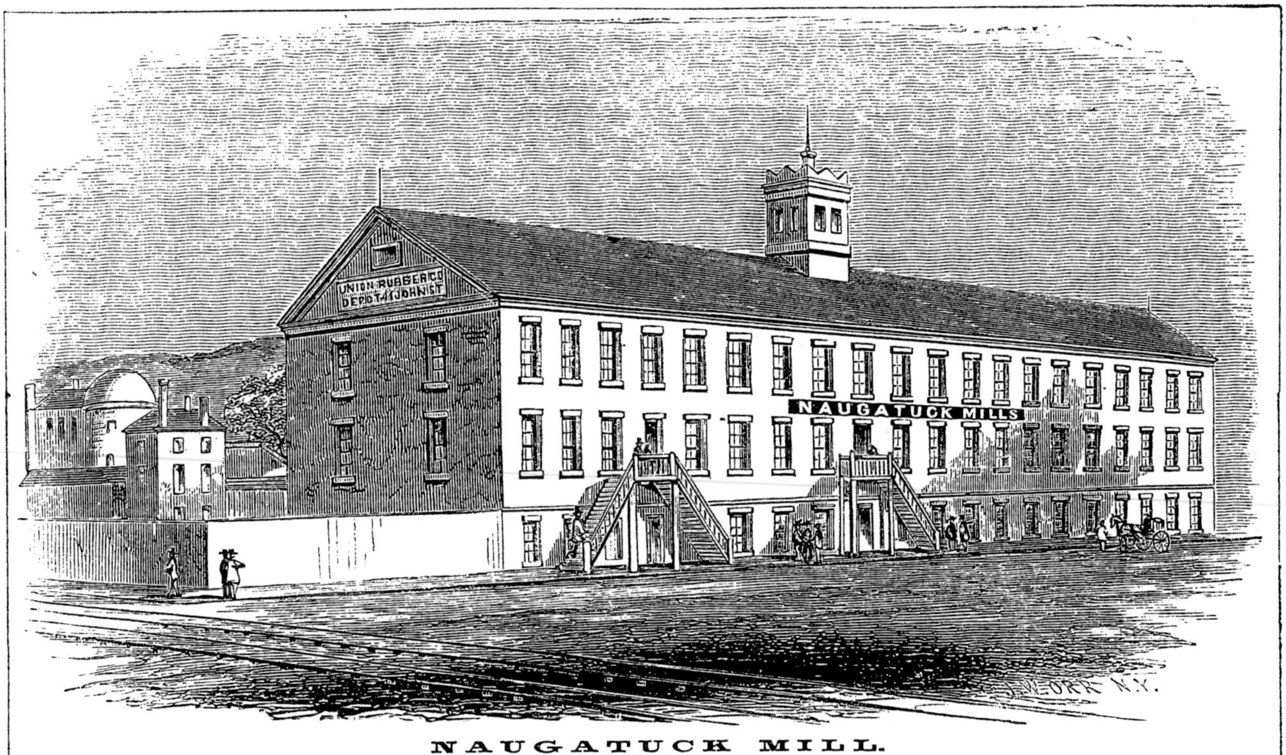

NAUGATUCK MILL.

Illustration shows the mill at Naugatuck, Conn, *owned by the Union India Rubber Company. This firm was licensed by Goodyear and was the primary manufacturer of Army and Navy rubber goods during the mid-19th century period which included the Mexican and Civil Wars.*

SECTION 9

ARTIFACT DATING AND PRESERVATION

"It may be a challenge to establish the exact age of an artifact, but there are some guidelines."

Author's opinion

DATING ARTIFACTS

India-rubber, gutta-percha, and hard rubber items can be difficult to pinpoint to a specific date or period, as the manufacture of many of these items continued to at least the end of the century. Raincoats, ponchos, syringes, combs, canes, and other items known to have been made and used during the Civil War are seen in catalogs and advertisements of the 1880s and 1890s.[1] Some novelty and advertising items of this later period are at times mistakenly represented as being of the Civil War era.[2]

The following guidelines are based on a combination of extensive research, actual documented artifacts, and, to a great extent, the opinion of the author.

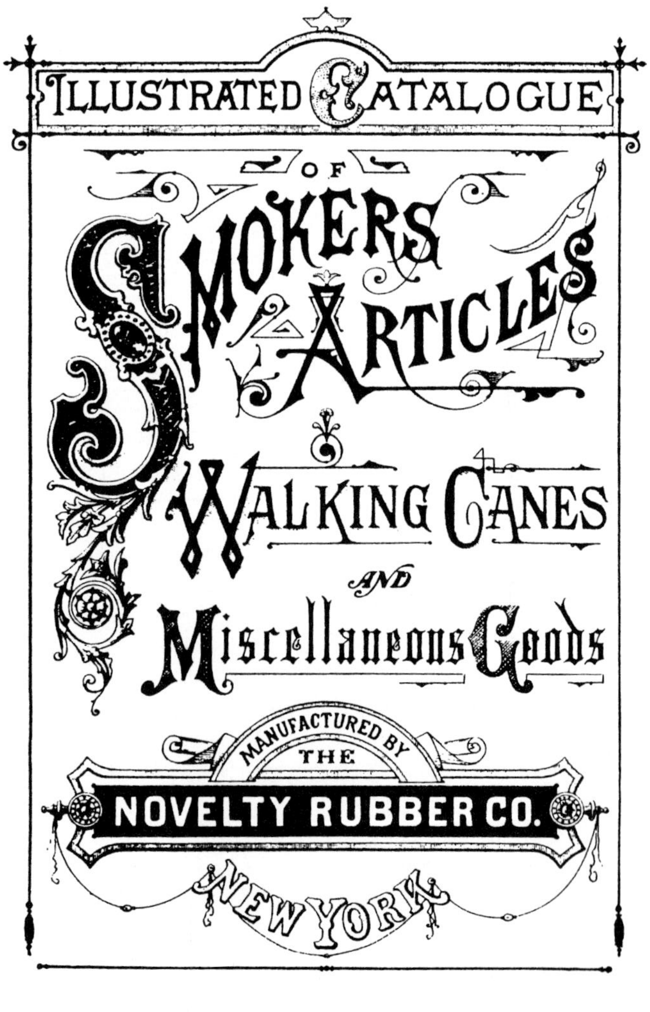

India-rubber and Gutta-percha

Soft India-rubber items made prior to Charles Goodyear's vulcanization patent of June 15, 1844 are almost non-existent. Most were deemed useless due to their inflexibility and were discarded or destroyed. Since licensees under the vulcanization patent were required to mark their products with the Goodyear name and patent date, items legally manufactured in this country are so marked. Several Goodyear patent dates are found on surviving artifacts, including all or part of the June 15, 1844 date of the original patent (U. S. Patent #3,633), the December 25, 1849 date of the first reissue, and the June 15, 1858 date of the patent extension. Most items manufactured in the Civil War years bear the 1849 and 1858 dates, along with the manufacturer's name. Dates of other Charles Goodyear patents, such as his 1848 patent for making hollow rubber balls and toys (U.S. Patent #5,536 dated April 25, 1848) are also seen on some items.

Some articles made by the Rubber Clothing Company are stamped with several 1858 and 1859 patent dates for recycled or restored waste rubber that appear to have been assigned or transferred from the Beverly Rubber Company.[3]

Items without patent dates may have been imported or produced illegally by manufacturers such as Horace Day, although at least one known item bears his name.[4] It is also possible that the markings, molded into items like boot and shoe soles or stamped in indelible red or black ink on the canvas side of clothing, blankets, etc., have worn away from use. Because of public opinion and sentiment against the Goodyear family, it is unlikely that the patent markings were used on many India-rubber items produced after the patent expired on June 15, 1865.[5]

In most cases, manufacturer's markings alone may not be good indicators of a specific date or period, as many leading companies of the 1850s and 1860s were in business long after the Civil War. Trademarks were issued to Goodyear's India Rubber Glove Manufacturing Company in 1879, Boston Rubber Shoe Company in 1894, and the Rubber Clothing Company in 1906. The Union India Rubber Company was issued a trademark for the label "Goodyear Storm Coat" in 1879.[6]

The Goodyear Tire & Rubber Company, incorporated in 1898 and named to honor Charles Goodyear, had no connection to any previous manufacturer at that time. In the 1970s, the modern-day company acquired the trademark rights to the Goodyear Rubber Company, which had been established in 1872.

Some surviving soft gutta-percha items are stamped with Rider and Murphy's patents of 1852 and 1854 (U.S. Patents #8,992 and #11,906), along with the manufacturer's name. Again, it is possible some gutta-percha articles were illegally manufactured or that the markings have worn off.

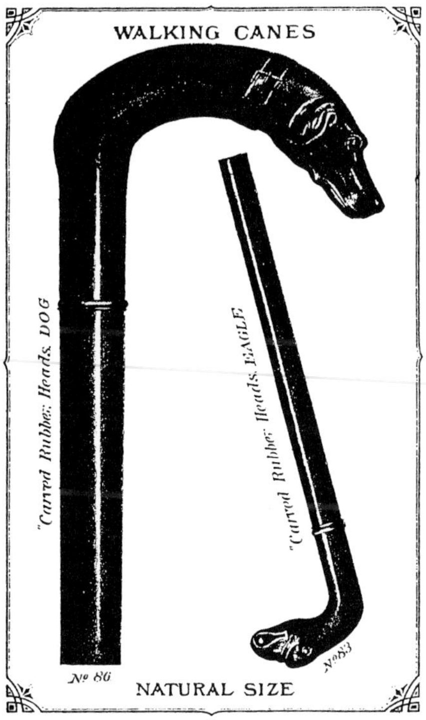

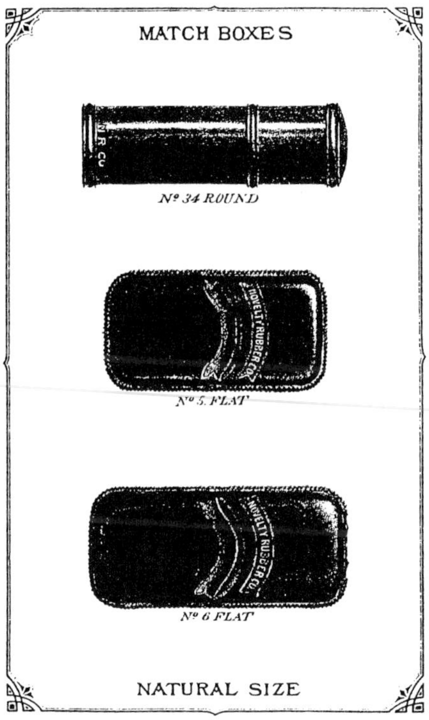

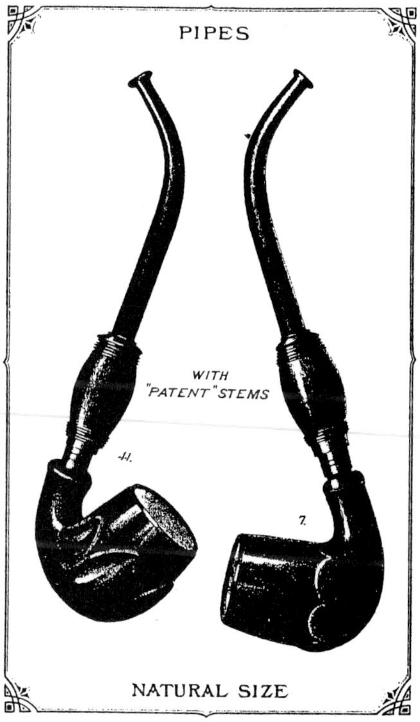

Previous Page and Above: ***Novelty Rubber Company*** *cover and pages from an 1876 catalog of hard rubber canes, match boxes, pipes, and other items. Note that these post-Civil War items are marked with only "Novelty Rubber Company" or "N. R. Co." and do not contain the Goodyear patent markings. The Novelty Rubber Company was a major producer of hard rubber items between 1853 and 1886. (Courtesy Paul Jung)*

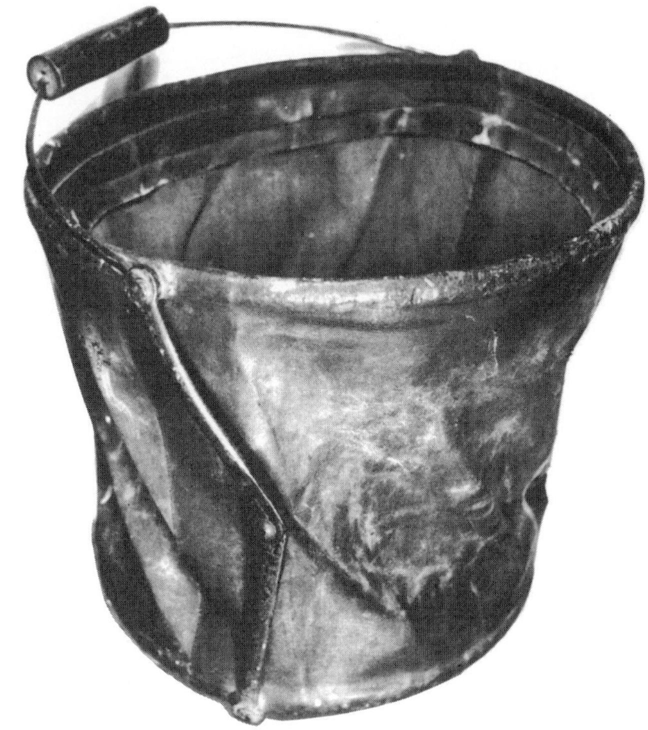

This folding rubber pail, likely a post-Civil War invention, is pictured in the 1878 catalog of the Brooklyn Rubber Depot. There are no patent or manufacturer's markings visible on the pail.

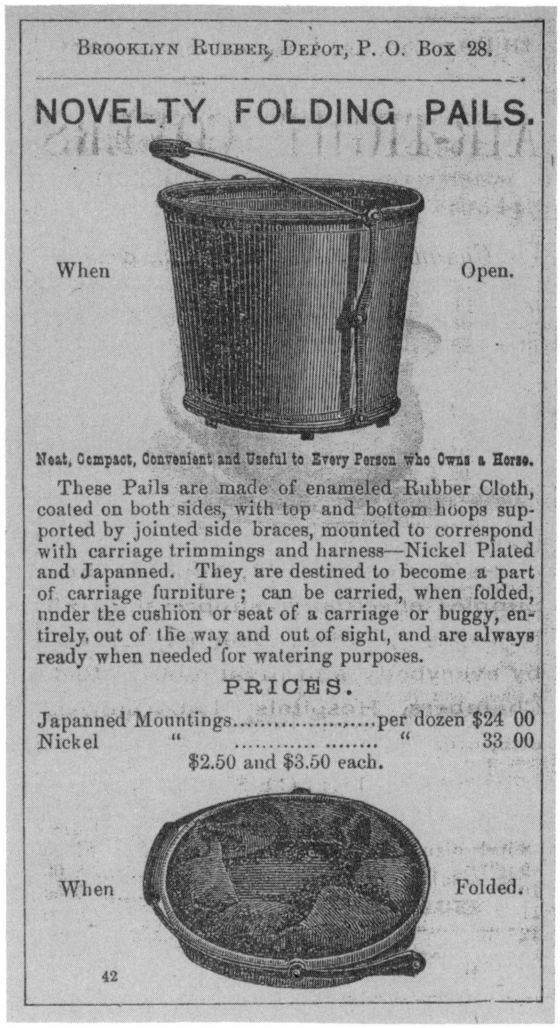

BROOKLYN RUBBER DEPOT, P. O. BOX 28.

NOVELTY FOLDING PAILS.

When Open.

Neat, Compact, Convenient and Useful to Every Person who Owns a Horse.

These Pails are made of enameled Rubber Cloth, coated on both sides, with top and bottom hoops supported by jointed side braces, mounted to correspond with carriage trimmings and harness—Nickel Plated and Japanned. They are destined to become a part of carriage furniture; can be carried, when folded, under the cushion or seat of a carriage or buggy, entirely out of the way and out of sight, and are always ready when needed for watering purposes.

PRICES.

Japanned Mountings	per dozen	$24 00
Nickel "	"	33 00

$2.50 and $3.50 each.

When Folded.

42

The lack of significant details on later gutta-percha military accoutrements or consumer goods in catalogs or advertising indicates that many of these applications likely diminished after the War.

Hard Rubber and Hard Gutta-percha

Early hard rubber artifacts were manufactured under Nelson Goodyear's patent of May 6, 1851. Some items also bear the 1849 date of Goodyear's filing for a caveat for patent. The hard rubber patent was extended on May 6, 1865 and expired on May 6, 1872. Many variations of manufacturer and patent markings are found on hard rubber items made between the 1850s and 1890s. While there was no requirement to mark items with the patent date after May 6, 1872, the India Rubber Comb Company continued to use the "Goodyear 1851" marking after 1872.

Dating of items with "Goodyear 1851" and "Goodyear's Pat. [or P=T] 1851" markings can be troublesome, as similar markings have been noted on Civil War era and post-war artifacts. However, items marked with the full patent date of May 6, 1851 almost certainly date before May 6, 1872 and likely before May 6, 1865.[7]

Some hard rubber whiskey flasks manufactured by the India Rubber Comb Company, as well as knife handles made by other manufacturers, bear the Goodyear name and "1865" date of the patent extension. Other items have a marking "Goodyear's Patent May 6, 1851 Extended." These articles were likely made between the patent extension date of May 6, 1865 and the expiration date of May 6, 1872.

Markings of other hard rubber patents, such as A. G. Day's patent of August 10, 1858, can be found on certain surviving artifacts. Some items may carry more than one patent date. The Niles hard rubber collapsible cups have the name and "June 5, 1860" date associated with Peter Niles' collapsible cup patent boldly molded into the lid and may also

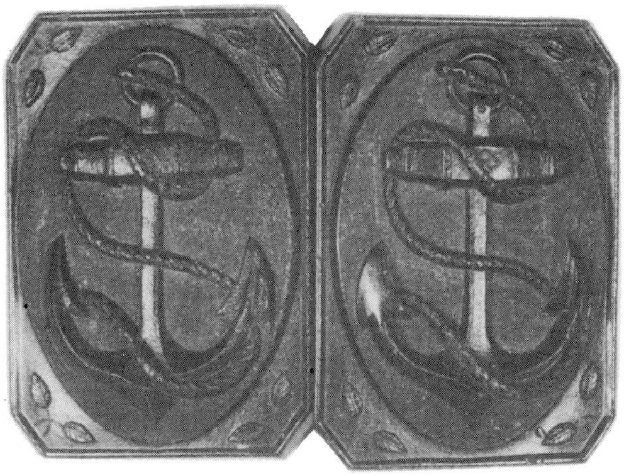

Composition two-piece Navy belt buckle. *There are no markings on this composition Naval-style belt buckle. The material and the stamped brass hooks on the back are similar to buckles made for the Centennial in 1876. (Courtesy Norm Flayderman)*

bear the "Goodyear's Patent May 6, 1851" markings in tiny letters on the top section of the cup.

Many artifacts of the 1850s carry the names of very early manufacturers, such as the A. G. Day Caoutchouc Company, Beacon Dam Company, or Poppenhusen and Koenig Company, the old American Hard Rubber Company (not to be confused with the new American Hard Rubber Company founded in 1898). However, a large number of these early items, generally brown in color and oftentimes without manufacturer's markings, bear only the marking "Goodyear's Patent May 6, 1851," (sometimes preceded and followed by "xxx" or "***"). The letters, tiny and possibly faint from wear, are almost always found around the upper "collar" of the item—be it a syringe, ink bottle, folding cup, or powder flask. The old American Hard Rubber Company is thought to be the maker of these early items without a maker's name.

One of the most frequently encountered names on Civil War period and post-war artifacts is the Novelty Rubber Company (N. R. Co.), founded in 1853. Many Novelty Rubber Company items, such as canes, shaving soap boxes, and Union Navy overcoat buttons, are marked with the company name and "Goodyear's Patent May 6, 1851." Others, like hard rubber eagle buttons and political buttons of the 1872 campaign, are marked "Goodyear's P=T 1851," possibly due to limited space. Articles made by Novelty Rubber Company and other manufacturers' hard rubber products without the Goodyear name and patent date were likely manufactured after the patent expired in 1872. The Novelty Rubber Company continued to make canes, match boxes, pipes, and other articles before ceasing operations in 1886.[8]

Rubber grain bag or navy duffel bag. *Rubber flaps with grommets at top close with the drawstring or stay open to increase the volume of the bag. Marked "Trade Mark - Aetna Rubber Works." Lack of Goodyear patent markings and research on the Aetna Rubber Works indicate that this item is likely from the early 1870s.*

Hard rubber knitting ring marked: *"The Butler Hard Rubber Co. Patented Feb. 15, 1898." Medical syringes and other items are also seen with the Butler company's markings. The Butler Hard Rubber Company was formed in 1882 and was included in the 1898 merger that formed the new American Hard Rubber Company.*

Hard rubber was manufactured in Germany *as early as the 1850s. But this comb box, with "Made in Germany" on the label, more likely held combs made in the 1890s.*

As mentioned earlier, the markings of the India Rubber Comb Company (I. R. C. Co.) can also be found on many items from the Civil War and post-War periods. While some of their earlier products are marked with the company name and "Goodyear's Patent May 6, 1851," many others carry their name and "Goodyear 1851." The latter marking appears to have been used by this company prior to 1865, but it is frequently noted on combs of the post-Civil War period. In fact, the words "Goodyear 1851" became a registered trademark for hard rubber combs made by the India Rubber Comb Company's successor, the new American Hard Rubber Company of New York on December 11, 1906 (U.S. Trademark #58,361). The specification notes that the marking had been in use for at least 10 years.[9]

Articles made by the Butler Hard Rubber Company of Butler, New Jersey and the Goodrich Hard Rubber Company of Akron, Ohio are definitely post-Civil War, despite the occasional appearance of a questionable "Patented 1860" marking on some artifacts.[10] The Goodrich Hard Rubber Company first appeared in the *Akron City and Summit County Directory* in 1888.[11] The publication showed an affiliation with The B. F. Goodrich Company, which had been formed in 1870.[12] The Butler Hard Rubber Company was established in 1882.[13] These two companies merged with the India Rubber Comb Company in 1898 to form the new American Hard Rubber Company.[14]

Hard vulcanized gutta-percha artifacts were also being manufactured through the 1800s, as

Hard rubber G. A. R. souvenir canteen. *1 inch diameter. Soldier motif with "Grand Army of the Republic Veteran 1861-1866" on face. Reverse reads: "26th Encampment GAR 1892 Washington, D. C."*

Hard rubber pipe with screw-on briar bowl. *The hard rubber stock and stem is an octagon-barreled rifle with the inscription "Army and Navy." This and similar pipes patterned after rifles and pistols were likely made in the 1880s.*

A smoking gun resulted *when tobacco was burned in these revolver pipes. Briar bowls screwed onto the hard rubber smoking pipes. (See pipe mold in Section 3.)*

This hard rubber personal treatment kit *includes a hard rubber male gonorrhea syringe, a glass vial with cork stopper, and a hard rubber recepticle with screw-on cap. Absence of patent markings suggests an imported item or production after the 1872 expiration of Goodyear's patent.*

evidenced by product labels and advertisements for "India Rubber and Gutta Percha" articles.[15] While some dates on artifacts represent patents for "...other vulcanizable gums," no patent dates directly linked to hard gutta-percha have been noted by the author on any artifacts.

Reproductions and Post-Civil War Items

Fortunately, this area of collecting has not been deluged with intentional fakes, although reproductions exist and some post-War items are sometimes offered as Civil War items. Collectors should be aware of the following:

Rubber Blankets and Ponchos: Reproductions of rubber blankets and ponchos are sold as kits or as completed articles for re-enactors. Rubber blankets and ponchos were also made for the Spanish American War. Most, but not all, post-Civil War blankets and ponchos have nickel grommets instead of the brass grommets of the Civil War era.

Eagle "Sharpshooter" Buttons: Reproductions were recently introduced for re-enactors. The poor quality makes them easily distinguishable from originals. No backmarks have been observed on the reproductions.

"The Military" comb *with "Goodyear 1851" on reverse, both in gold letters. This is another example of "Goodyear 1851" appearing on a late 19th century item.*

Hard rubber match boxes *advertised in the Novelty Rubber Company's 1876 Illustrated Catalogue of Smokers Articles, Walking Canes and Miscellaneous Goods. These markings date from the Civil War period but were usually accompanied by Goodyear's patent. (Courtesy Paul Jung)*

Navy Buttons: Hard rubber Navy buttons with only a large upright anchor on the face and "A. H. R. Co." on the reverse are late 19th to early 20th century, according to Albert's button book.[16] They would have been made by the new American Hard Rubber Company after 1898.

Navy Combs: A number of original hard rubber combs marked "U.S. Navy" on the face and "Goodyear's Patent 1851" on the reverse were part of the stores on board the sunken U.S.S. *Cairo*. Boat and cargo were recovered in 1964. The Ace Comb Company, a successor to The India Rubber Comb Company and American Hard Rubber Company, made a number of reproductions and distributed them as promotional items. "Ace" hard rubber combs are still made today by Goody Products, Inc.

Questionable Patent Markings: Personal items, syringes, etc. made of hard rubber or compositions sometimes contain a branded or stamped marking, "Patented 1860" or "PAT 1860." Research has turned up no connection between these items and any patent or design patent issued that year. Some of the items with these markings also bear the name of the Butler Hard Rubber Company, which indicates the items were produced after 1882. Any Civil War period patent markings without the name of the patentee or a full patent date that can be traced should be viewed with skepticism.

(Also see **Sections 1-5**.)

Intricate hard rubber watch chain *shown with a hard rubber advertising token. Absence of patent or manufacturer's markings make the watch chain difficult to date. Token advertises "John Thornley's - Philadelphia" and on the reverse, "All kinds of India Rubber and Gutta Percha goods." Believed to date from the 1860s.*

CLEANING & PRESERVING ARTIFACTS

India-rubber and Gutta-percha

While many India-rubber and gutta-percha artifacts of the mid-1800s are still in excellent condition today, some items are very stiff and flaking badly. The problems may be traced to an inferior raw gum, a poor vulcanization process, or an alternative process such as solarization. The deterioration may also be due to the conditions in which the items have been stored.

The aging stability of rubber compounds is influenced by such factors as heat, light, and atmospheric conditions. Tests have shown that exposure to air and ozone—a major component of air pollution—is the biggest cause of deterioration in rubber. When rubber is exposed to ozone, a chemical change similar to rusting in metal occurs, causing the rubber to become dry and brittle.[17] Improvements in resistance to attack by oxygen are accomplished by incorporating antioxidants and antiozonants into modern rubber products.[18]

The author has carefully treated stable, pliable, vulcanized India-rubber goods with a compound such as ArmorAll© Protectant to help preserve their flexibility. Gutta-percha and unvulcanized items, usually found stiff and brittle today, do not respond well to this treatment and may streak from this application.

Based on this limited experience with a few items in good condition, no firm recommendation can made for treatment of surviving artifacts. In the author's opinion, any articles that are flaking badly should be left untreated and stored or displayed in one permanent position. An hour of exposure to the warm sun is often all that is required to soften any rolled or folded stiff articles that must be rearranged for display.

Hard Rubber

Hard rubber's resistance to chemicals and acids was recognized immediately as an advantage over other contemporary materials. Recovered hard rubber relics display the same resistance to sand, soil, and water. Most hard rubber artifacts—even those buried or underwater for almost a century and a half—are still in excellent condition.

Items can be cleaned carefully with soap and water, but may discolor slightly. Polishing with a compound like ArmorAll© Protectant can enhance the appearance of a hard rubber item. But exercise caution! The polished article becomes very slippery and easy to drop.

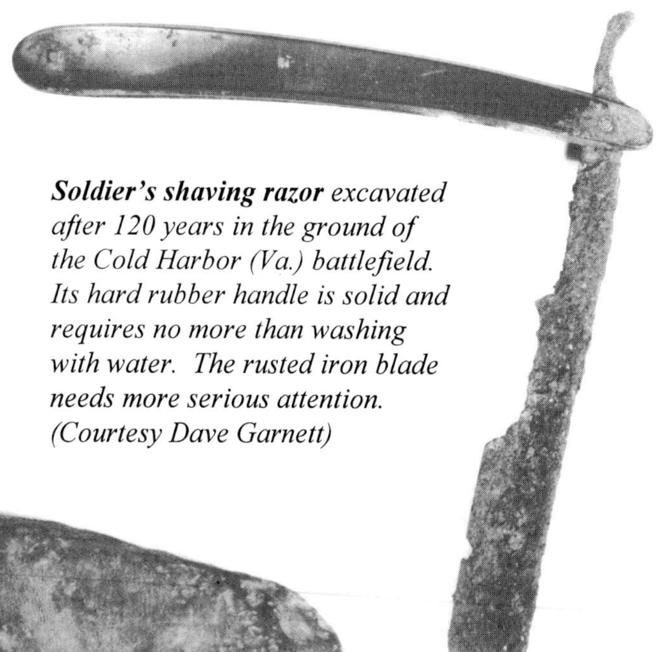

***Soldier's shaving razor** excavated after 120 years in the ground of the Cold Harbor (Va.) battlefield. Its hard rubber handle is solid and requires no more than washing with water. The rusted iron blade needs more serious attention. (Courtesy Dave Garnett)*

***India-rubber overshoe** dating from the 1850s in rough condition as it was found in a pre-Civil War dumpsite along the Hudson River. The rubber retains its flexibility and should have the dirt removed with gentle soap and water. Later, store carefully away from light, and either leave as is, or coat with ArmorAll© Protectant as described above.*

Bishop Gutta Percha Company. 181 Broadway, N.Y. Founded by S. T. Armstrong in 1847 as the American Gutta Percha Co. The company name was changed in 1853 when Samuel C. Bishop succeeded Armstrong as general agent. Made rubber, gutta-percha, and hard gutta-percha products. In the late 1860s and 1870s, the company listed insulators, submarine wire, and combs among their products.

Bishop, Samuel C. See Bishop Gutta Percha Co.

Boston & Lynn India Rubber Company. India-rubber cloths and blankets.

Boston Belting Company. Boston. Founded 1843. Rubber belting. Succeeded the Naugatuck India Rubber Co. Still in business and issued a trademark in 1889.

Boston Elastic Fabrics Company. Boston. 1865. Rubber thread.

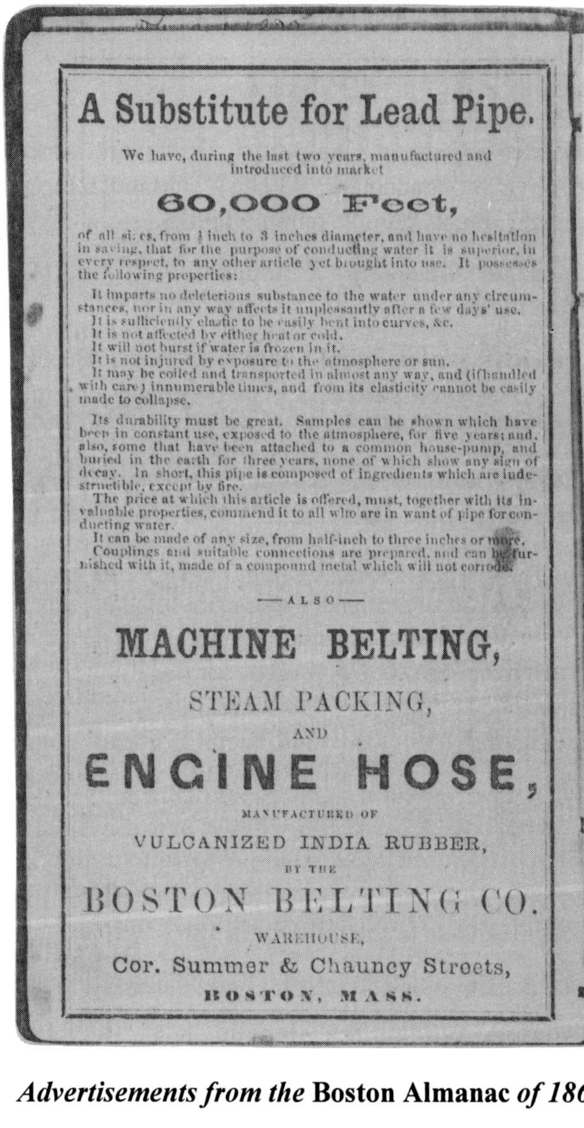

Advertisements from the* Boston Almanac *of 1862.

Boston Rubber Shoe Company. Incorporated as Malden Manufacturing Co. in 1853. Name changed in 1855. Rubber shoes. Manufactured rubber blankets for the Union Army under a license from the Union India Rubber Co. Still in business and issued a trademark in 1894.

Boston Steam Factory. Pitts Street, Boston. 1835-1836. Rubber aprons, caps, shoes, clothing.

Breeden, B. F. & Bros. 245 Pearl and 20 Cliff St., N.Y. 1855. Agent for Ford Rubber Co.

Candee, L. & Co. Started in New Haven, Conn. in 1842 by Leverett Candee. One of the first to receive a Goodyear license for making rubber shoes. Still in business at the turn of the century.

Car Spring Hose and Tubing Company. N.Y. 1855. Gutta-percha railroad car springs, hoses and tubing under Rider and Murphy's patents.

Chaffee, Edwin. Roxbury, Mass. Early rubber pioneer; founder and part owner of Roxbury Rubber Co. (1833). Designed the first rubber compounding machine, called "The Monster," which weighed 30 tons.

Church & Crittenden. Rubber Shoes.

Clapp, Charles M. & Co. Also known as C. M. Clapp & Co. 183 and 185 Devonshire Street and 42 to 48 Arch Street, Boston. Appeared in the 1874 Boston City Directory in an advertisement linking the company to National Rubber Co., Aetna Rubber Co., and Goodyear Rubber Co. The ad contains a title "Established 1865" and a note "Proprietors-Goodyear Rubber Co., Store 24 Milk Street, Boston."

Colchester Rubber Company. See Hayward, Nathaniel

Congress Rubber Company. 28 Cortlandt, N.Y. Shirred goods. Along with Horace Day, named as a defendant in a lawsuit filed by Charles Goodyear in 1855.

Continental Gutta Percha Company. 24 Pine Street, N.Y. 1867.

Cutler & Warner. Springfield, Mass. 1842. Founded by Horace Cutler and Solomon C. Warner. Manufactured shoes, suspenders, trusses, umbrella tops, vest springs and other elastic goods under Goodyear's patents of the 1830s. All goods were marked "Goodyear's Patent." Cutler provided Horace Day with Goodyear's vulcanization process in 1842.

Davenport, Henry. 265 Broadway, N.Y. Advertised in the Pittsburgh Gazette of October 19, 1854. "Manufacturer and dealer in every description of Goodyear's patent rubber fabrics, fancy rubber goods and toys.

Davidson Syringe Company/Davidson Rubber Company. Davidson Syringe Co., Charlestown Bridge, Charlestown Square from 1863 to 1866. Davidson Rubber Co., Boston, from 1872 to 1900. India-rubber syringes and general articles.

Davis, George N. & Bro. 152 & 154 Congress Street, Boston. A full page advertisement for this company appeared in the *Historical Picture Gallery,* published in 1856. Articles manufactured by eleven India-rubber and hard rubber companies were detailed and illustrated.

Davis, S. D. 23 Courtland Street, N.Y. Submitted proposal for India-rubber blankets in 1862.

Day, Horace. New Brunswick, N.J. Rubber pioneer who began experimenting with rubber and turpentine in 1837 when he was only 14 years old. Held several patents and manufactured a variety of rubber and hard rubber goods. Became the most notorious of the rubber "pirates" who illegally produced rubber goods without a license. Day was the defendant in "The Great India Rubber Case-Goodyear vs. Day" of 1852. Proposed "Day's Shelter Tents" to the Union Army in 1862.

DeForest, William. Brother-in-law of Charles Goodyear. Involved in early rubber industry. Loaned Goodyear sizable amounts of money to continue work. As personal friend, helped manage family finances, even after Goodyear's death.

Eagle India Rubber Company. Easton, Mass. and Woburn, Mass. India-rubber goods. Later became the Eagle India Rubber Co. and again later Newell and Humphrey. Organized around Nathaniel Hayward's operation in 1835 and bought by Charles Goodyear in 1838 for experimenting.

Easthampton Rubber Thread Company. 1863. Rubber thread.

Farley, Clagstone & Co. St. Louis. Circa 1850s. Listed in "The Saint Louis Guide" of 1859: "...who have for many years been engaged in the manufacture and sale of rubber goods, have very recently opened a depot for the sale of every variety of the goods of which India Rubber is a whole or component part."

Firestone Tire & Rubber Company. Akron, Ohio. Established in 1900.

Ford, J. R. & Co. New Brunswick, N.J. Founded 1843. Produced shirred goods and rubber shoes. Later became Meyer Rubber Co.

Framington India Rubber Company. Framington, Mass. 1835-1836.

Frisbie, Myron J. 248 Pearl, N.Y. 1855. India-rubber shoes.

Gay, Gamaliel. Agent for the Naugatuck India Rubber Co. His name appears on military proposals for "Goodyear's Patent Metallic Rubber" goods from 1845 to 1848.

Goodrich, B. F. Company. Akron, Ohio. Established in 1870 as Goodrich, Tew and Co. The first rubber company west of the Alleghenies. The founder, Dr. Benjamin Franklin Goodrich, had been a stockholder in the Hudson River Rubber Co. at Hastings-on-Hudson, N.Y. in 1869.

Goodrich, Isaac J. Stationer. State Street, Boston. Advertised Gum Elastic shoes, India Rubber balls, and overshoes in 1824.

Goodyear, Charles. Persistent inventor who discovered vulcanization in 1839 and patented the process on June 15, 1844 (U.S. Patent #3,633). Spent the next sixteen years pursuing improvements and new applications for India-rubber. Died in 1860.

Goodyear, Charles, Jr. Son of Charles Goodyear. Patented process for making artificial teeth of hard rubber. Inventor of the "Goodyear Welt" shoe improvement.

Goodyear Elastic Fabrics Company. Circa 1860s. N.Y. Advertised "India Rubber Army Blankets" in a wartime newspaper.

Goodyear, Henry B. Brother of Charles Goodyear. Credited with several inventions. Was the businessman of the family and managed the estates of brothers Charles and Nelson. (Also see Naugatuck India Rubber Co.)

Goodyear India Rubber Stopple Company. Boston, Mass. Was assigned patents for various stopples in the 1850s and 1860.

Goodyear Manufacturing Company. Formed in 1845 and only in existence a short time while Charles Goodyear was using an idle plant for making experimental rubber goods.

Goodyear Metallic Rubber Shoe Company. Naugatuck, Conn., 1843-1890s. One of the first plants to manufacture rubber shoes under a Goodyear license. Started in 1843 as Samuel J. Lewis and Co. and became Goodyear (also advertised as Goodyear's) Metallic Rubber Shoe Co. in 1845.

Goodyear, Nelson. Brother of Charles Goodyear and patentee of the hard rubber process in America (U.S. Patent #8,075 dated May 6, 1851). Died in 1852.

Goodyear Rubber Company. New York. Established in 1872. Rubber footwear. May have been affiliated with the Rubber Clothing Co. Still in business and issued a trademark in 1893. (Also see Charles M. Clapp & Co. and Goodyear Tire & Rubber Co.)

Goodyear Rubber Emporium. 159 Broadway, N.Y. India-rubber goods.

Goodyear Tire & Rubber Company. Akron, Ohio. Formed in 1898 and named to honor Charles Goodyear. No direct relationship with any of the earlier rubber companies. Acquired the trademark rights to Goodyear Rubber Co. in the 1970s.

Goodyear's India Rubber Glove (Manufacturing) Company. Formed as the Litchfield India Rubber Co. in Litchfield, Conn. by Samuel Lewis in 1844. The name changed when the company moved to Naugatuck, Connecticut Later moved to New York. An early Goodyear licensee manufacturing rubber gloves and other goods. Still in business and issued a trademark in 1879.

Goodyear's India Rubber Packing Company. Appeared in an advertisement published in the *Historical Picture Gallery* in 1856. "Steam Belting, packing, hose, etc." Listed as Goodyear Rubber Packing Co. at 98 Broadway, N.Y. in *Wilson's Business Directory of New York City,* 1855.

Goodyear's Rubber Belting and Packing Company. Philadelphia. In 1861, supplied rubber hose, pipes and packing for gas generators used by Professor Lowe's Balloon Corps.

Greacen, John. 96 Broadway, N.Y. Early 1850s. Licensee under Charles Goodyear's vulcanization patent. Manufactured what Goodyear called "deck scrubs," and what sailors called "squeel gees," of gum-elastic sponge for cleaning decks.

Gutta Percha and Rubber Manufacturing Company. 197 Broadway and 530 W. 25th Street, N.Y. 1865.

Gutta Percha Manufacturing Company. 844 W. 25th St. and 139 Broadway, N.Y. In business from 1860 to 1864. Clothing, hospital sheeting, cloths, hose, belting and packing.

Haering, R. & Company. 1182 Broadway N.Y. 1858-1859.

Hale, Alfred Rubber Company. Boston. 1837.

Hale, Holman J. 16 Beckman, N.Y. 1860-1862.

Hartshorn, Isaac & Company. 1840s through at least 1855. Founded by J. W. Clark of Boston, Massachusetts and Charles Jackson of Providence, R.I. Shoes and shirred goods.

Hayward, Daniel. Patented one of the first successful processes for recycling waste India-rubber in 1854.

Hayward, Nathaniel and Hayward Rubber Company. Nathaniel Hayward was actively involved with the rubber industry and was instrumental in steering Goodyear towards the use of sulphur in the vulcanization process. He operated the Eagle India Rubber from 1835 to 1838. In 1843, he started the Nathaniel Hayward Co. in Lisbon, Connecticut and manufactured rubber shoes. In 1847, he moved the company to Colchester, Connecticut and changed the name to Hayward Rubber Co. In consideration of royalties paid to Charles Macintosh & Co., Hayward was granted sole right to sell its rubber footwear in the British market. In 1888, the company name was changed to the Colchester Rubber Co. (Also see Eagle India Rubber Company and Haywardville Rubber Co.)

Haywardville Rubber Company. Haywardville, Mass. Started by Nathaniel Hayward near Stoneham, Mass. in 1858, when he purchased a rubber mill that had been operated by Elisha Converse since 1840. Known to the people in the area as "Red Mills." Made buckets, shoe soles, etc. of recycled India-rubber under Daniel Hayward's 1854 patent.

Hodgman, Daniel. Store at 27 Maiden Lane, N.Y. Had U.S. Patent #2504 for "Overshoe." Advertised heavily in 1848: "Manufactured from Goodyear's Metallic Rubber and would retain their brilliance of gloss, not stiffen in cold weather and are extremely light and durable." Also had a license for rubber door-springs.

Hood, J. M. In 1833, displayed boots made in New York and sent to South America to be varnished with liquid rubber as it came from the tree.

Hotchkiss & Pritchard. Waterbury, Conn. Woven elastic webs.

Howes Hyatt and Company. N.Y. Entered a bid on March 1, 1862 for India-rubber blankets. Bid noted an affiliation with the Union India Rubber Co.

Hudson Manufacturing Company. 1850. Gutta-percha goods. Bought by Bishop Gutta Percha Co. in 1853.

Hudson River Rubber Company. See B. F. Goodrich Co.

Huthinson & Runyon. Newark, N.J. 1843. Shirred goods.

India Rubber Belt Company. N.Y. India-rubber goods.

Johns & Crosley. 510 Broadway, N.Y. Appeared in Nov. 23, 1861 advertisement in *Household Journal.* "Sole manufacturers" of "gutta-percha cement roofing."

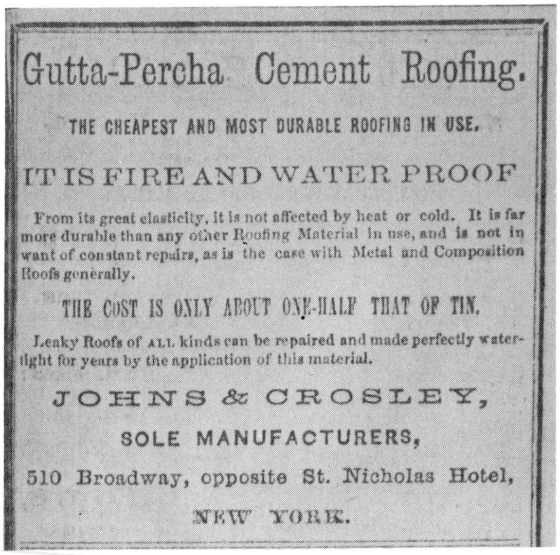

***Advertisement for roofing** by Johns & Crosley in* The Household Journal, *November 23, 1861.*

Joslin India Rubber Company. N.Y. Assigned patents for treating and vulcanizing rubber in 1859.

Judson, William. 37-38 Park Row, N.Y. Submitted proposal for India-rubber blankets in 1862.

Lambertville Rubber Company. 1860. Still in business and issued a trademark in 1894. Rubber footwear and coats.

Lewis, Samuel J. & Company. See Goodyear Metallic Rubber Shoe Co.

Lewis, Thomas. See Naugatuck India Rubber Co.

Litchfield India Rubber Company. See Goodyear's India Rubber Glove Manufacturing. Co.

Lynn Printing Company. Lynn, Mass. 1834-1837. Rubber-coated cloth.

Malden Manufacturing Company. 1853. Later became the Boston Rubber Shoe Co.

Meyer Rubber Company. New Brunswick, N.J. 1861. Started as J. R. Ford Co. in 1843. Rubber footwear.

Montreal Rubber Company. Appeared in an advertisement published in the *Historical Picture Gallery* in 1856. "Metallic rubber shoes and boots, for export."

Nashawannock Manufacturing Company. Easthampton, Mass. 1851. Gum-elastic webbing, suspenders, etc.

National India Rubber Company. Providence, R.I. 1840. Rubber footwear.

National Rubber Company. See Charles M. Clapp & Co.

Naugatuck India Rubber Company. Naugatuck, Conn. Founded 1844. Henry B. Goodyear was superintendent; Thomas Lewis was president. Maintained friendly relationship with Charles Goodyear and enjoyed some exclusive rights to manufacture India-rubber goods. Made suspenders, shoes, elastics, rubberized clothing. Was succeeded by the Boston Belting Co.

Newark India Rubber Manufacturing Company. Newark, N.J. In business from 1848 to 1876. Rubber goods including shoes and blankets. Had early Goodyear licenses for rubber shoes, team whips and buggy whips. Manufactured rubber blankets for the Union Army under a license from the Union India Rubber Co. Store at 59 Maiden Lane, N.Y.

New Brunswick Rubber Company. New Brunswick N.J. 1844. Rubber shoes. Also had a stores shown at 100 Liberty, 105 Cedar, Nassau Street and Maiden Lane, New York.

New England Car Spring Company. 104 Broadway, N.Y. 1840s-1860s. Railroad car springs.

New England India Rubber Company. 1835-1836. Coats and hats.

New England India Rubber Hat Company. Washington Street, Boston. India-rubber hats covered with plush.

New York Belting & Packing Company. N.Y. 1846. Rubber clothing. Succeeded the firm of Josiah Tomlinson of Sandy Hook, Conn., who has been granted a Goodyear license for rubber clothing. Still in business and issued a trademark in 1892.

New York Gutta Percha Company. 60 Beckman Street, N.Y. 1857. Moved to 171 Broadway in 1862.

New York India Rubber Company. Assigned patents in 1854 for manufacturing hollow rubber toys, balls, etc. Shown as New York Rubber Co. in *Wilson's Business Directory of New York City,* 1855.

North American Gutta Percha Company. 1854. Located at 144 Broadway and W. 25th & Tenth Ave., N.Y. in 1858. Moved to 153 Broadway in 1864.

Onderdonk & Letson. New Brunswick, N.J. Founded 1843. Shirred goods, rubber shoes.

Phoenix Rubber Company. A subsidiary of the Union India Rubber Co. Manufactured rubber blankets for the Union Army under the license from Union India Rubber.

Providence Rubber Company. Circa 1850s-1860s. India-rubber goods. Made vulcanized India-rubber blankets for the Union army without a license and was ordered to cease.

Providence Rubber Shoe Company. Established in 1840 at Providence, R.I. by George Bourn and David Winslow (company name unknown). Became Bourn & Brown in 1847, Brown, Bourn & Chaffee in 1851 and in 1861 was known as the Providence Rubber Shoe Co.

Richards, George H. Patent for "Fluid Rubber for Waterproofing Articles" in 1831.

Rider, W. & Brothers. Fourth Avenue, N.Y. Bought a Goodyear license to manufacture mail bags in 1847 and received a government contract for them. In 1848, they received another license for the manufacture of army and navy equipment, druggists goods and other articles. Manufactured a line of gutta-percha items in the 1850s. William, Emory, and John Rider assisted Charles Goodyear financially during his trying times. William Rider was President of the North American Gutta Percha Co. John Rider became the president of the New York Rubber Co.

Robbins, Eli. Lexington, Mass. 1835. Coats and hats. Sold out to the New England India Rubber Company in 1835.

Rogers & Wyckoff. 60 Cortlandt, N.Y. 1855. India-rubber goods.

Roxbury India Rubber Company. Formed by Edwin Chaffee, Charles Davis, Sr., Lemuel Blake and Luke Baldwin in Roxbury, Mass. in 1833. John Haskins was also involved with Luke Baldwin. In business until the early 1840s, the company manufactured coats, hats, shoes, life-preservers, mail-bags and other items. Charles Goodyear worked with Roxbury on many experimental items.

Rubber Clothing Company. Started approximately 1860. Beverly, Mass. Businesses in New York, Boston, Philadelphia, Cincinnati, and Chicago. General and military rubber goods. Still in business and issued a trademark in 1906. May have been affiliated with the Goodyear Rubber Co.

Salem India Rubber Company. Salem, Mass. Circa late 1830s. India-rubber apron cloth, caps, hats, life-preservers, and belting.

Smith, Stephen. N.Y. Held early patent for "India Rubber Boots and Shoes" (U.S. Patent #503 dated Dec. 7, 1837).

Smith, William. 57 Ann St., N.Y. *Wilson's Business Directory of New York City,* 1855, noted, "Manufacturer of patent elastic silk and cotton gores for Congress Boots, ladies' elastic dress bands, and plain of figured elastic Garters."

Suydam, David L. 1844. License for shirred goods.

Tobey, W. B. Providence, R.I. In 1862, submitted a bid for a new design of blanket with the India-rubber wholly covered by flocks of fur.

Trotter, Jonathon. Brooklyn, N.Y. 1848. First president of Union India Rubber Co. License for Army and Navy goods, druggist's goods and other sundry articles. (An identical license was issued to W. Rider Brothers.)

Tryon, D. & Co. 275 Pearl, N.Y. 1855. India-rubber shoes.

Tuttle Manufacturing Co. 32 Dey, N.Y. 1855. India-rubber shoes.

Union India Rubber Company. Naugatuck, Conn., New York. 1848. Goodyear license for wearing apparel and druggists' sundries such as syringes, water bottles bandages and air cushions. Due to the transfer of several licenses, the Union India Rubber held exclusive license for manufacturing Army and

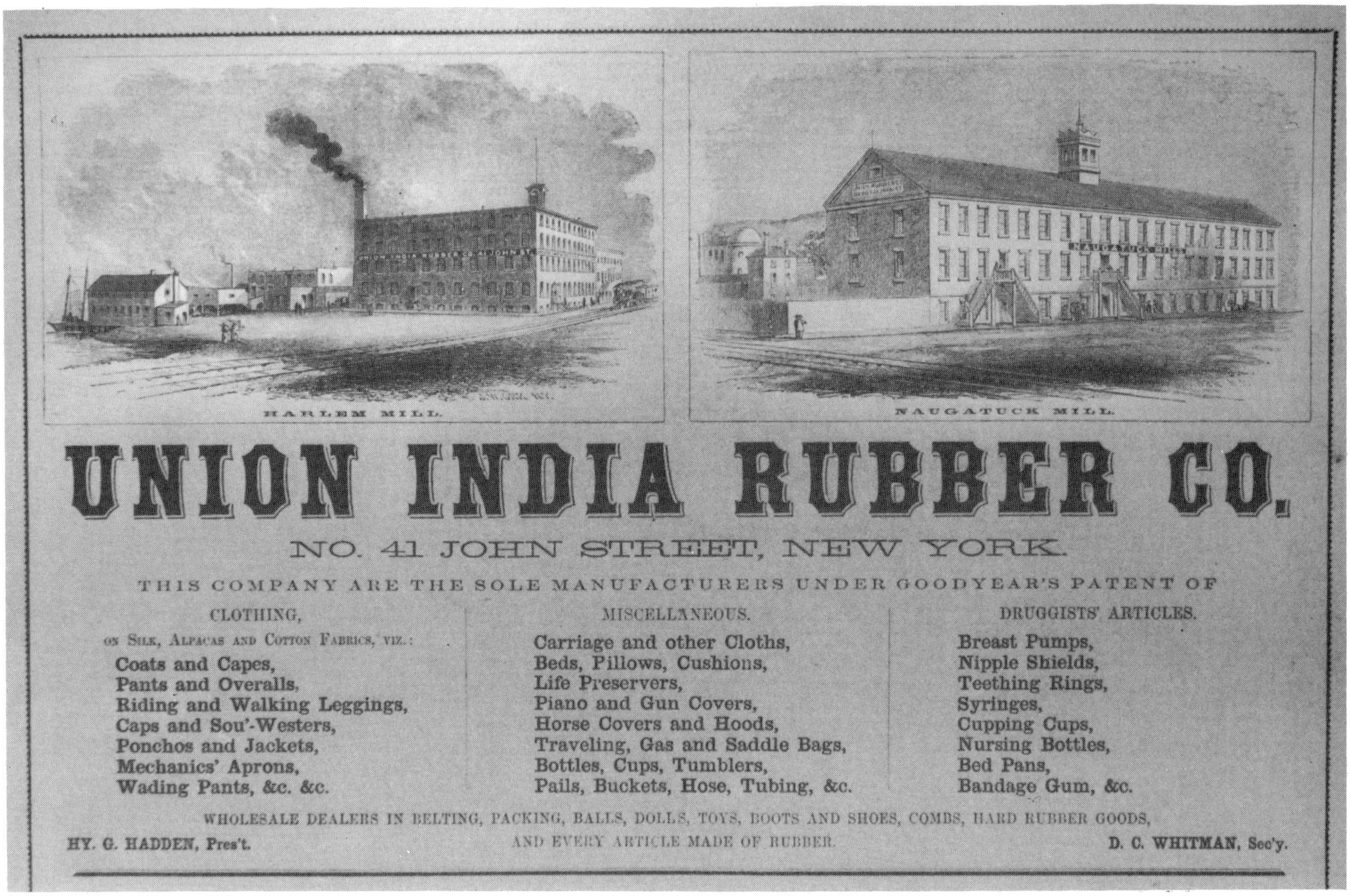

***Broadside of the Union India Rubber Company** dating from the 1860s. Shown are their two manufacturing plants and a list of the various rubber goods they made under Goodyear's patents.*

Navy military goods and manufactured most of the rubber blankets contracted for by the Union Army. Sub-contracted the manufacture of many blankets to the Phoenix Rubber Co., the Boston Rubber Shoe Co., and the Newark India Rubber Manufacturing Co. Still in business in 1879 and issued a trademark for a label reading "Goodyear Storm Coat."

United States Belting and Packing Company. Pearl Street, N.Y. 1857-1858.

United States Rubber Company. 1892. A consolidation of many existing rubber companies including the early firms of L. Candee and Goodyear Metallic Rubber Co.

United States Vulcanized Gutta Percha Belting and Packing Company. 66 Liberty, N.Y. 1857-1858.

Vulcanized Belt & Packing Company. 45 Courtland Street, N.Y. 1854-1855.

Wagner, F. C. 91 Liberty Street, N.Y. Submitted proposals for India-rubber blankets in 1862.

Ward & Sinclair. 102 Broadway, N.Y. 1855. India-rubber goods.

Warren, Ackerman & Company. Assigned a license to "...make and sell rubber goods in the same manner the Union India Rubber Company" by Union India Rubber on May 1, 1862.

Westbrooke & Company. White & Elm, N.Y. 1854-1855.

Whitney, D. M. 3rd Avenue, New York. Submitted proposal for India-rubber blankets in 1862.

Confederate States

Brands & Korner. Columbus, Ga. India-rubber cloth for military equipment.

Camp, D. B. . Savannah, Ga. 1860. India-rubber goods, military work. Formerly Robinson & Camp.

Eagle Manufacturing Company. Columbus, Ga. 1850. Also known as Eagle Mills. Manufactured textiles and cloth suitable for uniforms and tent-cloth for the Confederate Army. May have also manufactured India rubber cloth.

England

Arnold Company. 1852 catalog lists India-rubber bandages, knee-caps, etc.

Cornish & Company. 1833. India-rubber thread. Later became Elastic Thread & Web Manufacturing Co.

Cow, P. B. & Company, Ltd. 1836. India-rubber goods.

Dunlop Rubber Company, Ltd. Started in 1896 as Byrne Brothers India Rubber Company, Ltd. Name changed to Dunlop Rubber Co. in 1900.

Glass, Elliot & Company. Circa 1850s. Rubber-covered cables.

Gutta Percha Company. 1843. Gutta-percha goods. Founded by Thomas Hancock and H. Beverly.

Hancock, Thomas. English rubber pioneer who patented the vulcanization process in England after seeing samples of Goodyear's rubber and noting the presence of sulphur. Likely the true discoverer of the hard rubber process.

Heinke, C. E. Company. 1852. Rubber goods.

Ingram, J. G. 1847. Rubber goods, toy balloons.

Leyland & Birmingham Rubber Company. 1862. Rubber goods.

Macintosh, Charles & Company. 1826-1851. Rubber and hard rubber goods. Charles Hancock's manufacturing company. Issued trademarks in the United States as C. Macintosh & Co. in 1885 and 1890.

Mandleberg, J. & Company, Ltd. Circa 1850s. Rubber goods.

Moulton, Stephen C. Showed samples of Goodyear's vulcanized rubber to British rubber pioneer Thomas Hancock, leading him to perfect his process and patent it in England. Obtained a British license for waterproof fabrics, shoes, and saddlery under Goodyear's patent. Awarded a medal for "India Rubber Goods" at the Great Exhibition in 1851.

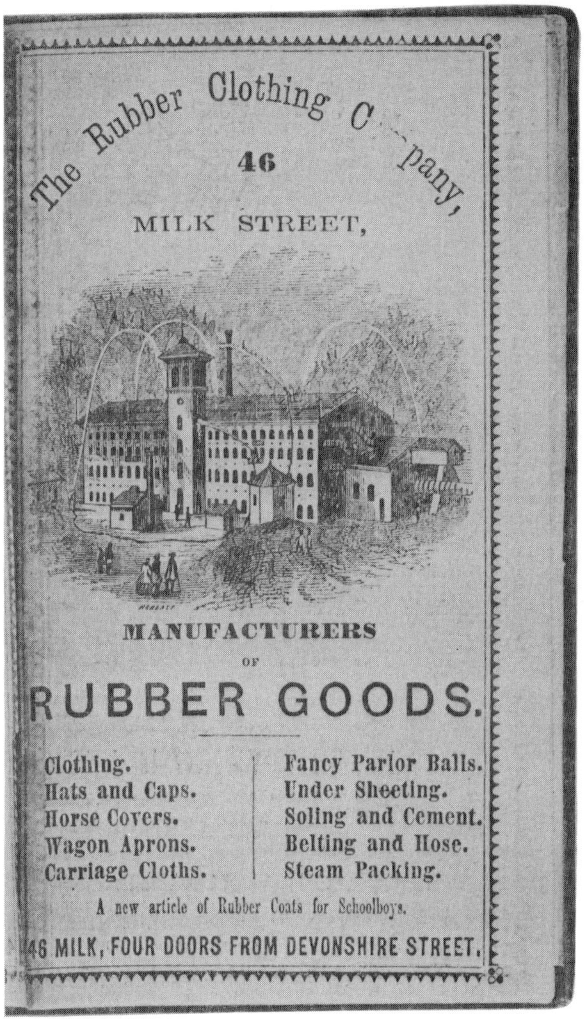

The Rubber Clothing Company, *as advertised in the* Boston Almanac *of 1861.*

Newall, R. S. & Company. Circa 1850s. Rubber-covered cables.

Siemens Brothers. 1864. Rubber-covered cables.

Silver & Company. 1844. India-rubber goods and cables.

Spencer, Moulton G. & Company. 1848. India-rubber goods.

Van Guens, M. M. Et Cie. Circa 1840s. Rubber-lined fire hose.

Warne, William & Company, Ltd. 1837. Rubber goods.

Scotland

North British Rubber Company, Ltd. 1855. Established by Henry L. Norris and Spencer T. Parmelee. Rubber footwear manufactured under Goodyear's patent. Also known as Norris & Co.

Liverpool Rubber Company, Ltd. 1859. Rubber footwear manufactured under Goodyear's patent.

HARD RUBBER AND HARD GUTTA-PERCHA

United States

American Hard Rubber Company (New). Formed in 1898 with the merger of the India Rubber Comb Co., Goodrich Hard Rubber Co., and Butler Hard Rubber Co.

American Hard Rubber Company (Old). Began in 1851 in Connecticut. The plant was moved to College Point, Long Island, N.Y., in 1861. The listed business address was 63 Maiden Lane, N.Y. from 1859 into the 1860s. Listed at 66 Cliff St., N.Y. in 1866 business directories. (Also see Poppenhusen, Conrad.)

Beacon Dam Company. Established in 1853 at Beacon Falls, Connecticut. Hard rubber buttons, syringes, photographic supplies, etc. Purchased by Conrad Poppenhusen in 1859 and moved to College Point, Long Island, New York. In 1861, it likely became part of the American Hard Rubber Co. when it was also moved to that location.

Butler Hard Rubber Company. Butler, N.J. Formed in 1882 and merged into the new American Hard Rubber Co. in 1898. Hard rubber syringes are seen with this name, some with the design name "Royal." Issued trademark in 1884 for "Fancy Hard Rubber Articles."

Caoutchouc Company. Appeared in an advertisement published in the *Historical Picture Gallery* in 1856. "Pen and pencil cases, pen handles, etc."

Day, A. G. Caoutchouc Company. Store in New York. 1850s. Hard rubber goods. Founded by Austin Goodyear Day, nephew of Charles Goodyear and no relation to Horace Day. Held his own patent for making hard rubber (U.S. Patent #21,122 dated August 10, 1858).

Goodrich Hard Rubber Company. Akron, Ohio. First appeared in the *Akron City and Summit County Directory, 1888-1889.* Affiliated with The B. F. Goodrich Company (founded in 1870). Merged into the new American Hard Rubber Co. in 1898.

Goodyear Dental Vulcanite Company. Boston. 1860s.

Goodyear, Nelson. Patented the hard rubber process (U.S. Patent #8,075 dated May 6, 1851).

Hard Rubber Cutlery Company. Circa 1850s. Knives and cutlery with vulcanized hard rubber handles.

India Rubber Comb Company. N.Y. Operated by this name from 1856 to 1898, when it merged and became part of the new American Hard Rubber Co. Made hard rubber combs, buttons, syringes, inkwells and many other articles. (Also see Poppenhusen, Conrad.)

India Rubber Handle Cutlery Company. N.Y. Circa 1850s. Knives and cutlery with vulcanized hard rubber handles.

Meyer, H. A. Partner in Meyer & Poppenhusen. Introduced hard rubber manufacture to Germany in 1856.

Meyer, L. Otto P. Relative of H. A. Meyer. Made radical changes in the hard rubber manufacturing process. Obtained several United States patents from 1853 to 1856.

Meyer & Poppenhusen. College Point, Long Island, N.Y. 1852-1854. Partnership of H. A. Meyer and Conrad Poppenhusen. Goodyear hard rubber licenses for artificial whalebone, combs and whipstocks

New York Gutta Percha & India Rubber Vulcanite Company. Circa 1850s. Hard rubber goods.

Novelty Rubber Company. Established 1853 in Beacon Falls, Connecticut. Incorporated and moved to New Brunswick, N.J. in 1855. Maintained a store in New York from 1856. Ceased operations in 1886. Hard rubber buttons, canes and many other articles. Manufactured military buttons for the Union Navy and Army, most notably those for Navy overcoats and those used by Berdan's Sharpshooters.

Poppenhusen, Conrad. Hard rubber pioneer and partner in several early hard rubber companies. One of the wealthiest and most influential men in the early rubber industry. Was a partner in Meyer & Poppenhusen and acquired the Beacon Dam Co. in 1859. He was also operating as the India Rubber Comb Co. and Poppenhusen & Koenig in 1856. Became president of the (old) American Hard Rubber Co. in 1861.

Poppenhusen & Koenig. See Poppenhusen, Conrad.

Vulcanite Optical Instrument Company. 1866. Began in 1866 from the dissolved original Bausch & Lomb Co. In the late 1850s, John Jacob Bausch began making eyeglass frames from hard rubber. In 1866, made hard rubber eyeglass frames and mountings for magnifiers under a contract or license from the India Rubber Comb Co.

England

India Rubber, Gutta Percha & Telegraph Works Company, Ltd. 1850. Hard rubber goods.

Macintosh, Charles & Company. 1826-1851. India-rubber and hard rubber goods.

Scottish Vulcanite Company, Ltd. 1861. Hard rubber goods.

COMPOSITIONS

United States

Critchlow, A. P. & Co. See Florence Mfg. Co.

Florence Manufacturing Company. Florence, Mass. Began in 1854 as A. P. Critchlow & Co. Manufactured articles from Alfred Critchlow's shellac-based composition. Name changed to Littlefield, Parsons & Co. in 1858 and to Florence Manufacturing Co. in 1866.

Holmes, Booth and Hayden. Waterbury, Mass. 1857.

Key, Frederick C. and Son. Philadelphia.

Littlefield, Parsons & Co. See Florence Mfg. Co.

Wadham's Manufacturing Company. Stoughton, Mass. 1858.

SOURCES:

Akron City and Summit County Directory—1888-1889 (Akron: N. H. Burch & Company, 1888)

Glenn D. Babcock, *History of the United States Rubber Company* (Bloomington: Bureau of Business Research, Graduate School of Business, Indiana University, 1966)

P. W. Barker, *Charles Goodyear—A Biography* (Godfrey L. Cabot, Inc., Boston, Mass., 1840).

Bausch & Lomb, "About History" (http://www.bausch.com/About/chronology.html)

Boston Almanac—1862.

Boston City Directory—1874.

Bruce S. Bazelon and William F. McGuinn, *A Directory of American Military Goods Dealers & Makers 1785-1915* (Manassas, VA: Published by the Authors, 1990)

Bruce S. Bazelon and William F. McGuinn, *Supplement to A Directory of American Military Goods Dealers & Makers 1785-1915* (Manassas, VA: Published by the Authors, 1995)

G. Leander Bishop, A *History of American Manufacturers from 1608 to 1860* (Edward Young & Co., Philadelphia, 1868).

Commissioner of Patents, *Annual Report of the Commissioner of Patents*, various years.

Chauncey M. Depew, *1795-1895 One Hundred Years of American Commerce* (D. O. Haynes & Co., New York, 1895).

Eighth Census of the United States, 1860.

The B. F. Goodrich Company, Community Relations Release. 1977.

Handbook of Hard Rubber and Plastics (New York: American Hard Rubber Company, 1951).

Institution of the Rubber Industry, *History of the Rubber Industry* (Cambridge, England: W. Heffer, 1952)

Lawrence C. Wassmer, *Butler, New Jersey in Story and Pictures* (Butler: The Butler Argus, 1951).

Wilson's Business Directory of New York City—1855. (New York: John F. Trow, 1855).

Zellie, Carole, & Martha Coons, *Stoneham, Massachusetts: A Shoe Town*, (Stoneham, Mass: Stoneham Historical Commission, 1981)

Various newspapers and periodicals.

Author's notes down through the years.

SECTION 11

UNITED STATES PATENTS

"The Patent System has added the fuel of interest to the fire of genius."

Abraham Lincoln

The United States patent records are an invaluable source of information for the historian, collector, and antique or Civil War buff. Perusing the endless creative inventions is educational as well as entertaining.

Patents were issued from 1790 to 1836 by name and date only. These patents were subsequently numbered, with the patent number preceded by an "X," to distinguish them from patents issued and numbered under the Patent Act of July 4, 1836.[1] Under the Act of 1836, patents were issued for fourteen years, with extensions of seven years possible with authorization.[2] Under the Patent Act of 1861, the term of the patent was increased to 17 years and the option of extension withdrawn. The Patent Act of 1842 authorized design patents that would be issued for a term of seven years.[3] In all cases, patents could also be reissued for technical changes, corrections, or additions.[4]

Under these Acts, inventors had to submit patent drawings and models for any new or improved invention.[5] The Patent Act of 1842 stipulated that items manufactured under the patent had to be marked as patented and include the date of the patent in order to protect the invention.[6] Some surviving articles contain multiple patent markings. For example, the hard rubber "Telescopic Drinking Cup," with the June 5, 1860 date of Peter Niles' U.S. Patent #28,597 may also bear the May 6, 1851 date of Nelson Goodyear's patent for the hard rubber process (U.S. Patent #8,075).

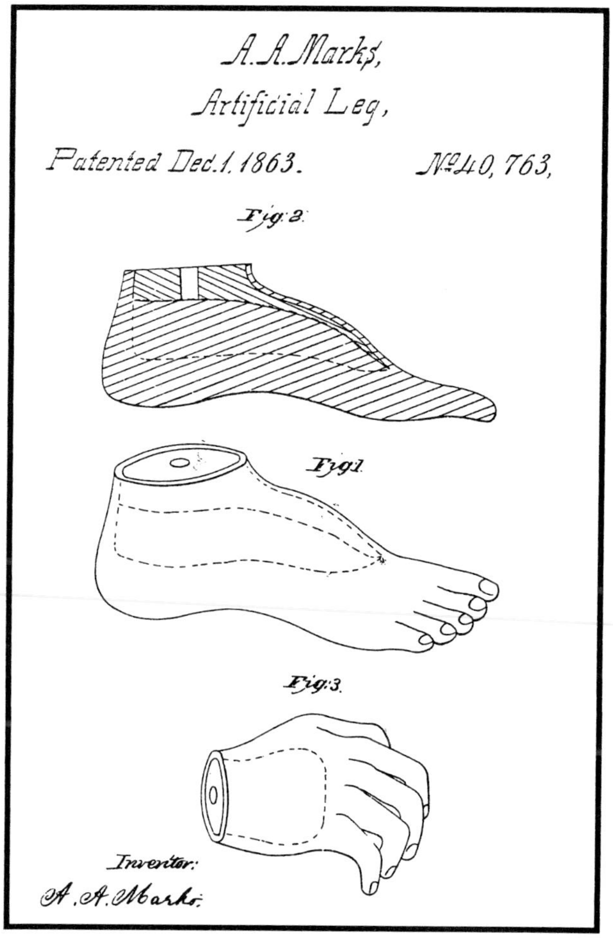

Amasa Marks' artificial limbs, *made of sponge rubber, hard rubber, and wood. Patented December 10, 1861 (U.S. Patent #40,763).*

Some patent numbers and dates found on articles may be dates of reissues, extensions, or design patents. In a few cases, patent dates on an article may not match the issue date and may represent the date the patentee filed for the patent. Some patent details may also be from foreign patents.

It should be noted that many patented processes and products were never used or manufactured commercially. As in modern times, ingenious ideas still required financial backing, an agreeable manufacturer, and a willingness to buy on the part of the public or the military.

The following lists include selected patents issued between 1813 and 1869 that relate to India-rubber, gutta-percha, hard rubber, compositions and waterproof materials. Most patents detailed were issued between 1844 (the year of Charles Goodyear's vulcanization patent) and 1865 (the end of the Civil War). A few significant and interesting patents from the earlier and later years are shown.

The first patent list, arranged chronologically by patent number and date, provides the patentees and a short summary of the patent. The second list is a cross-reference, arranged alphabetically by subject. This provides the researcher several alternatives for locating a specific patent or artifact.

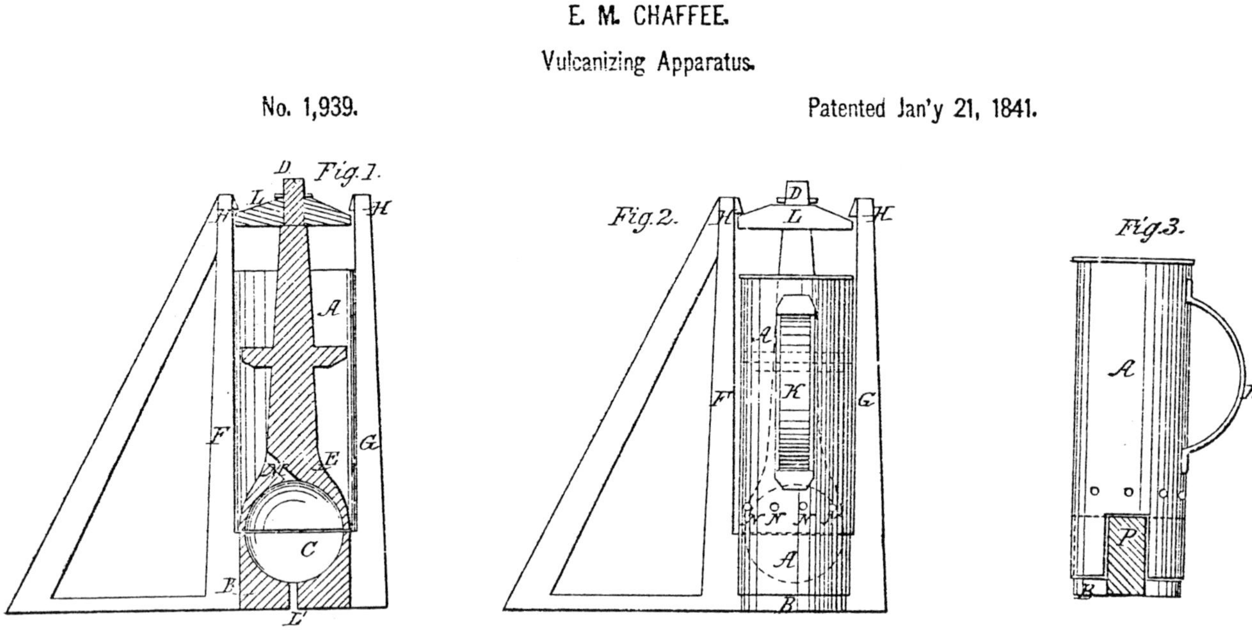

Vulcanizing apparatus. *Chaffee's 1841 patent, which pre-dated Charles Goodyear's 1844 vulcanization patent, specified a method of manufacturing toy balls by heating small particles of India-rubber and pressing the softened substance into molds.*

NOTES - United States Patents

[1] Kenneth W. Dobyns, *The Patent Office Pony* (Fredericksburg, VA: Sergeant Kirkland's Museum and Historical Society, Inc., 1994) p. 111.
[2] U.S. Department of Commerce, *The Story of the United States Patent Office—1790-1956* (Washington: U.S. Government Printing Office, 1956) p. 6.
[3] *Ibid.,* p. 8.
[4] Cooper-Hewitt Museum-The Smithsonian Institution, *American Enterprise: Nineteenth-Century Patent Models*, (Washington: The Smithsonian Institution, 1984) p. 15. [Kendall J. Dood, "Why Models?," *Journal of the Patent Office Society*.]
[5] *Ibid.,* pp. 14-16.
[6] *The Story of the United States Patent Office—1790-1956,* p. 9.

INDIA RUBBER, GUTTA PERCHA, AND RELATED U.S. PATENTS 1813-1868

PATENT	DATE	PATENTEE	DESCRIPTION
1913X	4/29/13	Hummel, J. F.	Shoes, waterproofing with gum-elastic varnish
7301X	11/24/32	Palmer, Warren C., M.D	Bathing cot, gum-elastic
7921X	12/31/33	Chaffee, E. M.	Mail & travelling bag, India-rubber
----	6/7/34	Goodyear, Charles	Pumps, air
8207X	6/7/34	Chaffee, E. M.	Boots, India-rubber
8731X	3/30/35	Goodyear, Charles	Pumps, air
9074X	9/9/35	Goodyear, Charles	Cement, Gum-elastic
240	6/17/37	Goodyear, Charles	Caoutchouc, dressing and bleaching (The "Acid Gas Process")
503	12/7/37	Smith, Stephen C.	Boots & shoes, India-rubber, mfg. of
849	7/24/38	Goodyear, Charles	Gum-elastic, mfg. (Covering overshoes with India-rubber)
1,090	2/24/39	Hayward, Nathaniel, assr. to Chas. Goodyear	Caoutchouc, mode of preparing with sulphur for the manufacture of various articles
1,634	6/17/40	Lane, John F., Col., U.S. Army (Deceased)	Pontoon Bridge (Floats coated with India-rubber) Filed by E. Huntington
1,841	10/30/40	Dupont, J. H. & T.Hyatt	Boots with gum-elastic gores (Gaiters)
1,939	1/21/41	Chaffee, E. M.	Balls, mfg. from caoutchouc or India-rubber
3,461	3/9/44	Goodyear, Charles	Fabrics, India-rubber (Corrugated goods of India-rubber, etc.)
3,462	3/9/44	Goodyear, Charles	India-rubber, machine for making (As described in Patent 3,461)
3,633	6/15/44	Goodyear, Charles	India-rubber (fabrics) This is the vulcanization patent. Reissued 12-25-49, extended 6-15-58, reissued 11-20-60, expired 6-15-65)
3,782	10/9/44	Tyer, Henry G. & John Helm	India-rubber, cutting, machine for
3,788	10/12/44	Day, Horace	India-rubber goods, corrugated and shirred, machine for making
4,005	4/22/45	Goodyear, Nelson	Fabrics, indurated India-rubber (Combining with emery, sand, grit,etc.)
4,047	5/13/45	Goodyear, Nelson	India-rubber, mfg. of (Combining of fibrous substances with the gum; adding sand, emery, etc. to rubber making a slip-resistant covering for floors, steps, shoe soles, etc.)
4,073	6/7/45	Tyer, Henry G. and John Helm	Threads, India-rubber, machine for cutting
4,099	6/16/45	Goodyear, Charles	India-rubber fabrics
4,131	8/4/45	Pratt, Elijah	Artificial nipples
4,138	8/9/45	Pratt, Elijah	Nursing bottle with artificial nipple
4,150	8/16/45	Day, Horace	Hose, combination (Textile tube and India-rubber tube combined)
4,356	1/15/46	Day, Horace	Boat, India-rubber, portable
4,357	1/15/46	Harmon, J. W.	India-rubber, softening before grinding (Use of turpentine or naptha)
4,827	10/24/46	Fuller, William	Springs & bumpers, India-rubber (Use on railway carriages)
4,984	2/27/47	Stanton, Henry	Carriages, bouyant (Combining of floats made of India-rubber cloth for transporting wagons, etc)
5,019	3/13/47	Tassie, Madeline	Shirt bosoms (Use of yoke and elastic strap for false bosom)
5,069	4/17/47	Ely, William F.	India-rubber, preparing
5,196	7/17/47	Gilbert, Charles J. & Gamaliel Gay	Fabrics, India-rubber
5,233	8/7/47	Story, Robert & Thomas Hopper	Shoes, India-rubber (With attached leather sole and heel)
5,271	9/4/47	Thomas, James	India-rubber, preparing
5,536	4/25/48	Goodyear, Charles	Hollow articles of India-rubber (Toys, spheres, etc.)
5,539	4/25/48	Durant, Charles F.	Gutta-percha & caoutchouc, dissolving

PATENT	DATE	PATENTEE	DESCRIPTION
5,543	5/2/48	Day, Horace	Fabrics, gutta-percha, preparing in imitation of patent leather
5,589	5/23/48	Bewley, Henry	Syringes & tubes, etc., flexible gutta-percha
5,590	5/23/48	Hancock. Charles	Bands or belts, gutta-percha, mfg.
5,592	5/23/48	Brooman, Richard A.	Articles, gutta-percha, mfg. by moulding, stamping, embossing
5,593	5/23/48	Keene, Charles	Boots, shoes, etc., gutta-percha combined with other fabrics
5,606	8/1/48	Ray, Fowler M.	Springs, India-rubber and pneumatic (Air-tight India-rubber chambers
5,904	11/7/48	Day, Horace & Frances Hayward	Suspenders and shoulder braces
5,952	12/5/48	Bassford, Abraham	Billiard tables, cushions for (India-rubber tube filled with gas)
6,053	1/23/49	Schnebly, Wm. & Thos.	Boats, life, self-inflating and folding
6,066	1/30/49	Tyer, H. G. & John Helm	India-rubber, mfg. of (zinc & sulphur)
6,231	3/27/49	Ray, Fowler M.	Springs, caoutchouc
6,363	4/17/49	Starr, Eben T.	Boats, flexible; Divisions between tubes
6,581	7/10/49	Seely, Samuel J.	Life-preserver hammock
6,786	10/16/49	Goodyear, Nelson	Suspenders, elastic cords for
7,189	3/19/50	Hayward, Frances D. & John C. Bickford	Cloth, India-rubber, process for rolling
7,196	3/19/50	Pridham, John, assr. to Horace Day	India-rubber, mfg. with oxide of tin
7,207	3/26/50	Bell, Hugh	Balloons and their appendages (Elastic pipes, etc.-British Pat. 11-23-48)
7,238	4/2/50	Fessenden, Abijah	Filtering & drinking tube, pocket
7,251	4/2/50	Ray, Fowler M.	Springs for cars, etc., India-rubber
7,643	9/17/50	Armstrong, Samuel F. & Charles J. Gibbert	Gutta-percha, process of working
7,650	9/17/50	Day, Horace & Richard McMullin	Hose, India-rubber (Flexible)
7,706	10/18/50	Ray, Fowler M.	Springs, vulcanized India-rubber combined with metal
7,816	12/3/50	Trotter, Jonathon F.	India-rubber, vulcanizing with zinc
7,880	1/1/51	Trotter, Jonathon F.	India-rubber, mfg. (Use of zinc)
8,016	4/1/51	McCurdy, David	India-rubber, mfg. (Use of potash)
8,051	4/22/51	Reynolds, James	Tubing and covering wire with gutta-percha, machines for
8,070	4/29/51	Mt. Storms, William	Vessels, flexible hose or float for supporting
8,075	5/6/51	Goodyear, Nelson	India-rubber, mfg. (Hard rubber. Reissued 5-18-58, extended 5-6-65, expired 5-6-72)
8,100	5/20/51	Day, Horace	Shoes, India-rubber
8,180	6/24/51	Armstrong, Samuel T.	Hollowware, gutta-percha (Moulding while soft)
8,449	10/21/51	Meech, Charles L.	Wind instrument, mouthpiece for (Made of elastic and waterproof material)
8,992	6/1/52	Rider, John	Gutta-percha, mfg. for processing (Extended 5-31-66)
9,173	8/3/52	Jones, John O.	Horseshoe, elastic (Sheet of vulcanized rubber between 2 plates)
9,246	9/7/52	Bronner, Frederick	India-rubber, preserving (Applying campeche or muriate of soda to the sap)
9,277	9/21/52	Pond, Daniel F.	Crayon rubber (For applying and blending colors in drawing)
9,319	10/12/52	Goodyear, Charles	Bat cloth, India-rubber, modes of making (Nearly waterproof, but not; Patented in England June 8, 1850)
9,570	2/1/53	Solis, Richard	India-rubber, mfg. of (Re-manufacturing)

PATENT	DATE	PATENTEE	DESCRIPTION
9,668	4/12/53	Goodyear, Charles, assr. to C. G. & Robert Haering	Gutta-percha and India-rubber, mfg. (Pat'd in England March 4, 1851)
9,695	5/3/53	Hinsdale, Richard L.	Exercising machine, elastic
9,790	6/14/53	Kingsley, John L.	Stereotype plates, moulding (Gutta-percha or India-rubber)
9,870	7/19/53	Westbrook, Leonardo	Gutta-percha, stereotype compositions
9,891	7/26/53	Armstrong, S. T., assr. of Henry Lee Norris	India-rubber, preserving in liquid state (Adding liquor of Ammonia to gum; Patented in England 7-26-53 and in France 2-24-53)
10,008	9/13/53	Chilcott, John & Robert Snell	Soles for boots & shoes, India-rubber (With leather lining and border)
10,106	10/11/53	Goodyear, Charles	Metals, covering with gutta-percha & caoutchouc (Pat. in England 3-4-51)
10,280	11/29/53	Bussell, Erastius T.	Springs, combined India-rubber & steel
10,339	12/20/53	Meyer, L. O. P.	Vulcanizing caoutchouc compounds
10,426	1/17/54	Davison, F.	Pump, saliva (For dentistry)
10,429	1/17/54	Hyatt, E. C. & C. Meyer	Boot & shoe soles, gutta-percha or India-rubber, mfg. of
10,586	2/28/54	Meyer, L. O. P.	India-rubber & other gums, vulcanizing
10,689	3/28/54	Ellsworth, D. S. Goodyear, assr. to New York India Rubber Co.	India-rubber, processes for treating (Hollow articles, balls, etc. Filling with water, which evaporates and turns to steam during vulcanization)
10,738	4/4/54	Goodyear, Charles	Gums, vulcanizable, treating
10,741	4/4/54	Meyer, L. O. P.	Gums, vulcanizable, treating
10,742	4/4/54	Mattson, Morris	Syringe, enema (With flexible tube)
10,838	4/25/54	Towle, Simeon	Syringe eye-bath (India-rubber bulb & cap form. Rim flap fits to eye)
10,977	5/30/54	Murphy, John	Gutta-percha, processes for treating
11,135	6/20/54	Needham, Orwell H.	Breast pump (Flexible & bulb tubes)
11,171	6/27/54	Minthorn, Daniel	Tube, inhaling (Flexible tube into medicine bottle for vapors)
11,608	8/29/54	Hayward, Daniel	India-rubber, mfg. of (Remolding worn-out India-rubber goods)
11,716	9/26/54	Brown, Edward	Goods, elastic (Rubber sheet between 2 thicknesses of stocking work)
11,758	10/3/54	Peck, Samuel	Daguerreotype cases, composition (Shellac and sawdust or fibrous material with color pressed through hot rollers)
11,762	10/3/54	Mabie, John	Pencil, improved (Mechanical; some had hard rubber bodies)
11,831	10/24/54	Herriet, Julius	Stereotype, gutta-percha or India-rubber (With pulverized graphite, soapstone, etc. heated and moulded)
11,897	11/7/54	Marcy, E. E.	Vulcanizing elastic gum
11,906	11/7/54	Rider, William E. & John Murphy	Gutta-percha, etc., desulphurizing (Use of hydrogen during vulcanization)
11,947	11/14/54	Pease, Julius A.	Overshoes, India-rubber (With ribbed or corrugated inner surface to allow air to circulate between it and shoe)
12,144	1/2/55	Tyer, Henry G. & John Helm	Cloth, India-rubber, process for making
12,212	1/9/55	Morey, Charles	Hard rubber compounds, reworking (Heating scrapings, dust, powder, etc. of hard rubber
12,334	1/30/55	Tyer, Henry G. & John Helm	Cloth, India-rubber, process for making
12,450	2/27/55	Stevenson, Joseph	Bed-boat or life-preserver (Mattress stored in berths and used as bed but convertible to boat; with India-rubber belts or aprons)
12,738	4/17/55	Converse, William F.	Springs, India-rubber

PATENT	DATE	PATENTEE	DESCRIPTION
12,753	4/17/55	Porter, Rufus & Jon. Bradley, assrs. to J. D. B. & G. Dennison	Tube pump, elastic (India-rubber)
12,983	5/29/55	Sigismund, Beer, assr. to Lewis & S.B Feuchtwanger	India-rubber, devulcanizing
13,272	7/17/55	Parmelee, S. T.	Soles, attaching metal to India-rubber
13,315	7/24/55	Larcher, E. B.	Boats, gutta-percha, making
13,410	8/7/55	Halvorson, Halvor,- assr. to H. Barnes	Daguerreotype cases, composition (Gum shellac and fibrous materials, heated)
13,420	8/14/55	Bashnagel, Francis	Composition for treating India-rubber or gutta-percha
14,389	3/11/56	Cheever, John H.	Belting or banding, India-rubber, mfg.
14,451	3/18/56	Gleason, R., Jr.	Inkstand (With hollow image of India-rubber attached to inkstand by an air-tight joint)
14,457	3/25/56	Schrag, Phillip	Mould for earthen vessels (Used an India-rubber lining that peeled off after baking; prevented adhesion of clay to the mould)
14,657	4/15/56	Hayward, Nathaniel	Fabrics, India-rubber, mfg. of
14,669	4/15/56	Lyman, William H.	Whip socket (Sheet of India-rubber applied to whip stock to prevent dust and moisture from entering)
14,739	4/22/56	Sibley, L. D.	Rings, elastic (To prevent nocturnal emissions)
14,811	5/6/56	Hayward, Nathaniel	Cloth, India-rubber/elastic, repairing
14,814	5/6/56	Tyer, Henry &John Helm	Gutta-percha or India-rubber, uniting with cloth, then vulcanizing
14,929	5/20/56	Puffer, A. D.	Gutta-percha, lining metal pipes with
14,972	5/27/56	Reynolds, James	Gutta-percha, apparatus for cleaning
15,021	6/3/56	Spooner, Lysander	Chairs, etc., elastic bottoms for
15,067	6/10/56	Day, Austin G.	India-rubber, cleaning
15,086	6/10/56	Reynolds, James	Tubing, gutta-percha, mandrills for making
15,087	6/10/56	Reynolds, James	Gutta-percha, feed apparatus for working
15,439	7/29/56	Reynolds, James	Wire, covering with gutta-percha
15,531	8/12/56	Shaw, William	India-rubber or gutta-percha, treating
15,897	10/14/56	Provines, William	Uterine supporter, elastic (Can be inflated to any desired degree)
15,947	10/21/56	Howell, Jacob H.	Hose, India-rubber, modes of making
15,998	10/28/56	Torstrick, Henry	India-rubber, vulcanized, working over (Extracting unorganic matter)
16,013	11/4/56	Hayward, Nathaniel	Soles, India-rubber, catch for
16,069	11/11/56	Sault, T.	India-rubber, process for cleaning
16,215	12/9/56	Reynolds, James	Cord, gutta-percha, making
16,269	12/23/56	Davenport, Henry	Thread, India-rubber, machines for cutting
16,601	2/10/57	Millerd, G. B.	Cloth, India-rubber, preparing
16,822	3/17/57	Babbitt, B. T.	Enema-giving apparatus with flexible tube (Uses hydrostatic pressure)
16,955	3/31/57	Poppenhusen, Conrad & C. F. E. Simon, assrs. to Conrad Poppenhusen	Table knives, method of securing blade to balance nut in handle
16,956	3/31/57	Davidson, Charles H. & Harman E.	Syringe, enema, with elastic bulb and tubes
17,037	4/14/57	Haering, Robert	India-rubber, etc., vulcanizing (Use of clay pipe)
17,214	5/5/57	Haering, Robert	Gutta-percha, process for purifying

PATENT	DATE	PATENTEE	DESCRIPTION
17,295	5/12/57	Poppenhusen, C. & Ludwig Held	India-rubber, devulcanizing
17,649	6/23/57	Winslow, Charles	Cloth, Elastic & India-rubber, preparing
17,702	6/30/57	Smith, Gilbert	Cartridges, improved (Case of India-rubber cloth or vulcanized rubber)
18,183	9/15/57	Butcher, William & William A.	Paint, India-rubber, waterproof (Linseed oil and boiled crude rubber and coloring thinned with painter's oil)
18,274	9/29/57	Bunker, Charles J.	Life preserver shirt (India-rubber with chambers; can be worn inside or outside other garments)
18,281	9/29/57	Hinckley, Charles H.	Hose pipe, coating with India-rubber
18,630	11/17/57	Cotton, Francis	Upholstery with India-rubber springs
18,809	12/9/57	Gibson, Abram J.	Life preserver (Deep belt with waterproof, air-tight material and hollow floats and paddles made to fit hands)
19,172	1/19/58	Hall, Hiram L., assr. to Beverly Rubber Co	Rubber, vulcanized, restoring waste (Grinding, boiling and reusing)
19,269	2/2/58	Wales, Thomas C.	Boots & gaiters, waterproof
19,328	2/9/58	Thomas, John F. assr. to himself and Samuel Remington	Cane guns (Made by Remington Co. with hard rubber-covered barrel and handle)
19,783	3/30/58	Lipman, Hymen L.	Lead pencil & eraser combination(India-rubber eraser in one end)
20,055	4/27/58	Fisler, Thomas	Rulers, application of India-rubber (To prevent slipping)
20,242	5/11/58	Hall, Hiram L., assr. to Beverly Rubber Co.	Rubber, vulcanized, utilizing waste (Grinding and mixing with shellac, resin, asphalt, etc.)
20,678	6/22/58	Baschnagel, Francis, assr. to himself & Beverly Rubber Co.	Rubber, vulcanized, restoring waste (Heating)
20,938	7/20/58	Cuppers, Gustavus	Rubber goods, hard, mfg. of (Improvement in hardening process)
21,122	8/10/58	Day, Austin G.	Caoutchouc, treatment of (Hard rubber. Ext'd. 8/10/72, Reissued 1/14/73)
21,270	8/24/58	Newell, John W.	Fabrics, elastic
21,799	10/12/58	Mayall, T. J., assr. to himself & B. F. Cooke	Stair pad, elastic
22,061	11/9/58	Delano, T. A.	Life-preserving vests (Vest with float of India-rubber between exterior and interior lining that can be inflated in a few minutes)
22,217	11/30/58	Hall, Hiram L., assr. to Beverly Rubber Co., Beverly, Mass.	Rubber, vulcanized, restoring waste (Steam)
22,218	11/30/58	Mayall, Thomas, assr. to himself & G. N. Davis	Rubber, hard, mfg. of (Adding olive oil)
22,265	12/7/58	Hall, Hiram L., assr. to Beverly Rubber Co.	Rubber, vulcanized, mfg. of (Combining fibrous materials with waste rubber)
22,480	1/4/59	Bishop, Lewis	Saddle, elastic (Seat suspended on India-rubber springs)
22,560	1/11/59	Joslin, H. W.	India-rubber, treatment of
22,566	1/11/59	McBurney, C.	Pipe, hose, India-rubber (Semi-elastic)
22,627	1/18/59	Colt, Samuel	Canteen Gun-stock (Hollow stock of gutta-percha or other material and stopper with India-rubber packing)
22,854	2/8/59	Cheever, J. H.	Pipe, hose, India-rubber, mfg. of seamless
22,976	2/15/59	Putnam, C. S.	Caoutchouc, vulcanizing apparatus

PATENT	DATE	PATENTEE	DESCRIPTION
23,001	2/15/59	Porter, Lucius P., assr. to New York Rubber Co	Toy, elastic (Combination of reed or speaking device with hollow elastic toy made of caoutchouc or gutta-percha)
23,103	3/1/59	Mattson, M.	Caoutchouc, treatment of
23,151	3/8/59	Beins, H. W.	India-rubber, vulcanized, treatment of
23,479	4/5/59	McBurney, C.	Sole, India-rubber, for boots & shoes (With holes for reception of pegs to fasten)
23,504	4/5/59	Shaw, William W.	Lead pencil head, rubber (Eraser with socket that is diameter of pencil)
23,582	4/12/59	Johns, Wm. B., U.S. Army	Knapsack (Shelters the soldier and holds his kit. Sold as "The Tent Knapsack" by the North American Gutta percha Company)
23,717	4/19/59	Shannon, A.	India-rubber, treatment of
23,740	4/19/59	Baschnagel, F., assr. to Beverly Rubber Co.	India-rubber, vulcanized, restoring waste
23,773	4/26/59	Mayall, T. J.	India-rubber and gutta-percha, vulcanized, treatment of
23,779	4/26/59	Parmelee, S. T.	Belting, elastic
23,805	4/26/59	Baschnagel, F., assr. to Beverly Rubber Co.	India-rubber, restoring waste
23,855	5/3/59	Parmelee, D. D.	Caoutchouc and gutta-percha, vulcanizing
23,948	5/10/59	Roberts, E. & W. J. Demorest	India-rubber, vulcanizing, portable stove apparatus
24,221	5/31/59	Mayall, Thomas J.	Sink, water-tight (India-rubber or gutta-percha)
24,265	5/31/59	Messer, H.	India-rubber, cutting into threads
24,321	6/7/59	McBurney, C.	Blanket or apron, India-rubber; for printing
24,401	6/14/59	Parmelee, D. D.	Caoutchouc and gutta-percha, vulcanizing
24,426	6/14/59	Cox, J., assr. to Horace Day	India-rubber, cutting into threads
24,620	7/5/59	Epes, E. & John F. Ellery	Paints, waterproof (Addition of India-rubber, linseed oil, gum shellac and alcohol)
24,693	7/5/59	Beins, H. W., assr. to New England Car Spring Co.	Spring for cars, vulcanized rubber
24,695	7/5/59	Eaton, A. K., assr., through others, to the Joslin India Rubber Co.	India-rubber, vulcanizing
24,996	8/9/59	DeWolfe, A. G.	Gums, vulcanized
25,100	8/16/59	DeForest, T. B.	Spring for cars, India-rubber
25,110	8/16/59	Goodyear, Charles	Rubber cloth, porous, improvement in
25,111	8/16/59	Goodyear, Charles	Fabrics, India-rubber
25,160	8/16/59	Hall, H. W., assr. to Beverly Rubber Co.	India-rubber and gutta-percha, vulcanized, restoring waste
25,175	8/23/59	Burnet, William	Inkstand, with India-rubber body
25,186	8/23/59	Essex, J. J.	Syringe, elastic bulb
25,239	8/30/59	Boyd, J. C.	Hose tubing, elastic (Cotton hemp, woven and fastened with rivets to layer of India-rubber or gutta-percha)
25,271	8/30/59	Marcy, E. E.	Caoutchouc, curing and treating
25,272	8/30/59	Marcy, E. E.	Caoutchouc, curing and treating
25,273	8/30/59	Marcy, E. E.	Caoutchouc, curing and treating
25,324	9/6/59	DuBois, M. D.	Roofing cement, Mfg. of (Includes 2 gallons of India-rubber solution)

PATENT	DATE	PATENTEE	DESCRIPTION
25,894	10/25/59	Fridley, Wm. & Cornman, Frederick	Preserve cans, Improvement in. (India-rubber or gutta-percha gasket that sealed the mouth of a vessel and showed if the contents were unspoiled)
25,957	11/1/59	Dieffenbach, G.	India-rubber, hard compound of
26,172	11/22/59	Eaton, A. K.	India-rubber, vulcanizing
26,175	11/22/59	Engelhard, George, Rudolph Franz & Heinrich Hinemann	Caoutchouc and allied gums, improvement in
26,177	11/22/59	Gately, D. C.	Belting, rubber
26,178	11/22/59	Gately, D. C.	Belting, rubber, making
26,204	11/22/59	Richardson, F. B.	Syringe, elastic enema (India-rubber or gutta-percha)
26,233	11/22/59	Joslin, H. W. & A. K. Eaton, assrs. to Joslin India Rubber Co.	India-rubber, treatment of
26,264	11/29/59	Gately, D. C.	Belting, rubber
26,265	11/29/59	Gately, D. C.	Belting, rubber
26,276	11/29/59	Mayall, Thomas J.	Hose, water-proof, mfg. of (Coating with India-rubber or gutta-percha
26,286	11/29/59	Parmelee, D. D.	Goods, India-rubber, hollow-moulded, mfg. of
26,358	12/6/59	Marcy, E. E.	Fabrics, India-rubber
26,359	12/6/59	Marcy, E. E.	Fabrics, India-rubber
26,360	12/6/59	Marcy, E. E.	Fabrics, India-rubber
26,519	12/20/59	Parmelee, D. D., assr. to J. A. Greene	Articles, India-rubber, mfg. of
26,549	12/20/59	Lee, B. F.	Belting, India-rubber (Combined with fibrous material)
26,551	12/20/59	Parmelee, D. D.	Goods, India-rubber, hollow-moulded (Moulding around bags then exhausting air from the bags)
26,580	12/27/59	Gately, D. C.	Belting, India-rubber, mfg. of
26,603	12/27/59	Mayall, T. J.	Belting, India-rubber, making
26,658	1/3/60	Day, Horace	Shirred goods, mfg. of
26,669	1/3/60	Gately, D. C.	Belting, India-rubber
26,720	1/3/60	Trotter, J. T.	India-rubber goods, finishing felt
26,891	1/24/60	Cummings, George N.	Spectacle cases, tobacco boxes, etc., improved catch for (Appears on a variety of metal cases, at times used as "ditty boxes" by soldiers to hold rubber water filter or other articles)
26,916	1/24/60	Munger, George N. & George	School slates (Applying India-rubber to the bottom of wooden frame or unframed school slate to protect)
26,982	1/31/60	Gately, D. C.	Belting, rubber or gutta-percha, mfg. of
27,251	2/21/60	Wilmann, A.	Caoutchouc, composition formed of
27,331	2/28/60	Fuller, Joseph G	Composition for painting bottoms of vessels (Contains India-rubber)
27,337	2/28/60	LaForge, Samuel	Waterproof leather goods (Undressed skins, exposed to heat, coated with charged leather solution, then vulcanized)
27,440	3/13/60	Gately, D. C.	Belting, Elastic
27,471	3/13/60	Peatfield, James & Sanford	Belting, Elastic
27,505	3/13/60	Wright, A. H.	Chain, Elastic (India-rubber)-or surge spring for ship's cables
27,706	4/3/60	Eaton, A. K.	Caoutchouc, vulcanizing
27,798	4/10/60	Harris, C. T.	Caoutchouc, vulcanizing
27,837	4/10/60	Simon, C. F. C.	India-rubber, vulcanized, restoring waste

PATENT	DATE	PATENTEE	DESCRIPTION
28,150	5/8/60	Broad, S., assr. to David W. Felt	Life-saving and troop landing raft (India-rubber or gutta-percha floats)
28,388	5/22/60	Mayall, T. J.	Hose, rubber, mfg. of
28,389	5/22/60	Mayall, T. J.	Hose tubing, rubber
28,420	5/22/60	Trotter, J. T. & I. F. Williams	India-rubber goods, finishing surface of
28,597	6/5/60	Niles, Peter	Drinking cup, telescopic-improved (Hard India-rubber or metal)
28,744	6/19/60	Eaton, A. K.	India-rubber, treatment of
29,717	8/21/60	Richard, A. C.	India-rubber, waste, de-vulcanizing
29,914	9/4/60	Ramsay, J. W. & D. Wilson	Tubs, bathing, India-rubber, portable
30,181	9/25/60	Parmelee, D. D., assr. to Beverly Rubber Co.	India-rubber, vulcanized, restoring waste (reworking)
30,450	10/23/60	Allen, T. F.	Springs, India-rubber, for railroad cars
30,465	10/23/60	Curtiss, Charles	Bucket, improved collapsible (Frame composed of a bottom and one or more rims, covered with India-rubber cloth.)
30,675	11/20/60	Goodyear, H. B.	India-rubber tips (shoe), vulcanizing
30,787	11/27/60	Hayward, F. D.	Vulcanization of caoutchouc, mode of holding sockets of molds
30,807	12/4/60	Falke, O. & A. C. Richard	Caoutchouc, vulcanizing
31,146	1/22/61	Bauhoefer, Louis	Boots and shoes (Gum with pieces of cork in soles and heels for adhesion to ice)
31,195	1/22/61	Teubner, Augustus	Composition with gum-elastic to render garments air and water tight
31,240	1/29/61	Havemann, R. F.	Caoutchouc compounds, substitute for ivory and bone (Mixing oxyd of zinc to chlorine rubber compound as described in Pat. #26,175 dated 11-22-59)
31,241	1/29/61	Havemann, R. F.	Caoutchouc compounds, substitute for ivory and bone (Mixing lime, aqua-ammonia to chlorine rubber compound as described in Pat. #26,175)
31,286	2/5/61	Aiken, George	Boots & shoes, heel attachment for (Filling a groove in iron heel with India-rubber to stop noise of metal)
31,311	2/5/61	Goodyear, H. B.	Globes, India-rubber; method of relieving geographical outlines (Used as a globe and a toy ball)
31,391	2/12/61	Hutchison, Hiram	Combination of India-rubber with leather or cloth
31,453	2/19/61	Goldenblum, R. & F. Steiner	Composition for waterproofing leather (Fish oil and India-rubber dissolved in turpentine and petroleum, rosin pitch, beeswax, tallow, soap and lard)
31,552	2/26/61	Mayall, T. J.	Hose tubing, India-rubber
31,613	3/5/61	Mayall, T. J.	Brushes, India-rubber or gutta-percha
31,614	3/5/61	Mayall, T. J.	Hose, India-rubber or gutta-percha
31,622	3/5/61	Power, J. N.	Telegraph connections (Use of India-rubber or gutta-percha to waterproof)
31,626	3/5/61	Richardson, F. B.	Syringe, enema (Elastic bulb & pipe)
31,634	3/5/61	Stevens, E. G.	Syringe, enema
31,729	3/19/61	Meyer, Christopher	Cloth, applying caoutchouc to
31,730	3/19/61	Meyer, Christopher	Shoes, India-rubber (Rubber soles)
31,770	3/26/61	Asay, A. M. & J. L.	Teeth, hard rubber
31,843	3/26/61	Willis, Martha	Obstetrical bandages, elastic
31,885	4/2/61	Goulard, E.	Bouying vessels, apparatus (With India-rubber balloons)
32,224	4/30/61	Haller, William L.	Brush, scrubbing (India-rubber strips)
32,288	5/14/61	Hicks, James	Eraser, India-rubber with handle

PATENT	DATE	PATENTEE	DESCRIPTION
32,360	5/21/61	Goodyear, R. A. & L. A. Sprague	Skates (With a block of India-rubber to allow yielding motion)
32,361	5/21/61	Hall, Henry A.	Water filter, portable (With India-rubber flexible tube and mouthpiece)
32,575	6/18/61	Meyer, Christopher	Boots and shoes with rubber-coated uppers
32,591	6/18/61	Turner, G. L.	Spring, car (With rubber cylinders)
32,695	7/2/61	Hutchison, John	Boat (Wood completely covered with India-rubber cloth)
32,715	7/2/61	Williams, E. C.	Tent, portable , waterproof cloth
32,762	7/9/61	Fuller, W. C.	Springs, India-rubber, on gun carriage to neutralize shock
32,815	7/16/61	Arnold, Benjamin	Filter with elastic bulb
32,901	7/23/61	Simon, F.	Rubber, hard, in the mold; method of giving smooth surface
33,017	8/6/61	Hingher, Const., assr. to Novelty Rubber Co.	Stem for smoking pipe with condensing chamber to send moisture back to pipe
33,094	8/20/61	McBurney, C.	Rubber, utilizing waste vulcanized
33,105	8/20/61	Russell, Stephen H.	Canteen filler with India-rubber parts
33,118	8/20/61	Whitford, George	Washboard with India-rubber or gutta-percha ribbed plate
33,160	8/27/61	Pearson, Joseph W.	Pen holder, elastic (Rubber band around inkstand to hold pens when not in use)
33,234	9/10/61	Boardman, Henry	Syringe, vaginal
33,246	9/10/61	Franklin, Bradley	Vulcanizing caoutchouc-single chamber
33,303	9/17/61	Gately, D. C.	Vulcanizing or curing India-rubber or gutta-percha
33,304	9/17/61	Gault, John	Eraser & knife sharpener combination
33,344	9/24/61	Gebney, George	Air gun (With India-rubber tube)
33,354	9/24/61	Livingston, A. A.	Stirrup (With India-rubber footpiece to prevent foot from slipping out)
33,361	9/24/61	Newall, John	Elastic cloth, manufacture of
33,385	10/1/61	Frobsham, B. & M. Levett	Knapsack and bed combined
33,429	10/8/61	English, B. C.	Cartridge case, India-rubber (Improvement of Patent #17702. Adds metallic bushing to prevent breaking when fired)
33,436	10/8/61	Kinsley, Ed	Soles, flexible, India-rubber
33,464	10/8/61	Franklin, B. W., assr. to American Hard Rubber Co.	Vulcanizing caoutchouc, Improvement in
33,523	10/22/61	Flake, O. & Simon, E.	Vulcanizing rubber and other gums, arrangement of metallic plates for
33,528	10/22/61	Johns, William B	Convertible cloak and tent
33,583	10/29/61	Hanlon, Wm. & George	Gymnastics-portable apparatus with elastic cords
33,597	10/29/61	Moulton, Stephen	Springs, India-rubber covered
33,809	11/26/61	Woodbury, S. D.	Carriage wheels, elastic (Interposing a layer of India-rubber in wheel
33,861	12/3/61	Whitney, Washington	Clothes wringer (India-rubber covered rollers to regulate pressure)
33,910	12/10/61	Weiss, F. W.	Cloak, military (India-rubber coat with inflatable compartment convertible to mattress)
34,001	12/24/61	Mayall, T. J.	Mats for floors (India-rubber or gutta-percha with vertical projections to clean boots, etc.)
34,004	12/24/61	Morrison, S. G.	Canteen (With gum-elastic cone or cape that acts as a funnel when inverted or may be removed and used as a cup)
34,140	1/14/62	Goodyear, Henry	Gaiters, India-rubber with a backing of knit web/textile fabric
34,223	1/21/62	White, Samuel	Teeth, artificial (Hard rubber)

PATENT	DATE	PATENTEE	DESCRIPTION
34,227	1/21/62	Ames, Nathan, assr. to Goodyear India-Rubber Stopple Co.	Stopple (Variations of rubber, rubber with wood or metal core, and with coating of gutta-percha)
34,428	2/18/62	Jenckes, T. A.	Waterproof fabric (India-rubber combined with wood, silk, fur, etc.)
34,429	2/18/62	Kelsey, G. R.	Buckles, Improved (This date stamped on buckles on gutta-percha knapsacks made by the Gutta Percha Manufacturing Company)
34,430	2/18/62	Kinsley, Rhodolphus	Tompion for firearms (Wooden pin split in two parts with a spring of metal or rubber inserted between to force sides against the bore)
34,437	2/18/62	Lyon, Henry	Boot or shoe sole (Ground cork mixed with gum-elastic dissolved in oil or turpentine. Placed inside boot or shoe to waterproof)
34,453	2/18/62	Washburn, O. C.	Oil cloth, composition for making (A coating paste of oil sweetmeats, boiled oil, white vitriol, carbonate of ammonia, lamp-black, light daubs, and water, applied in several coats)
34,458	2/18/62	Browne, J. W., assr. to J. M. Van Wagner	Moulding, weatherstrip (Felt, rubber, or other elastic substance secured to a rigid backing)
34,409	2/28/62	Craig, Henry	Microscope (Manufactured in hard rubber and brass models)
34,682	3/18/62	Godfrey, B. D.	Boots and shoes (Double India-rubber sole)
34,705	3/18/62	Tracy, Albert-US Army	Tompion, wood or elastic material
34,947	4/15/62	Ditto, Joseph	Waterproof cloth coating (Made of India-rubber and residual gum)
34,983	4/15/62	Ray, George	Penholder with rubber tube which acts as a support for fingers
35,056	4/22/62	Boisset, Pierre, assr. to B. Antogini & self	Boot or shoe heels, India-rubber (With points of metal to prevent wear)
35,057	4/22/62	Boisset, Pierre, assr. to B. Antogini & self	Boots & shoes (Incorporating a nut in India-rubber sole so upper can be screwed on)
35,196	5/6/62	Allen, C. F.	Springs, car bumper & draw-head
35,231	5/13/62	Goodyear, Chas. Jr.	Casters from vulcanized compositions
35,292	5/20/62	Bogia, M. C.	Sword or dagger handles, India-rubber (Give firm grip & prevent abrasion)
35,418	5/27/62	Haskins, John, assr. to S. P. Blake	Nipple protector, India-rubber
35,544	6/10/62	Ritner, Michael	Projectiles of rifled ordnance, sabot for (A hollow sabot of vulcanized rubber to reduce concussion and expand by pressure of the gases)
35,599	6/17/62	Falke, Oscar	Rubber, hard, compound (Addition of sulphuret of antimony, sulphite of soda & India-rubber, gutta-percha or other vulcanizable gum to remove offensive odor
35,723	6/24/62	White, Thomas	Limbs, artificial (Legs with lower part made of gutta-percha or vulcanized rubber)
35,854	7/8/62	Greene, J. F.	Fabric, waterproof
35,855	7/8/62	Greene, J. F.	Fabric, waterproof, machine to mfg.
35,992	7/29/62	Burt, Lorin	Rubber cloth, figured, mfg. of
36,003	7/29/62	Hall, H. M.	Shawl pins (Hook with straight shank and button on head; Some were made with Union Eagle buttons and used with ponchos)
36,081	8/5/62	Ernst, John	Bayonet guard, removable (With elastic ball for practice)
36,406	9/9/62	Haering, Robert	Composition substitute for horn, hard rubber, etc. (Linseed oil, asphalt, sulphur, gutta-percha)
36,685	10/14/62	Brecht, T. C.-US Army & Sigesmond, S. B., assrs. to John Kuliuski & J. H. Housewright	Combined cloak, tent, bed (Used as a hammock, mattress, or life preserver)

PATENT	DATE	PATENTEE	DESCRIPTION
36,743	10/21/62	Whittemore, Joshua	Crutches (With India-rubber socket and buffer)
36,872	11/4/62	Timby, T. R.	Barometer (Bottom & cover of India-rubber to prevent expansion of mercury
36,879	11/4/62	Bird, J.N., assr. to Horace Day	Projectile wads, India-rubber (Act as a cushion between projectile and pressure of gases to prevent windage which is forced into grooves of projectile)
36,933	11/11/62	Putnam, Abel, Jr.	Spring hook for fastening garments (i.e. India-rubber blankets, etc. Some were made with Union Eagle buttons)
37,035	12/2/62	Day, Horace	Ordnance breeches, India-rubber (To lessen pressure of gases on gun parts)
37,112	12/9/62	Sault, Thomas	Wire-cable, covering with gutta-percha
37,192	12/16/62	Alden, Henry, assr. to New York Rubber Co.	Hose and flexible tube, India-rubber or gutta-percha
37,222	12/23/62	Ernst, John	Bayonet scabbard & guard (India-rubber ball secured on bayonet by a socket)
37,231	12/23/62	Hull, Liverus	Caoutchouc, treating (Adding carbon spirits and chloride of sulphur)
37,262	12/23/62	Wait, W. B., assr. to Jos. Fairbanks & self	Feed-bag for horses (With elastic mouth or head cap)
37,287	1/6/63	Goodyear, Chas. Jr.	Boot & shoe soles, India-rubber with piece of wire cloth for form & rigidity
37,446	1/20/63	Hull, Liverus	Machine for cutting caoutchouc
37,677	2/17/63	Colvin, L. O.	Breast pump (India-rubber pump/piston)
37,866	3/10/63	Hull, Liverus	Caoutchouc or India-rubber, improved (Use of chloride sulphur/carbon spirits
37,919	3/17/63	Neefus, Peter	Poncho (With air-tight compartment in corner to act as pillow when inflated. Inflatable margins to act as pillows
37,936	3/17/63	Clark, Augustus, assr. to Rubber Clothing Co.	Knapsack (With combination of shoulder straps, guides, loops, and fasteners)
38,005	3/24/63	Simpson, Edwin, assr. to Jared Post & self	Composition for waterproofing fabrics (Gutta-percha and campene or naptha)
38,006	3/24/63	Simpson, Edwin, assr. to Jared Post & self	Fabrics, mfg. of flocked, waterproof (Gutta-percha and campene or naptha)
38,072	3/31/63	White, Samuel S.	Teeth, Artificial (Recesses in teeth so vulcanite projects into them and becomes firmly attached)
38,073	3/31/63	White, Samuel S.	Teeth, Artificial (Recesses in teeth so vulcanite projects into them and becomes firmly attached)
38,245	4/21/63	Seely, Samuel J.	Spring, rubber
38,369	5/5/63	Bagnoll, Wm. R.	Caoutchouc, apparatus for curing
38,442	5/5/63	Bond, F., assr. to John Murray	Tourniquet (Endless band or tube of rubber - with or without pads)
38,615	5/19/63	Weeks, Theodore	Boot & shoes soles, India-rubber
38,674	5/26/63	Goodyear, Chas. Jr.	Straps, elastic rubber (For aprons or carriages with ornamented edge to imitate appearance of stitched leather)
38,686	5/26/63	Lewis, G. H.	India-rubber, printing and ornamenting
38,701	5/26/63	Southward, E. F.	Knapsack (Combination knapsack and waterproof blanket with yoke and straps. Could be attached to poles, muskets, etc. to form litter)
38,703	5/26/63	Steinberg, David	Teeth, artificial (Insertion of bars to strengthen India-rubber)
38,785	6/2/63	Elmer, William	Artificial leather (Using caoutchouc or gutta-percha)

PATENT	DATE	PATENTEE	DESCRIPTION
38,813	6/9/63	Davis, Abbot R.	Balls, springs, etc. (Sponge pressed into desired shape, dried and covered with India-rubber)
38,892	6/16/63	Faber, Eberhard	Pencil, eraser & stamp
38,895	6/16/63	Flanagan, Francis	Hats (Straw, palm leaf, etc. covered with India-rubber, hatter's varnish and felt or fabric)
38,897	6/16/63	Goodyear, Chas. Jr.	Whip sockets, vulcanized India-rubber
39,208	7/14/63	Baxter, Charles F.	Bottle stopper, elastic
39,237	7/14/63	Mayall, Thomas	Hose (Woven from threads saturated with India-rubber or gutta-percha)
39,238	7/14/63	Mayall, Thomas	Hose, elastic
39,267	7/21/63	Alger, Francis	Explosive shells, removable chamber for (Tube or pouch of India-rubber)
39,310	7/21/63	Russell, Clement	Springs, railroad car (Rubber disks)
39,434	8/4/63	Tripp, George	Dental plates, vulcanized rubber base with gold or other metal covering
39,452	8/4/63	Rawson, Smith E. G.	Testicle supporter (Pouch with elastic straps, rubber belt or band, etc.)
39,476	8/11/63	Fuller, John	Teeth, artificial (Rubber, metal and porcelain)
39,538	8/11/63	McClelland, J. A.	Dental plates with India-rubber base
39,594	8/18/63	Stevens, E. M., assr. to Alfred B. Ely	Boots & shoes, stiffening (India-rubber and fibrous material)
39,662	8/25/63	Lockwood, H. D.	Syringes, elastic (Enema)
39,665	8/25/63	Merryman, Frederick	Mitten for divers, waterproof (With elastic sleeve to exclude water)
39,667	8/25/63	Mossman, Sarah	Cap, military with waterproof cape & cover that folds away
39,678	8/25/63	Russell, Dwight	Powder injector (With elastic bulb and flexible probe)
39,754	9/1/63	Ross, Joseph	Inkwell (Float of cork covered with rubber or gutta-percha which guides the pen and reduces evaporation)
39,865	9/8/63	Simpson, Edwin, assr. to Jared Post & self	Boots & shoes, waterproof (Using process described in Patent #38,005)
39,897	9/15/63	Devlan, Patrick	Composition for filling projectiles (8 lb. paper pulp, 5 lb. plaster of Paris, 1/2 lb. caoutchouc dissolved in rosin oil; molded while soft around projectile. Used to form the expandable ring which, in the act of firing, is pressed into the rifle grooves)
39,958	9/15/63	Rolph, Reuben	Holdback for carriages (Hook with piece of India-rubber tube which bulges against point of hook to retain ring)
40,114	9/29/63	McKeen, T. C.	Diving apparatus (Pipe with air knapsack; enough air for several hours)
40,192	10/6/63	Singer, J.	Injecting & douching instrument (Shows hard rubber stop-cock)
40,319	10/20/63	Beadle. John	Breast pump, elastic
40,407	10/27/63	Hayward, C. H. & D. E.	Treating waste rubber
40,408	10/27/63	Hayward, Daniel	Boot & shoe soles, India-rubber
40,409	10/27/63	Hayward, Daniel	Boot & shoe heel, India-rubber stiffener or counter
40,421	10/27/63	Mayall, Thomas J.	Pontoon (Waterproof fabric)
40,491	11/3/63	Mayall, Thomas J.	Restoring waste rubber
40,580	11/10/63	Syrcher, John	Billiard cushion (Horn & India-rubber)
40,720	11/24/63	Wilmott, George	Tompion for firearms, expanding (Ring or packing of vulcanized rubber)
40,742	12/1/63	Davison, John	Cork, air-tight (Application of layer of gutta-percha or similar material)
40,763	12/1/63	Marks, A. A.	Limbs, artificial (Sponge rubber and hard rubber and wood)
41,067	1/5/64	Hamilton, Edward, assr. to Henry Goodyear & self	Bottle stopper, elastic
41,260	1/12/64	Howard, Frank, assr. to Vyse & Sons, N.Y.	Hat, rendering underside waterproof by suitable preparations

PATENT	DATE	PATENTEE	DESCRIPTION
41,284	1/19/64	Denton, Drake W.	Coffins & other articles, mfg. of (Asphaltic compound: Pitch or coal tar with plaster and sand or India-rubber)
41,292	1/19/64	Gengembre, Phillip W.	Button (with unique shank surrounded by a tube of India-rubber that acts as a spring in securing the button to the fabric)
41,347	1/19/64	Alden, Henry, assr. to New York Rubber Co,	Moulds for vulcanizing rubber
41,418	2/2/64	Abbott, Granville of U.S. Army	Knapsack-Hammock (Waterproof material)
41,453	2/2/64	Taylor, Frederick	Flocking India-rubber and gutta-percha goods
41,481	2/9/64	Carr, Peter F.	Gun barrels, implement for cleaning bores of (India-rubber plunger coated with emery)
41,495	2/9/64	Eastman, H. G.	Penman's assistant (Pen or pen holder with a metal or hard rubber hand support to conform to shape of hand)
41,719	2/23/64	Perry, Edward L.	Boat & pontoon (Waterproof & elastic materials) Antedated 2-7-64
41,879	3/8/64	Webb, Benjamin	Boots & shoes, rubber
41,940	3/15/64	Pratt, E. L. & John Thompson	Carriage wheel, elastic (Band of rubber around wheel which acts as cushion)
42,228	4/5/64	Sanders, Benjamin	Cushion for piercing instruments, elastic (India-rubber enables pointed end to be self-retracting from substances into which they may have been driven)
42,415	4/19/64	Streeter, L. R.	Roller, elastic (Caoutchouc)
42,423	4/19/64	Ware, Preston, Jr.	Soles, rubber & composition-moulds
42,427	4/19/64	Zeno, Sigismund	Bottle, nursing (With flexible tube between bottle & mouthpiece)
42,533	4/26/64	Falke, O. & P. Shrag	Fabric, manufactured from caoutchouc, India-rubber or gutta-percha
42,632	5/10/64	Ayling, H. A.	Caoutchouc manufacture (Using carbon spirits & chloride of sulphur)
42,769	5/17/64	Holmes, D. E.	Penholder & ink eraser
42,883	5/24/64	Shephard, F. M.	Tablet, India-rubber
42,900	5/24/64	Harris, Isaac	Tubes, mfg. of flexible & other tubes (India-rubber and vulcanite)
42,937	5/31/64	Gardner, E. M.	Cushion, India-rubber (For horses feet)
42,942	5/31/64	Hackert, Julius	Compound for artificial ivory (Vinegar and nitric ether with shavings of bone, ivory, horn, etc. heated and moulded)
42,953	5/31/64	Liddell, R.	Waterproof fabric (Linseed oil, black lead & plumbago with cloth)
42,972	5/31/64	Taylor, Frederick P.	Tubing, India-rubber or gutta-percha
43,009	6/7/64	Cummings, John	Gum & palate, artificial (Vulcanite)
43,158	6/14/64	Shepard, F. M., assr. to W. A. Shepard & self	Syringe, India rubber with hard rubber fittings. (Lightweight and non-corrosive)
43,316	6/28/64	Lewis, George H.	Spring, car (Air-tight India-rubber shell filled with water or other liquid)
43,347	6/28/64	Terrell, John	Teeth, artificial (Vulcanite base)
43,377	6/28/64	Pagenstecher, Albert	Water defenses as protecting armor of vessels (With large India-rubber pockets filled with water)
43,454	7/5/64	Fickey, Fred. Jr., assr. to William Fickey	Pipe stem, tobacco (With 3 chambers)
43,526	7/12/64	Rowe, Philip	Pipe, pocket tobacco (Carried in pocket while smoked; Flexible rubber tube and rubber stem)
43,572	7/19/64	Hawkins, Clark & Frank Wilder	Overshirt (India-rubber body, close shirt collar and lapel with buttons. flannel lining)
43,620	7/19/64	Beach, Wm., assr. to F. Chamberlain & self	Cravats, vulcanized rubber (Neckties & cravats of hard rubber. Ornamented and ventilated by perforations)

PATENT	DATE	PATENTEE	DESCRIPTION
43,666	8/2/64	Burchardt, C. T.	Inlaying & ornamenting rubber articles
43,679	8/2/64	Fuller, Orange M.	Bandage apparatus for cooling parts of the body. (Bandage for head made of rubber tubing so cold water running through it will cool head)
43,762	8/9/64	Eliaers, A.	Button or stud, elastic (To protect chair legs, etc.)
43,902	8/23/64	Dodge, William C.	Calf feeder (With India-rubber teat)
43,907	8/23/64	Foley, D. D.	Pipe stem, tobacco (With air space for reception of saliva)
43,931	8/23/64	Southworth, Israel	Comb, toilet (2 parts, identical; secured at rear ends in a handle)
44,160	9/13/64	Burr, Melbourne	Scrubber (India-rubber strip folded into wood handle)
44,541	10/4/64	McDermut, John	Dental apparatus for vulcanizing (For making hard rubber teeth)
44,654	10/11/64	Perry, Edward L.	Horse Cover with louvers for ventilation (India-rubber, gutta-percha or other waterproof material)
44,832	10/25/64	Bullen, Wm. S., assr. to Wm. McIntire & self	Scrubber-mop (India-rubber scrubber)
44,843	10/25/64	Smith, Austin	Injector for hair (With elastic bulb or bellows)
44,905	11/1/64	Williams, William	Breech-loading ordnance (Block of India-rubber between rear of chamber and breech pin to receive recoil)
44,926	11/8/64	Arnold, L. L. & Francis X. Hazman	Cigarettes, Instrument for making (Frame & rollers with India-rubber band)
44,961	11/8/64	Landfear, William	Device to prevent slipping on ice (Elastic band with gritty substance to be worn around boot or shoe)
45,020	11/15/64	Cleu, John F.	Tompion cap for muzzles of ordnance exhausted of air (Wood with muzzle-cap of India-rubber)
45,073	11/15/64	Perry, E. L.	Life raft (Elastic cylinder and waterproof sacks)
45,112	11/15/64	Albertson, A.	Bottle stopper (India-rubber disc and spring)
45,373	12/6/64	Beard, Josiah & Moses Fairbanks, assrs. to Automatic Bottle Closing Co.	Bottling still liquids (Using a ball of India-rubber)
45,523	12/20/64	Robertson, D. F.	Bandage, catamenial (Elastic)
45,586	12/22/64	Cleu, John F.	Breech for ordnance, elastic (Shoulder of needle is faced with India-rubber to act as a gas check)
45,641	12/27/64	Schermerhorn, Thos. & Joseph Anderson	Tompion, expanding, for firearms (With elastic bands)
45,709	1/3/65	Gordon, T. E.	Crutch (With India-rubber spring)
45,766	1/3/65	Staunton, J. Galusha	Boxes and cases, air-tight. (Inside coated with paraffin or gum. Thread or welt of rubber on joints of boxes) Issued 5-18-65
45,794	1/3/65	Hicks, F. L., assr. to Ben & P. Laurence	Inkstand (With elastic bottom & turning rod to raise ink when required)
45,909	1/17/65	Colburn, G. F. J.	Comb, folding (With graduations to use a rule.)
46,082	1/31/65	Colburn, G. F. J.	Comb (With case that slips on as handle)
46,168	1/31/65	Wilcox, H. W.	Bucket, folding (Parts connected by joints with India-rubber or waterproof cloth attached to them.)
46,346	2/14/65	Fowler, N. C.	Teeth, artificial, using vulcanized rubber.
46,347	2/14/65	Fowler, N. C.	Teeth, artificial, using vulcanized rubber.
46,608	2/28/65	Simpson, E. L.	Hose, belting, packing, mfg. of India-rubber for (As in Patent #46,610)
46,609	2/28/65	Simpson, E. L.	India-rubber, hard, mfg. of (As in Patent #46,610)

PATENT	DATE	PATENTEE	DESCRIPTION
46,610	2/28/65	Simpson, E. L.	India-rubber, gutta-percha, etc., process of mfg. (Boiled oil with oxide or carbonate of lead added and heated to rid material of odor and deleterious properties of vulcanized rubber)
46,611	2/28/65	Simpson, E. L.	Waterproof fabric (Boiled oil, sulphur, etc. as in Patent #46,610)
47,457	4/25/65	Rice, G. S.	Articles of hard rubber (Lockets, etc)
47,477	4/25/65	Webster, William	Packing rings, India-rubber (For sealing fruit cans)
47,540	5/2/65	Gilbert, J. P., M.D.	Medical instrument with rubber nipple for curing piles
47,705	5/16/65	Doty, G. W.	Bottles, siphon. (Bottle & cork with 2 India-rubber tubes)
48,015	5/30/65	Crosby, T., assr. to American Water-proof Cloth Co.	Carpets, uniting fabrics by a coating or sheet of India-rubber
48,762	7/11/65	Crosby, T., assr. to American Water-proof Cloth Co.	Carpets, uniting fabrics by a coating or sheet of India-rubber
48,992	7/25/65	Parmelee, D. D.	Hard rubber, process using chloride of sulphur and bisulphide of carbon
48,993	7/25/65	Parmelee, D. D.	Hard rubber, process using chloride of sulphur and bisulphide of carbon
48,996	7/25/65	Percival, George	Rim, India-rubber (Attaches to bottom of tumblers, etc. to prevent breaking)
49,220	8/8/65	Bock, Louis & Al F. Wheeler	Artificial cork (Compound of wood dust of fiber and India-rubber)
49,527	8/22/65	Jenkins, Nathaniel	Valve, rubber (For faucets)
49,969	9/19/65	Brown, F. C.	Dental packing, rubber (Machinery)
50,457	10/17/65	Colvin, L. O.	Breast-pump (India-rubber valve and hard rubber rim)
50,461	10/17/65	Dibble, William H.	Dental elastic suction cup & tube (Holds jaw open and tongue away from mouth)
50,981	11/14/65	Wright, Nelson	Photographic bath (Box lined with rubber to preserve silver solution)
51,001	11/21/65	Ballard, William	Teeth, artificial with vulcanized plates
51,036	11/21/65	Forsyth, J. B.	India-Rubber, curing
51,161	11/28/65	Ensign, C. A.	Binding for India-rubber fabrics
51,207	11/28/65	Moffitt, J. R.	Moulding of rubber
51,331	12/5/65	Marquard, F.	Rubber, white (Curing with ammonia & mixing with phosphate of lime)
51,332	12/5/65	Marquard, F.	Rubber, white (Washing with hot water & mixing with phosphate of lime)
51,642	12/19/65	Woolworth, I.	Dental rubber, mixing with tints
51,697	12/26/65	Clough, S. C.	Spring, rubber (For wagons)
51,715	12/26/65	Hartman, John Jr.	Car brake, automatic (India-rubber sheet to remove animal from track and actuate brake)
51,778	12/26/65	Davidson, Herman E.	Syringes, elastic (Syringe & flexible pipe in one piece)
51,846	1/2/66	Malcolm, James	India-rubber and gutta-percha compound, colored
55,764	6/19/66	Eastabrook, Alanson. C., assr. to J. S. Parsons & Geo. A. Scott	Brush (Bristles project through perforated plate and are embedded in cement. Seen on "Florence" brushes and mirrors)
59,154	10/23/66	Somes, Daniel E.	Composition for ornamental articles (Contains sulphur, soapstone, clay, caoutchouc, gums, varnishes, etc.)
60,597	12/18/66	Walker, J. H.	Umbrella, India-rubber covered
65,231	5/31/67	Jennings, J. H.	Horses, rubber boot for
79,719	7/7/68	Alden, H. A.	Baseball, India-rubber
81,082	8/18/68	Halsey, W. H.	Watch-cases and lockets, hard rubber

DESCRIPTION	PATENT	DATE
Goods, India-rubber, hollow	26,551	12/20/59
Gum-elastic, mfg. (Overshoes)	849	7/24/38
Gum & palate, artificial	43,009	6/7/64
Gums, vulcanizable, treating	10,738	4/4/54
Gums, vulcanizable, treating	10,741	4/4/54
Gums, vulcanized	24,996	8/9/59
Gun barrels, cleaning bores	41,481	2/9/64
Gutta-percha & caoutchouc	5,539	4/25/48
Gutta-percha and India-rubber	9,668	4/12/53
Gutta-percha or India-rubber	14,814	5/6/56
Gutta-percha, cleaning	14,972	5/27/56
Gutta-percha, desulphurizing	11,906	11/7/54
Gutta-percha, feed apparatus	15,087	6/10/56
Gutta-percha, lining metal pipes	14,929	5/20/56
Gutta-percha, mfg.	8,992	6/1/52
Gutta-percha, process of working	7,643	9/17/50
Gutta-percha, purifying	17,214	5/5/57
Gutta-percha, stereotype	9,870	7/19/53
Gutta-percha, treating	10,977	5/30/54
Gymnastics, portable apparatus	33,583	10/29/61
Hard rubber, process	48,992	7/25/65
Hard rubber, process	48,993	7/25/65
Hard rubber, reworking	12,212	1/9/55
Hat, rendering waterproof	41,260	1/12/64
Hats (Straw, palm leaf, etc.)	38,895	6/16/63
Holdback for carriages	39,958	9/15/63
Hollow articles of India-rubber	5,536	4/25/48
Hollowware, gutta-percha	8,180	6/24/51
Horse cover with louvers	44,654	10/11/64
Horses, rubber boot for	65,231	5/31/67
Horseshoe, elastic	9,173	8/3/52
Hose (Woven from threads)	39,237	7/14/63
Hose and flexible tube	37,192	12/16/62
Hose pipe, coating with rubber	18,281	9/29/57
Hose tubing, elastic	25,239	8/30/59
Hose tubing, India-rubber	31,552	2/26/61
Hose tubing, rubber	28,389	5/22/60
Hose, belting, packing	46,608	2/28/65
Hose, combination	4,150	8/16/45
Hose, elastic	39,238	7/14/63
Hose, India-rubber	7,650	9/17/50
Hose, India-rubber	31,614	3/5/61
Hose, India-rubber, mfg.	15,947	10/21/56
Hose, rubber, mfg. of	28,388	5/22/60
Hose, water-proof, mfg.	26,276	11/29/59
India-rubber & gutta-percha	23,773	4/26/59
India-rubber & gutta-percha	51,846	1/2/66
India-rubber & other gums	10,586	2/28/54
India-rubber (de-vulcanizing)	29,717	8/21/60

DESCRIPTION	PATENT	DATE
India-rubber (Hollow articles)	10,689	3/28/54
India-rubber (Re-manufacturing)	9,570	2/1/53
India-rubber (Vulcanization pat.)	3,633	6/15/44
India-rubber fabrics	4,099	6/16/45
India-rubber goods, finishing	28,420	5/22/60
India-rubber goods, finishing felt	26,720	1/3/60
India-rubber goods, shirred, mfg.	3,788	10/12/44
India-rubber or gutta-percha	15,531	8/12/56
India-rubber tips (shoe)	30,675	11/20/60
India-rubber, restoring waste	23,740	4/19/59
India-rubber, restoring waste	27,837	4/10/60
India-rubber, reworking	30,181	9/25/60
India-rubber, cleaning	15,067	6/10/56
India-rubber, cleaning	16,069	11/11/56
India-Rubber, curing	51,036	11/21/65
India-rubber, cutting into threads	24,265	5/31/59
India-rubber, cutting into threads	24,426	6/14/59
India-rubber, cutting, machine	3,782	10/9/44
India-rubber, devulcanizing	12,983	5/29/55
India-rubber, devulcanizing	17,295	5/12/57
India-rubber, etc., vulcanizing	17,037	4/14/57
India-rubber, gutta-percha, etc.	46,610	2/28/65
India-rubber, hard compound	25,957	11/1/59
India-rubber, hard, mfg.	46,609	2/28/65
India-rubber, machine for making	3,462	3/9/44
India-rubber, mfg.	4,047	5/13/45
India-rubber, mfg.	6,066	1/30/49
India-rubber, mfg.	7,196	3/19/50
India-rubber, mfg.	7,880	1/1/51
India-rubber, mfg.	8,016	4/1/51
India-rubber, mfg. (Hard rubber)	8,075	5/6/51
India-rubber, ornamenting	38,686	5/26/63
India-rubber, preparing	5,069	4/17/47
India-rubber, preparing	5,271	9/4/47
India-rubber, preserving	9,246	9/7/52
India-rubber, preserving	9,891	7/26/53
India-rubber, restoring waste	23,805	4/26/59
India-rubber, restoring waste	25,160	8/16/59
India-rubber, softening	4,357	1/15/46
India-rubber, treatment of	22,560	1/11/59
India-rubber, treatment of	23,151	3/8/59
India-rubber, treatment of	23,717	4/19/59
India-rubber, treatment of	26,233	11/22/59
India-rubber, treatment of	28,744	6/19/60
India-rubber, vulcanizing	7,816	12/3/50
India-rubber, vulcanizing	23,948	5/10/59
India-rubber, vulcanizing	24,695	7/5/59
India-rubber, vulcanizing	26,172	11/22/59
India-rubber, working over	15,998	10/28/56

DESCRIPTION	PATENT	DATE
India-rubber (Remolding)	11,608	8/29/54
Injecting & douching instrument	40,192	10/6/63
Injector for hair	44,843	10/25/64
Inkstand	14,451	3/18/56
Inkstand (Elastic bottom)	45,794	1/3/65
Inkstand (India-rubber body)	25,175	8/23/59
Inkwell (India-rubber float)	39,754	9/1/63
Inlaying & ornamenting rubber	43,666	8/2/64
Knapsack-Hammock	41,418	2/2/64
Knapsack	37,936	3/17/63
Knapsack ("Tent Knapsack")	23,582	4/12/59
Knapsack (With blanket)	38,701	5/26/63
Knapsack and bed combined	33,385	10/1/61
Lead pencil & eraser	19,783	3/30/58
Lead pencil head, rubber (Eraser)	23,504	4/5/59
Life-preserver hammock	6,581	7/10/49
Life-preserving vests	22,061	11/9/58
Life-saving & troop landing raft	28,150	5/8/60
Life preserver (Deep belt)	18,809	12/9/57
Life preserver shirt	18,274	9/29/57
Life raft (Elastic cylinder & sack)	45,073	11/15/64
Limbs, artificial	35,723	6/24/62
Limbs, artificial (Sponge rubber)	40,763	12/1/63
Machine for cutting caoutchouc	37,446	1/20/63
Mail & travelling bag	7921X	12/31/33
Mats for floors	34,001	12/24/61
Medical instrument to cure piles	47,540	5/2/65
Metals, covering	10,106	10/11/53
Microscope (hard rubber)	34,409	2/28/62
Mitten for divers, waterproof	39,665	8/25/63
Mould for earthen vessels	14,457	3/25/56
Moulding of rubber	51,207	11/28/65
Moulding, weatherstrip	34,458	2/18/62
Moulds for vulcanizing rubber	41,347	1/19/64
Nipple protector, India-rubber	35,418	5/27/62
Nursing bottle (artificial nipple)	4,138	8/9/45
Obstetrical bandages, elastic	31,843	3/26/61
Oil cloth, composition for making	34,453	2/18/62
Ordnance breeches, India-rubber	37,035	12/2/62
Overshirt (India-rubber body)	43,572	7/19/64
Overshoes, India-rubber	11,947	11/14/54
Packing rings, India-rubber	47,477	4/25/65
Paint, India-rubber, waterproof	18,183	9/15/57
Paints, waterproof	24,620	7/5/59
Pen holder, elastic (Rubber band)	33,160	8/27/61
Pencil, eraser & stamp	38,892	6/16/63
Pencil, improved (Mechanical)	11,762	10/3/54
Penholder & ink eraser	42,769	5/17/64
Penholder with rubber support	34,983	4/15/62

DESCRIPTION	PATENT	DATE
Penman's assistant	41,495	2/9/64
Photographic bath	50,981	11/14/65
Pipe stem, tobacco	33,017	8/6/61
Pipe stem, tobacco	43,454	7/5/64
Pipe stem, tobacco	43,907	8/23/64
Pipe, India-rubber, semi-elastic	22,566	1/11/59
Pipe, India-rubber, seamless	22,854	2/8/59
Pipe, pocket tobacco	43,526	7/12/64
Poncho (With air-tight pillow)	37,919	3/17/63
Pontoon (Waterproof fabric)	40,421	10/27/63
Pontoon Bridge	1,634	6/17/40
Powder injector (Elastic bulb)	39,678	8/25/63
Preserve cans (Rubber gasket)	25,894	10/25/59
Projectile wads, India-rubber	36,879	11/4/62
Projectiles, composition for	39,897	9/15/63
Projectiles, sabot	35,544	6/10/62
Pump, saliva (For dentistry)	10,426	1/17/54
Pumps, air	----	6/7/34
Pumps, air	8731X	3/30/35
Restoring waste rubber	40,491	11/3/63
Rim, India-rubber (For tumblers)	48,996	7/25/65
Rings, elastic	14,739	4/22/56
Roller, elastic (Caoutchouc)	42,415	4/19/64
Roofing cement	25,324	9/6/59
Rubber cloth, figured	35,992	7/29/62
Rubber cloth, porous	25,110	8/16/59
Rubber goods, hard, mfg.	20,938	7/20/58
Rubber, hard (Smooth surface)	32,901	7/23/61
Rubber, hard, compound	35,599	6/17/62
Rubber, hard, mfg.	22,218	11/30/58
Rubber, restoring waste	19,172	1/19/58
Rubber, restoring waste	20,678	6/22/58
Rubber, utilizing waste	33,094	8/20/61
Rubber, vulcanized (Waste)	22,265	12/7/58
Rubber, white	51,331	12/5/65
Rubber, white	51,332	12/5/65
Rubber, restoring waste	22,217	11/30/58
Rubber, utilizing waste	20,242	5/11/58
Rulers (With India-rubber)	20,055	4/27/58
Saddle, elastic (Rubber springs)	22,480	1/4/59
School slates	26,916	1/24/60
Scrubber-mop	44,832	10/25/64
Scrubber (India-rubber strip)	44,160	9/13/64
Shawl pins (Poncho hooks)	36,003	7/29/62
Shirred goods, mfg. of	26,658	1/3/60
Shirt bosoms (Elastic strap)	5,019	3/13/47
Shoes, India-rubber	8,100	5/20/51
Shoes, India-rubber	5,233	8/7/47
Shoes, India-rubber (Soles)	31,730	3/19/61

DESCRIPTION	PATENT	DATE
Shoes, waterproofing	1913X	4/29/13
Sink, water-tight	24,221	5/31/59
Skates (With a block of rubber)	32,360	5/21/61
Sole, India-rubber, for boots	23,479	4/5/59
Soles for boots & shoes	10,008	9/13/53
Soles, attaching to India-rubber	13,272	7/17/55
Soles, flexible, India-rubber	33,436	10/8/61
Soles, India-rubber, catch for	16,013	11/4/56
Soles, rubber & composition	42,423	4/19/64
Spectacle cases, tobacco boxes	26,891	1/24/60
Spring for cars, India-rubber	25,100	8/16/59
Spring for cars, vulc. rubber	24,693	7/5/59
Spring hook for garments	36,933	11/11/62
Spring, car (Rubber cylinders)	32,591	6/18/61
Spring, car (Shell with liquid)	43,316	6/28/64
Spring, rubber	38,245	4/21/63
Spring, rubber (For wagons)	51,697	12/26/65
Springs & bumpers, India-rubber	4,827	10/24/46
Springs for cars, India-rubber	7,251	4/2/50
Springs, caoutchouc	6,231	3/27/49
Springs, car bumper	35,196	5/6/62
Springs, for railroad cars	30,450	10/23/60
Springs, gun carriage	32,762	7/9/61
Springs, India-rubber	12,738	4/17/55
Springs, India-rubber & metal	7,706	10/18/50
Springs, India-rubber & steel	10,280	11/29/53
Springs (Air-tight chambers)	5,606	8/1/48
Springs, India-rubber covered	33,597	10/29/61
Springs, railroad car (Disks)	39,310	7/21/63
Stair pad, elastic	21,799	10/12/58
Stem for smoking pipe	33,017	8/6/61
Stereotype plates, moulding	9,790	6/14/53
Stereotype, Gutta-percha	11,831	10/24/54
Stirrup (India-rubber footpiece)	33,354	9/24/61
Stopple (Rubber with wood etc.)	34,227	1/21/62
Straps, elastic rubber (Aprons)	38,674	5/26/63
Suspenders and shoulder braces	5,904	11/7/48
Suspenders, elastic cords for	6,786	10/16/49
Sword or dagger handles, rubber	35,292	5/20/62
Syringe eye-bath	10,838	4/25/54
Syringe, elastic bulb	25,186	8/23/59
Syringe, elastic enema	26,204	11/22/59
Syringe, enema	31,634	3/5/61
Syringe, enema (Elastic bulb)	16,956	3/31/57
Syringe, enema (Elastic bulb)	31,626	3/5/61
Syringe, enema (Flexible tube)	10,742	4/4/54
Syringe, (Rubber & hard rubber)	43,158	6/14/64
Syringe, vaginal	33,234	9/10/61

DESCRIPTION	PATENT	DATE
Syringes & tubes, etc., flexible	5,589	5/23/48
Syringes, elastic (Enema)	39,662	8/25/63
Syringes, elastic (One piece)	51,778	12/26/65
Table knives, securing handle	16,955	3/31/57
Tablet, India-rubber	42,883	5/24/64
Teeth, artificial	39,476	8/11/63
Teeth, artificial	46,346	2/14/65
Teeth, artificial	46,347	2/14/65
Teeth, Artificial	38,072	3/31/63
Teeth, Artificial	38,073	3/31/63
Teeth, artificial (Hard rubber)	34,223	1/21/62
Teeth, artificial (Strengthening)	38,703	5/26/63
Teeth, artificial (Vulcanite base)	43,347	6/28/64
Teeth, artificial (Vulcanized plate)	51,001	11/21/65
Teeth, hard rubber	31,770	3/26/61
Telegraph connections	31,622	3/5/61
Tent, portable , waterproof cloth	32,715	7/2/61
Testicle supporter	39,452	8/4/63
Thread, India-rubber, machines	16,269	12/23/56
Threads, India-rubber, machine	4,073	6/7/45
Tompion cap for muzzles	45,020	11/15/64
Tompion for firearms	34,430	2/18/62
Tompion, expanding, for firearms	40,720	11/24/63
Tompion, expanding, for firearms	45,641	12/27/64
Tompion, wood or elastic	34,705	3/18/62
Tourniquet (Rubber band)	38,442	5/5/63
Toy, elastic	23,001	2/15/59
Treating waste rubber	40,407	10/27/63
Trunks, valises, boxes	87,467	3/2/69
Tube pump, elastic	12,753	4/17/55
Tube, inhaling (For vapors)	11,171	6/27/54
Tubes (India-rubber & vulcanite)	42,900	5/24/64
Tubing and covering wire	8,051	4/22/51
Tubing, gutta-percha, mfg.	15,086	6/10/56
Tubing, India-rubber	42,972	5/31/64
Tubs, bathing, India-rubber	29,914	9/4/60
Umbrella, India-rubber covered	60,597	12/18/66
Upholstery (Rubber springs)	18,630	11/17/57
Uterine supporter (Inflatable)	15,897	10/14/56
Valve, rubber (For faucets)	49,527	8/22/65
Vessels, Float for supporting	8,070	4/29/51
Vulcanization of caoutchouc	30,787	11/27/60
Vulcanizing caoutchouc	10,339	12/20/53
Vulcanizing caoutchouc	33,246	9/10/61
Vulcanizing caoutchouc	33,464	10/8/61
Vulcanizing elastic gum	11,897	11/7/54
Vulcanizing or curing	33,303	9/17/61
Vulcanizing rubber & other gums	33,523	10/22/61

DESCRIPTION	PATENT	DATE
Washboard (With ribbed plate)	33,118	8/20/61
Watch-cases and lockets	81,082	8/18/68
Water defenses-armor of vessels	43,377	6/28/64
Water filter, portable	32,361	5/21/61
Waterproof cloth coating	34,947	4/15/62
Waterproof fabric	34,428	2/18/62
Waterproof fabric	42,953	5/31/64
Waterproof fabric	46,611	2/28/65
Waterproof leather goods	27,337	2/28/60
Whip socket	14,669	4/15/56
Whip sockets	38,897	6/16/63
Wind instrument, mouthpiece for	8,449	10/21/51
Wire, covering with gutta-percha	15,439	7/29/56
Wire-cable, covering	37,112	12/9/62

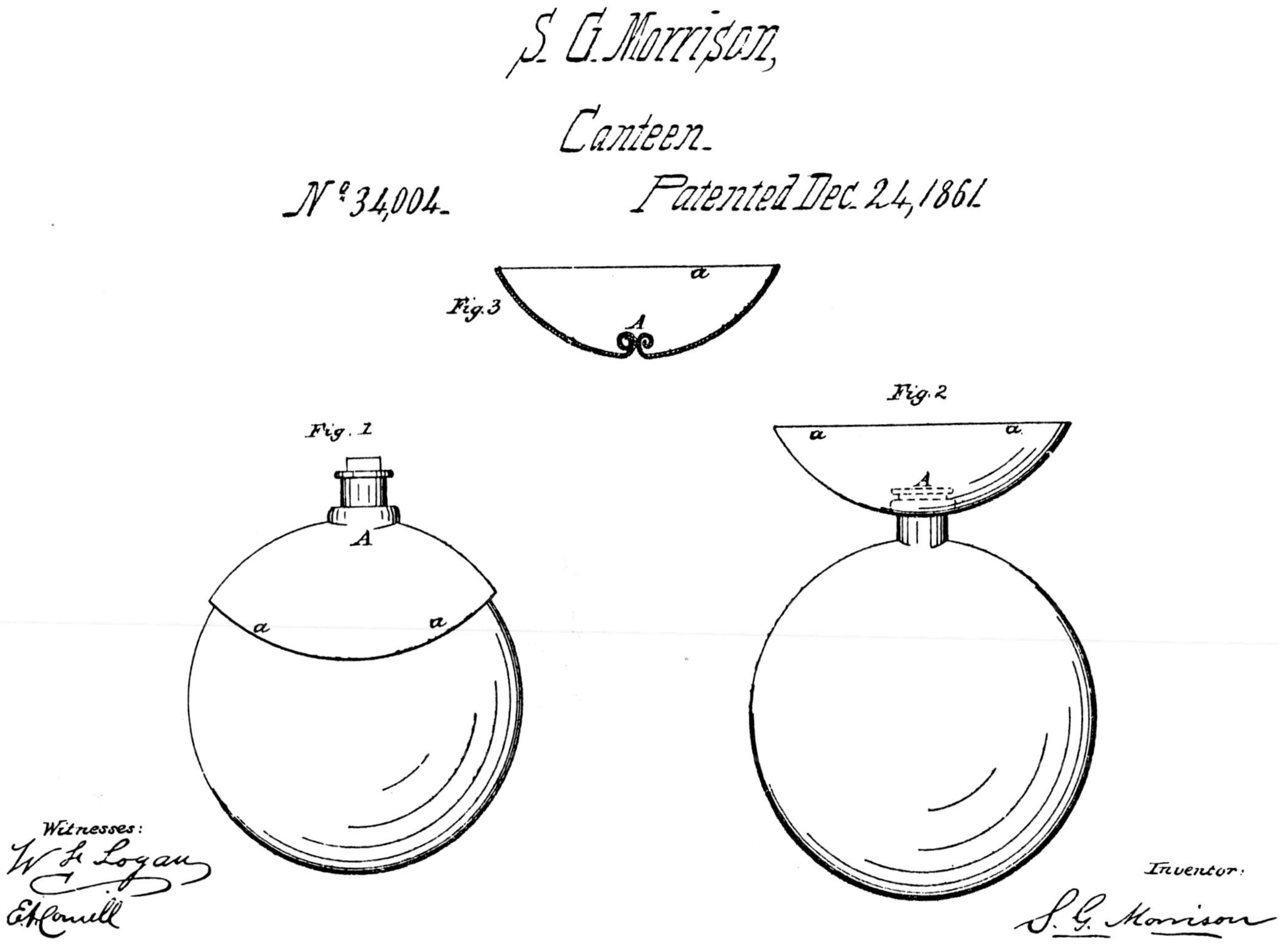

***S. G. Morrison's improved canteen** consisted of an India-rubber or gutta-percha "cape" that fit over the spout and the upper part of the canteen. When lifted up and inverted, the cape formed a funnel to facilitate filling the canteen.*

In Conclusion

The mid-nineteenth century and the Civil War era were just the beginning of the modern-day rubber and plastics industries. Catalogs and advertisements of the 1880s and 1890s show that many early items continued to be made and used through the remainder of the century. The primary uses for gutta-percha were in dentistry, golf balls, and insulated cables. Hard rubber and hard vulcanized gutta-percha were still in use for a variety of personal items but were being replaced with celluloid and various other plastic and plastic-like materials. Noted exceptions include flexible and durable hard rubber combs, manufactured throughout the 20th century.

Rubber (still called India-rubber or "gum" by some well into the 20th century) boots, garments, covers, medical supplies, and other commercial and industrial products continued to grow in popularity. The rubber industry's output in 1895 was ten times what it was in 1860. The sale of rubber footballs, baseballs, and associated equipment exceeded $1,000,000 annually in 1895. The annual value of rubber boots and shoes was $29,000,000 and the output of mechanical goods was valued at $20,000,000. All told, the industry employed 150,000 people in this country and manufactured $75,000,000 worth of goods annually.[1]

But the pneumatic tire, a revived invention of the late 1840s, would be the application that created the giants of the rubber industry. In 1895—before the automobile revolutionized transportation—over 6,000,000 pounds of crude rubber was used just for bicycle tires.[2] As demand for pneumatic automobile tires developed, modern-day companies like The Firestone Tire & Rubber Company, The B. F. Goodrich Company, The Goodyear Tire & Rubber Company, the United States Rubber Company, and a host of other large and small companies responded to meet the needs of a changing world.

The modern-day rubber industry and the evolution to synthetic molded materials had its basis in the innovative products and processes of the nineteenth century. Surviving artifacts and their rich history, coupled with the widespread acceptance of their modern descendants, provide a true and lasting testimony to the creativity, hard work, and persistence of the early pioneers of the industry.

NOTES - In Conclusion

[1] Chauncey M. Depew, *1795-1895 One Hundred Years of American Commerce* (D. O. Haynes & Co., New York), 1895, Vol. 2, pp. 501-503.
[2] *Ibid.*, p. 501.

Bibliography

MANUSCRIPT COLLECTIONS

U.S. National Archives and Records Administration, Washington, D.C.:

Consolidated Correspondence File, 1794-1890, Letters Received by the Office of the Quartermaster General and Other Offices, (CCF) Entry 225, Records of the Quartermaster General (OQMG), Record Group 92 (RG 92).

Subjects: "Blankets," "Goodyear's Metallic Rubber Co.," "Gutta Percha," "Horace Day," "India-Rubber," "India Rubber Bags," "Knapsacks," "Life Preservers," "North American Gutta Percha Company," "Rubber," "Uniforms," "Union India Rubber Co."

Invoices of Stores to be Delivered to Units, June 8-Dec. 14, 1846, Vol. 33 of 52, Entry unknown, RG 92.

Letters Sent, Book "L," Commissary General of Purchases, Masterson Entry 35, Bound Volumes of Philadelphia Supply Agencies, 1795-1858, Entry 2117, RG 92.

Letters Sent (Clothing and Equipage), Vol. 18, Entry 999, RG 92.

Maj. G. H. Crosman, Letters Received, 1835-62, Entry 332, RG 92.

Orders for Army Supplies, No. 1, June 17, 1846 - Feb. 15, 1847, Entry unknown, RG 92.

Press Copies of Letters Sent Relating to Orders, Purchases, and Contracts, Aug, 1861 to Dec., 1861, Vol. 5, Entry 2195, RG 92, Philadelphia Branch, U.S. National Archives and Records Administration, Philadelphia, PA.

Register of Letters Received (Clothing and Equipage), Entry 1003, RG 92.

Records of the Office of the Secretary of War, Letters Sent, Vol. 41, Entry unknown, Record Group 107, (RG 107).

Records of the Chief of Ordnance, Record Group 156 (RG 156).

Correspondence Relating to Inventions, 1812-1870, Entry 994.

Letters Received, 1836, 1847, 1855, Entry 21.

Letters, Telegrams, and Endorsements Sent to Ordnance Offices and Military Storekeepers, 1839-89, Vol. 15, Entry 6.

Records of the Treasury Department, Quartermaster Contracts, 2nd Controller's Office, Entry 236, Record Group 217 (RG 217).

NEWSPAPERS AND PERIODICALS

Army and Navy Chronicle, 22 Sep. 1836.

Army and Navy Journal, 27 May 1865.

Harper's Weekly, 2 Nov. 1861.

Illustrated London News, 9 Aug. 1851, 6 Sep. 1851.

Journal of Practical Medicine and Surgery. No. 522, Article 6240, 2 July 1862.

Newark Daily Advertiser, 26 Dec. 1855.

Peterson's Ladies Magazine, Spring 1861.

Scientific American, 8 Nov. 1851, 1 Nov. 1856, 25 Oct. 1856, 31 Jan. 1857, 10 Oct. 1857, 24 Oct. 1857, 14 Aug. 1858, 4 Sep. 1858, 23 Oct. 1858, 19 Feb. 1859, 10 Sep. 1859, 24 Sep. 1859, 29 Oct. 1859, 22 Jan. 1860, 26 Jan. 1860, 8 Sep. 1860, 11 May 1861, 8 June 1861, 22 June 1861, 13 July 1861, 1 Feb. 1862, 22 Mar. 1862, 29 Mar. 1862, 27 Sep. 1862, 29 Nov. 1862, 27 Dec 1862, 10 Jan. 1863, 23 Feb. 1863, 18 July 1863, 30 Apr. 1864, 16 July 1864, 19 Nov. 1864, 8 April 1865, 1 July 1865, 12 Aug. 1865, 23 Sep. 1865.

United States Magazine of Science, Art, Manufactures, Agriculture, Commerce and Trade. Circa 1855. Vol. 1.

PUBLISHED MATERIALS

Akron City and Summit County Directory—1888-1889. Akron: N. H. Burch & Co., 1888.

Albaugh, William A. and Edward N. Simmons. *Confederate Arms.* New York: Bonanza Books, 1957.

Albert, Alphaeus H. *Buttons of the Confederacy; A Descriptive and Illustrated Catalog of the Buttons Worn by the Troops of the Confederates* [sic] *States of America, 1861-1865.* Boyertown, PA: Boyertown Publishing Company, 1963.

Albert, Alphaeus H. *Record of American Uniform and Historical Buttons.* Alphaeus H. Albert, Bicentennial Edition, 1775-1976. Boyertown, PA: Boyertown Publishing Company, 1977.

American Chemical Society. *A Centennial Volume of the Writings of Charles Goodyear and Thomas Hancock.* Boston: The American Chemical Society, 1939.

American Hard Rubber Company. *Handbook of Hard Rubber and Plastics.* New York: American Hard Rubber Company, 1951.

An Officer in the U.S. Army. *The Soldier's Guide - A Complete Manual and Drill Book for the Use of Volunteers, Militia, and The Home Guard.* Philadelphia: T. B. Peterson and Brothers, 1861.

Archer, William H., M. B. Asbell, W. B. Irby. *History of the Development of Anesthesia, Oral Surgery, and Hospital Dental Service in the United States of America.* Pittsburgh: N.p., 1971.

Ashton, T. J. *On the Diseases, Injuries and Malfunctions of the Anus and Rectum.* Philadelphia: Blanchard & Lea, 1860.

B. F. Goodrich Company, Community Relations Release, 1977.

Babcock, Glenn D. *History of the United States Rubber Company.* Bloomington: Bureau of Business Research, Graduate School of Business, Indiana University, 1966.

Barker, P. W. *Charles Goodyear.* Boston: Godfrey L. Cabot, Inc., 1940.

Bausch & Lomb, "About History." http://www.bausch.com/About/chronology.html.

Bazelon, Bruce S. and William F. McGuinn. *A Directory of American Military Goods Dealers & Makers 1785-1915.* Manassas, VA: Published by the Authors, 1990.

Bazelon, Bruce S. and William F. McGuinn. *Supplement to A Directory of American Military Goods Dealers & Makers 1785-1915.* Manassas, VA: Published by the Authors, 1995.

Bedford, Gunning S. *The Principles and Practice of Obstetrics.* New York: William Wood & Co., 1868.

Beem, Barbara and Ken. "Collectors May Find Their Future in Plastics." *Antique Week,* 12 Apr 1993.

Bell, John B. "The Craig Microscope." *Rittenhouse - Journal of the American Scientific Instrument Enterprise*, Vol. 8, No. 3, Spring 1994.

Bennion, Elisabeth. *Antique Medical Instruments.* Berkely, CA: University of California Press, 1979.

Berg, Paul. *Nineteenth Century Photographic Cases and Wall Frames.* Huntington Beach, CA: Huntington Valley Press, 1995.

Bishop, J. Leander. *A History of American Manufacturers from 1608 to 1860.* Philadelphia, Edward Young & Co., 1868.

Boston Almanac for the Year 1862. Boston: J. E. Tilton & Co., 1862.

Boston City Directory—1874.

Braithwaite, W. *The Retrospect of Practical Medicine and Surgery.* New York: W. A. Townsend & Co., 1864.

Brannt, William T. *A Practical Treatise on Caoutchouc and Gutta Percha.* Philadelphia: Henry Carey Baird & Co., 1883.

Brannt, William T. *India Rubber, Gutta-Percha, and Balata.* Philadelphia: Henry Carey Baird & Co., 1900.

Brown, Tony. *A Short History of Soccer.* http://www.innotts.co.uk/soccerstats/hist1.htm, 1996.

Bush, David R. "Johnson's Island Civil War Military Prison: Intrasite Micro-Chronology." Research report for Center for Cultural Resource Research, University of Pittsburgh, October, 1992.

Bush, David R. "The Unknown Soldier." *The Gamut.* Cleveland State University, No. 36, Summer 1992.

Carleton, J. Henry. *The Prairie Logbooks.* Lincoln: University of Nebraska Press, 1983.

Carrington, George D. Edited and translated by Ken Bauman. *Diary of George D. Carrington of Company B, 11th Illinois Infantry.* Original in Chicago Historical Society, N.p., N.d..

Civil War Letters of Major Fredrick C. Winkler, 1863. (http://www.agro/agri.umn.edu/~lemedg/wis26/26pgwk63/htm, 1997.

Coates, E. J. and F. C. Gaede. Unpublished Compilation of U.S. Army Quartermaster Department Contracts, 1861-1865. From NA, RG 217, Entry 236. (Production Draft No. 4, Dec., 1994)

Coggins, Jack. *Arms and Equipment of the Civil War.* Garden City, NJ: Doubleday & Co., Inc., 1962.

Concannon, Dale. *Golfing Bygones.* Buckinghamshire, UK: Shire Publications, Ltd., 1994.

Coons, Zellie, Carole, & Martha. *Stoneham, Massachusetts: A Shoe Town.* Stoneham: Stoneham Historical Commission, 1981.

Covey, E. N. "The Interdental Splint." *The Richmond Medical Journal*, Feb. 1866.

Cullum, George W. *Papers on Practical Engineering, No. 4, Description of a System of Military Bridges with India-Rubber Pontons.* New York: D. Appleton & Co., 1849.

Dammann, Gordon. *Pictorial Encyclopedia of Civil War Medical Instruments and Equipment.* Missoula, MT: Pictorial Histories Publishing Co., 1983.

Depew, Chauncy M. *One Hundred Years of American Commerce.* New York: D. O. Haynes & Co., 1895.

Dobyns, Kenneth W. *The Patent Office Pony.* Fredericksburg, VA: Sergeant Kirkland's Museum and Historical Society, Inc., 1994.

Dodd's Curiosities of Industry. (Published in England) N.d., circa 1851.

Dood, Kendall J. "Why Models?" Published in *American Enterprise: Nineteenth-Century Patent Models.* Washington: Cooper-Hewitt Museum, The Smithsonian Institution, 1984.

Doyle, Marian I. "Mirror, Mirror on the Wall, Who's the Fairest of Them All?" *Antique Week*, 7 Jan. 1994.

"Dredging Up the Past." *Monsanto Magazine,* Summer 1965.

Durkee, Silas. *A Treatise on Gonorrhea and Syphilis.* Philadelphia: Lindsay & Blakiston, 1864.

Eklund, Vern. "Another Visit to Littlefield, Parsons & Co." *Smith & Wesson Journal,* Vol. 24, Number 4, Winter 1990.

Eklund, Vern. "You Just Never Know." *The Doorknob Collector,* Number 45, Jan.-Feb. 1991.

Eklund, Vern. "'Gutta Percha' Pistol Cases." *Smith & Wesson Journal,* Vol. 21, Number 3, Fall 1987.

Encyclopedia Brittanica. Chicago: Encyclopedia Brittanica, 1943.

Executive Committee. *Transactions of the Connecticut State Agricultural Society for the Year 1855.* Hartford: Case, Tiffany and Company, 1856.

Fendelman, Helaine. "What is It? What is It Worth?" *Country Living,* July 1998.

Fisher, Henry Linn. *Rubber and Its Use.* London: Macmillan & Co., Ltd., 1941.

Flayderman, Norm. *Flayderman's Guide to Antique Firearms - 5th Edition.* Northbrook, IL: DBI Books, Inc., 1990.

Foote, Shelby. *The Civil War - A Narrative.* 3 vols. New York: Random House, 1986.

Furber, George G. *The Twelve Months Volunteer or Journal of a Private in the Tennessee Regiment of Cavalry in the Campaign, in Mexico. 1846-7.* Cincinnati: J. A. and U. P. James, 1850.

Goodyear, Charles. *Gum-Elastic and Its Varieties.* 2 vols. New Haven, CT: Goodyear, 1853 and 1855, Vol. I and II.

Goodyear's Manual, Catalog of goods issued by T. G. Thorne of the Brooklyn Depot for Goodyear's Rubber Goods, 1878.

Hancock, Thomas. *Personal Narrative of the Origin and Process of the Caoutchouc or India-Rubber Manufacture in England.* London: Longman, Brown Green, Longman's & Roberts, 1857.

Harran, Susan and Jim. "Jewelry was an Important Part of Mourning." *Antique Week*, 1 Dec. 1997.

Haydon, Frederick Stansbury. *Aeronautics in the Union and Confederate Armies.* New York: Arno Press, 1980, c1941.

Hole, Frank. *The Acadia - A Civil War Blockade Runner, Technical Report Number One.* Houston: Rice University, Department of Anthropology, August, 1974.

Holland, Keith V., Lee B. Manley, James W. Towart—Board of Editors. *The Maple Leaf - An Extraordinary American Civil War Shipwreck.* Jacksonville: St. Johns Archeological Expeditions, Inc., 1993.

Hunt's Merchant Magazine and Commercial Review. New York, W. & Jno. A. Wood, 1851, Vol. 24; 1860, Vol. 42.

Institution of the Rubber Industry. *History of the Rubber Industry.* Cambridge, England: W. Heffer, 1952

Johnson, David F. *Uniform Buttons - American Armed Forces 1784-1948.* 2 vols. Watkins Glen, NY: Century House, 1948.

Kay, Frederick George. *Pioneers of British Industry.* London: Rockliff Publishing Corp., Ltd., 1952.

King, Constance Eileen. *The Collector's History of Dolls.* New York: Bonanza Books, ©1977, 1981.

Krainik, Clifford and Michele with Carl Walvoord. *Union Cases-A Collector's Guide to the Art of America's First Plastics.* Grantsburg, WI: Centennial Photo Service, 1988.

Kramer, W. *The Aural Surgery of the Present Day.* London: The New Sydenham Society, 1863.

Lock, R. H. *Rubber and Rubber Planting.* New York: G. P. Putnam's Sons, 1913.

Lord, Francis A. "Underwater Time Capsule: The Wreck of the Maple Leaf." *North South Trader's Civil War*, Vol. 19, No. 3, May-June 1992.

Lord, Francis A. *Civil War Collector's Encyclopedia.* Harrisburg, PA: The Stackpole Co., 1965.

Lord, Francis A. *Civil War Sutlers and Their Wares.* Cranbury, NJ: Thomas Yoseloff, 1969.

Lufkin, Arthur Ward. *The History of Dentistry.* Philadelphia: Lea & Febiger, 1938.

Magune, Jack. Civil War List of "The House of Billy Yank." Worcester, MA, August, 1992.

Manches, C. *Dr. Manches Matrimonial Guide.* New York: C. Manches, 1867.

Marcot, Roy M. "The Uniforms & Field Equipment of the Berdan U.S. Sharpshooters." *North South Trader's Civil War*, Vol. 17, No. 6, 1990.

McKee, Alexander. *How We Found the Mary Rose.* New York: St. Martin's Press, 1982.

McKittrick, Rosemary. "Antiques." *Pittsburgh Post-Gazette - Week End,* 26 July 1996.

McLaughlin, James A. *Charles Goodyear of Woburn, Mass.* Woburn: 1931.

Merriam Webster's Collegiate Dictionary. Springfield, Mass: Merriam-Webster, Inc., 1993.

Montgomery, Florence M. *Textiles in America—1650-1870.* New York: W.W. Norton & Co., 1984.

Newark, New Jersey's Greatest Manufacturing Centre, Illustrated, Newark: The Consolidated Illustrating Company, 1894.

New Encyclopedia Britannica. Chicago: Encyclopedia Britannica, Inc., 1993.

North American Gutta Percha Company. *Catalog.* N.p. [1854]. Consolidated Correspondence File, Entry 225, Records Group 92, "North American Gutta Percha Company," National Archives.

Ohio State Historical Society. *Diary of Sergeant Thomas Booth*, Box 4, Booth papers, Ohio State Historical Society.

Payne, P. L. *Rubber and Railways in the Nineteenth Century - A Study of the Spencer Papers.* Liverpool: Liverpool University Press, 1961.

Pearson, Henry. *Crude Rubber and Compounding Ingredients.* New York: The India Rubber Publishing Company, 1920.

Peirce, Bradford K. *Trials of an Inventor: Life and Discoveries of Charles Goodyear.* New York: Carlton & Porter, 1866.

Petsche, Jerome B. *The Steamboat Bertrand - History, Evacuation, and Architecture.* Washington: Government Printing Office, 1974.

Reed, John H. *Report of the Quartermaster General, February 1st 1862.* Appended to the Adjutant General's Report for 1862 in Massachusetts Legislative Documents 1862. Boston. N.p. 1862.

Regli, Adolph C. *Rubber's Goodyear.* New York: Julian Messner, Inc., 1941.

Rieser, Raymond G. "The Elastic Weapons of R. E. Hodges." *Journal of the Society of Archer-Antiquaries*, Vol. 31, 1988 .

Riling, Ray. *The Powder Flask Book.* New York: Bonanza Books, 1953.

Rinhart, Floyd and Marion. *American Miniature Case Art.* Cranbury, NJ: A. S. Barnes, 1969.

Robinson, Julie. "Bakelite: The Material of a Thousand Uses." *Antique Week,* 31 July 1995.

Robinson, Julie. "John Wesley Hyatt and the Invention of Celluloid." *Antique Week,* 19 June 1995.

Savas, Theodore P. "Bulwark of the Beleaguered Confederacy - The Augusta Powder Works." *Civil War,* Sep.-Oct. 1991.

Singer, Charles Singer and E. Ashworth Underwood. *A Short History of Medicine.* England: Oxford University Press, 1962.

Skinner, Henry Alan. *The Origin of Medical Terms.* Baltimore, Williams & Wilkins Co., 1961.

Speert, Harold. *Iconographia Gyniatrica - A Pictorial History of Gynecology and Obstetrics.* Philadelphia: F. A. Davis Co., 1973.

"Sports." *Science 81,* July-Aug. 1981.

Springer, George F. *Concise History of the Camp and Field Life of the 122d Regiment, Penn'a Volunteers.* Lancaster, PA: The New Era Steam Book Print, 1885.

Sylvia, Steven and Michael J. O'Donnell. *Civil War Canteens.* Orange, VA, Moss Publications, 1990.

Tafel, Anthony. "The Leather Gaiter." *North South Trader*, Vol. 5, No. 2, Jan.-Feb. 1978.

Tannahill, Reay. *Sex in History.* New York: Stein and Day, 1980.

Tomes, John. *Dental Surgery.* London: J & A Churchill, 1859.

U.S. Department of Commerce. *The Story of the United States Patent Office—1790-1956.* Washington: U.S. Government Printing Office, 1956.

U.S. House, *Letter* [April 11, 1864] *from the Secretary of the Navy in Answer to resolutions of the House and Senate in Relation to the Operation of Armored Vessels Employed.* 38th Cong., 1st sess., 1863-1864. H Exe. Doc. 69. Serial 1193.

U.S. House, *Letter* [March 2, 1865] *from War Department, Washington, City to Speaker of the House of Representatives on Expenditures from the Contingent Fund of the War Department.* 38th Congress 2d sess., H. Exe. Doc. No. 85.

U.S. House. *Annual Report of the Secretary of War.* 38th Cong., 2d sess., 1864-5. H. Exe. Doc. No. 83. Serial 1230.

U.S. Patent Office, *Annual Report of the Commissioner of Patents*—1844-1872.

U.S. Patent Office, *Official Gazette*—1872-1906.

U.S. Quartermaster General. *Report of the Quartermaster General of the United States to the Secretary of War for the Year Ending June 30, 1865.* Washington: Government Printing Office, 1865.

U.S. Sanitary Commission. *Narration of Privations and Sufferings of U.S. Officers and Soldiers While Prisoners of War in the Hands of the Rebel Authorities.* Philadelphia: U.S. Sanitary Commission, 1864.

U.S. Surgeon General. *Medical and Surgical History of the War of the Rebellion.* Surgical Volume, Parts I-III. Washington: Government Printing Office, 1883.

U.S. War Department, General Orders No. 64, 29 November 1848.

U.S. War Department. *The War of the Rebellion: a Compilation of the Official Records of the War of the Union and Confederate Armies.* 128 vols. Washington: Government Printing Office, 1880-1901.

U.S. War Department. *The War of the Rebellion: a Compilation of the Official Records of the War of the Union and Confederate Armies; The Civil War CD-ROM.* Carmel, Indiana: Guild Press of Indiana, 1996.

U.S. War Department. *Regulations for the Army of the United States*-[May 1]*1861.* New York: Harper & Bros.

U.S. War Department. *Revised Regulations for the Army of the United States*-[Aug. 10]*1861.* Philadelphia: J. G. L. Brown.

Van Doren Stern, Phillip. *Soldier Life in the Union and Confederate Armies.* New York: Bonanza Books, 1961.

von Niemeyer, Felix. *A Textbook of Practical Medicine.* New York: D. Appleton & Co., 1876. Volume II.

Wassmer, Lawrence C. *Butler, New Jersey in Story and Pictures.* Butler: The Butler Argus, 1951.

Wells, J. Soelberg. *A Treatise on Diseases of the Eye.* Philadelphia: Henry C. Lea, 1873.

White, Samuel S. *A History of Dental and Oral Science in America.* Philadelphia: White, 1876.

White, Samuel S. *Catalog of 1862.* Philadelphia: Lippencott & Co., 1862.

"Why Dry Wipers Corrode Faster." *The Pittsburgh Press*: 22 June 1981.

Wiard, Norman. *Wiard's System of Field Artillery.* New York: Holman, 1863.

Wideman, John C. *The Sinking of the USS Cairo.* Jackson: The University Press of Mississippi, 1993).

Wilson's Business Directory of New York City—1855. New York: John F. Trow, 1855.

Wolf, Howard and Ralph. *Rubber.* New York: Covici-Friede, 1936.

Wolf, Ralph F. *India Rubber Man.* Caldwell, Idaho: The Caxton Printers, Ltd., 1939.

Woodward, Joseph J. *The Hospital Stewards Manual; for the Instruction of Hospital Stewards, Wardmasters, and Attendants, in their Several Duties; Prepared in Strict Accordance with Existing Regulations and the Customs of Service in the United States of America, and Rendered Authoritative by Order of the Surgeon-General.* Philadelphia: J. B. Lippincott & Co. c. 1862.

Yeldham, Stephen. *Homeopathy in Venereal Diseases.* London: H. Turner and Co., 1864.

Index

(Also refer to Section 10 - Manufacturers and Pioneers & Section 11 - United States Patents.)

About the Author

Mike Woshner began collecting Civil War artifacts in 1969. He became interested in the history of hard rubber after purchasing a navy soap box marked "Goodyear's Patent May 6, 1851" and soon became totally intrigued by this unique material. The search for information and artifacts soon encompassed rubber, hard rubber, and gutta-percha and has since consumed much of Mike's spare time. The resultant collection of artifacts made of these materials has been showcased in public displays and has earned fourteen awards at Civil War collector's shows, including five coveted "Best of Show" honors.

As a consultant for the *Maple Leaf* Project (St. John's Archeological Expedition, Inc.), the *Bertrand* Project (U.S. Department of the Interior), and the Johnson's Island Project (University of Pittsburgh Center for Cultural Resources), Mike identified and provided important references on various Civil War rubber, hard rubber, and gutta-percha artifacts recovered at the sites. He has authored fifteen articles in collector and historical publications, including *Antique Week, Collector's News, Military Collector and Historian,* and *North South Trader's Civil War.*

Mike conducts unique presentations on the fascinating history of the discovery, patents, and novel applications for rubber and the predecessors of plastic—hard rubber and gutta-percha. The educational and entertaining talks are creatively structured to suit any group, and include hands-on displays of many rare and unusual Civil War period military and civilian artifacts.

An ongoing learner, Mike welcomes comments and information regarding any related artifacts, applications, patents, or processes.

Mike Woshner
2306 Spokane Avenue
Pittsburgh, PA 15210
412-884-9299
mwoshner@bellatlantic.net